DANIEL HILLEL

The Natural History
of the Bible

AN ENVIRONMENTAL
EXPLORATION OF THE
HEBREW SCRIPTURES

Columbia University Press   New York

Columbia University Press
Publishers Since 1893
New York Chichester, West Sussex

Library of Congress Cataloging-in-Publication Data

Hillel, Daniel.
The natural history of the Bible : an environmental exploration
of the Hebrew scriptures / Daniel Hillel.
p. cm.
Includes bibliographical references and index.
ISBN 0–231–13362–6 (cloth : alk. paper)
1. Bible. O.T.—Criticism, interpretation, etc. 2. Human ecology in the Bible.
3. Human ecology—Biblical teaching. 4. Human ecology—Religious aspects—
Judaism. I. Title.
BS1199.N34H55   2005
221.9'1—dc22   2005050740

Columbia University Press books are printed
on permanent and durable acid-free paper.

Printed in the United States of America
c 10 9 8 7 6 5 4 3 2 1

The Natural History of the Bible

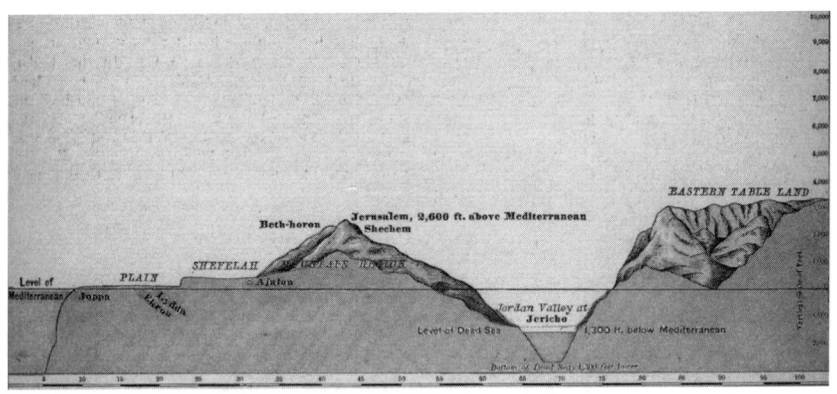

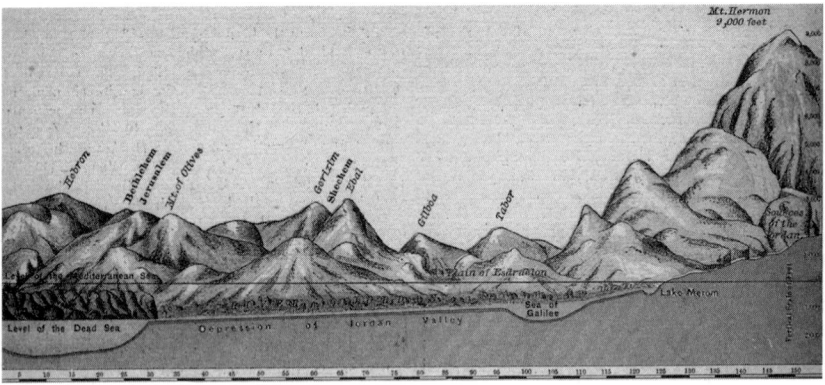

East–west (*above*) and north–south (*below*) physiographic transects of Canaan.

Dedicated to the gentle spirit and enduring legacy of my grandfather,
Haim Mordekhai Fromberg, *'alaiv hashalom*

CONTENTS

ACKNOWLEDGMENTS

W HEN I first chose the title of this book, I was unaware that it had been used before. Only later did I discover that the English naturalist and explorer Henry B. Tristram had published a book with the same title as long ago as 1873, under the aegis of the London-based Society for Promoting Christian Knowledge. His is indeed a beautiful book, elegantly written and richly illustrated, remarkable for its time and still fascinating to read. However, while describing the geography, climate, fauna, and flora of the Holy Land, as it was then known, Tristram made no attempt to define and interpret the cultural evolution of the Israelite people and of their Scriptures as influenced by the natural environment. Hence I tend to think of Tristram's book as "the natural history of the *land* of the Bible" (as, indeed, he defined it in his preface), rather than as the natural history of the Bible per se. So while I credit the pioneering explorer Henry Tristram with originating the title, I feel free to use it some 130 years later.

Several colleagues read parts of the manuscript at various stages of its development. Among them were Professor Michal Artzy of Haifa University; Professor Bill Dalesky of Hebrew University, Jerusalem; Professor Raymond Scheindlin of Jewish Theological Seminary, New York; Dr. Vivien Gornitz of the Goddard Institute, New York; Dr. David Arnow of the New Israel Foundation, New York; Professors Mary Evelyn Tucker and John Grim of Bucknell University, Lewisburg, Pennsylvania; and—last but certainly not least—author Bill McKibben. Others with whom I discussed different aspects of the subject include Professor Bustenay Oded of Haifa University; Professors Avinoam Danin and Moshe Greenberg of Hebrew University; author Evan Eisenberg; linguist Shira Hillel; Paul Gorman of the National Religious Partnership for the Environment, Amherst, Massachusetts; Professor Shmuel Bolozky of the University of Massachusetts, Amherst; and Dr. Alon Tal of the Arava Institute for Environmental Studies, Ketura, Israel. Each of them made helpful comments, but none of them is responsible for whatever shortcomings the book may yet contain.

An acknowledgment is also due to Jose Mendoza, who drew (and redrew) many of the illustrations.

Finally, I thank my literary agent, John Thornton, for guiding and sustaining me through difficult times, as well as my editors at Columbia University Press, Robin Smith, Wendy Lochner, and, most especially, Irene Pavitt (who is a treasure!), for their caring and astute handling of the manuscript and its final transmutation into a published book.

A NOTE ON TRANSLATION

A DIFFICULT DECISION I had to make at the outset was the choice of the translation to use in rendering passages of the Hebrew Bible into English. The classic King James Version (1611) is still the most familiar one to English-speaking readers of the Bible. Indeed, it is a magnificent translation, conveying much of the majestic tone and flavor of the original Hebrew. To many present-day readers, however, that language seems anachronistic and somewhat stilted. A more serious problem is that it is inaccurate in too many instances.

Several modern translations have attempted to rectify those flaws. Among the translations, one of the latest is that published by the Jewish Publication Society (2000), and it is the one I chose to use. This version, a work of meticulous scholarship, is perhaps less poetic than the classic translations, but it seems to be considerably more accurate. Still, there were places where I felt that even the JPS version fails to capture and convey the subtle meaning of the Hebrew text, especially as it pertains to the environment. Whenever I could offer a word or phrase to express what I understand to be the original intent or context, I inserted my own translation.

Another decision I had to make related to the name or names of the Deity. Instead of using the translated names (for example, "God" for Elohim, "The Lord" for Yahweh, "Lord of Hosts" for Yahweh Tsevaot), I chose to transliterate the actual names used in the Hebrew text. The context within which each of the names of the Deity was used and the progression toward the ultimate Yahweh thus become clearer.

Transliteration of the Hebrew Letters

The Latin letters used in the transliteration of Hebrew words appear in the following list

Hebrew Letter	English Pronunciation
א	*a*h, *eh*, *ee*, *aw*, *oo*, or ' (glottal stop)
ב	*b* as in *b*ig
כ	*v* as in *v*oice
ג	*g* as in *g*ood
ד	*d* as in *d*o
ה	*h* as in *h*ome
ו	*v* as in *v*oice or *w* as in *w*as
וּ וֹ	*u* as in *b*oot, *o* as in *o*n
ז	*z* as in *z*ebra
ח	*ḥ* as in German *ch*lor
ט	*t* as in *t*en
י	*y* as in *y*es or *i* as in *ti*er
כ	*k* as in *k*eep
ך, כ	*kh* as in Scottish lo*ch*
ל	*l* as in *l*ust
ם, מ	*m* as in *m*an
ן, נ	*n* as in *n*ever
ס	*s* as in *s*ea
ע	' (glottal stop, guttural)
פ	*p* as in *p*ost
ף, פ	*f* as in *f*ate, *ph* as in *ph*oto
ץ, צ	*ts* (*tz*) as in ca*ts*
ק	*q* as in Ira*q*
ר	*r* as in French t*r*oi or Italian *R*oma
שׁ, שׂ	*sh* as in *sh*ow, *s* as in *s*ee
ת, תּ	*t* as in *t*ea, *th* as in *th*ick

CHRONOLOGY

Prebiblical

Ninth to eighth millennia B.C.E.	Pre-Pottery Neolithic
Seventh to sixth millennia B.C.E.	Pottery Neolithic
Fifth to fourth millennia B.C.E.	Chalcolithic
Third millennium B.C.E.	Early Bronze

Biblical

Early second millennium B.C.E.	Middle Bronze
Mid- to late second millennium B.C.E.	Late Bronze
Twelfth to seventh centuries B.C.E.	Iron
Early to late sixth century B.C.E.	Assyrian–Babylonian
Late sixth to late fourth centuries B.C.E.	Persian

Postbiblical

Late fourth to first centuries B.C.E.	Hellenistic
First century B.C.E. to third century C.E.	Roman
Fourth to seventh centuries C.E.	Byzantine
Seventh to eleventh centuries C.E.	Arab

The Natural History of the Bible

PROLOGUE

A Personal Testament

It is a tree of life for those who grasp it,
and whoever holds on to it is happy.
Its ways are ways of gentleness
and all its paths are peaceful.

PROVERBS 3:18, 17

A BOOK ABOUT the Bible written by an environmental scientist? Some may find that an anomaly. In an age of fragmented and compartmentalized disciplines, we have come to expect that, with the possible exception of journalists, only those who are academically or professionally certified as bona fide specialists are qualified to write authoritatively on their respective disciplines. One may venture to deviate from that norm only at great risk of opprobrium.

Mainstream biblical scholars have written voluminously and continue to do so, and many of their works are profoundly insightful. But if conventional wisdom mandates that a book on the Bible be written by only a certified biblical scholar, I hereby challenge that stricture. To paraphrase Georges Clemenceau's dictum regarding war and generals, the Bible is too important to be left exclusively to those who are officially designated its academic gatekeepers. Rather, the Bible is such a prodigious mother lode of abiding cultural and historical interest that it can and should be reexamined by scholars of differing perspectives. Like a diamond with many facets, each of which exposes but a single aspect of the whole, the Bible should be viewed from various angles to appreciate more fully its timeless richness of content and nuances of meaning.

As an environmental scientist and an ecologist, my quest is to explore systems in which physical, biological, and human factors operate conjointly and interactively. My discipline attempts to reintegrate what specialists tend to segregate.

An ecologist's reading of the Bible is likely to differ significantly from that of a "standard" biblical scholar. The latter's main concern is generally with the text, whereas the former's is more likely to be with the *context*, the milieu, or (why beat around the burning bush?) the *original environment* of the Near East, within which the biblical events reportedly occurred. That environment was both diverse and labile. It included several domains that differed from one another in climate, topography, soils, vegetation, and human habitability. Defining those domains and determining the differing ways ancient societies evidently adapted to them are tasks that call for enlisting sources of knowledge additional to those brought to bear on the subject by traditional biblical scholarship. In particular, insufficient attention has been given to the ways the disparate ecological domains that the Israelites experienced in the course of their turbulent history influenced the development of their culture.

MY OWN INTEREST in the region in which the Bible evolved has been lifelong. Born in the United States, I was taken at an early age to Palestine, then in the very first stages of modernization. I spent part of my childhood in a pioneering settlement in the Valley of Jezreel, where, according to the Book of Judges, Gideon once drove off the pastoral nomads who encroached periodically on the laboriously cultivated fields of sedentary farmers.

Here, the ancient contest for survival has been waged between the archetypal figures of Cain and Abel, Isaac and Abimelech, and Jacob and Esau, since civilization began. And it was in this frontier between "the wilderness and the sown"[1] that I was first captivated by the region's environment, with its vivid contrasts: unlidded sun and parched earth, prolonged drought and sporadic rain, rugged mountains and flat plains, wind-whipped sands and soggy marshes, barren wasteland and fruitful fields.

I began to absorb the lore of the Bible from my grandfather Reb Haim Mordekhai Fromberg, who had left Charleston, South Carolina, in the late 1920s to live the life of the spirit and engage in good works in the Holy Land, for which he had always yearned. I dedicate this book to his memory. (Although I suspect that he, being a traditionalist, would have taken exception to some of the unorthodox ideas I express, I cherish the notion that he would have respected my viewpoint, however different from his, and would have been pleased by my abiding interest in a subject so dear to his heart.) His gentle voice and sonorous intonation still echo in my mind. As he related the age-old stories to me, and later as I learned to read them myself, the figures of the Bible came to life in my mind and stimulated my imagination. What the literary characters Natty Bumppo and Huck Finn were (in that pre-television era) to an American child; and what Ivanhoe, Robin Hood, and Jim Hawkins were to an English child; Abraham, Jacob, Joseph, and Gideon were to me. Samson was my Superman, while Joshua and David were the heroes I most longed to emulate.

In my early teens, whenever I could escape the confinement of school, I explored the countryside with my friends, using the Bible as guidebook. We seem to have been the last generation to witness the land as it was before the full onslaught of modernity. The trappings of the age of mechanization had not yet usurped the seemingly timeless landscape. (Asphalt highways, electric-power lines, smoke-spewing factories, sprawling cities, overflying airplanes—along with all the other intrusions of industrial development—were to begin changing the scenery only a few years later.) Although human habitation and exploitation over the millennia had certainly left their imprint on the land, it still appeared to retain its inherent character. Ancient villages, with their clustered rough-hewn houses and narrow alleys, still perched on hillsides; flocks of goats and sheep still grazed the brush-covered slopes; farmers still grooved the soil with spiked plows drawn by oxen and winnowed the harvested grain by heaving it into the air to let the wind blow away the chaff; women still carried water from wells and cisterns in jars balanced gracefully on their heads; and caravans of camels and donkeys laden with tradable goods still trudged along unpaved trails.

In the 1930s and early 1940s, the backdrop yet looked so authentic that many of the events described in the Bible could be seen in the mind's eye as though they were actual recollections. It seemed as though time had stood still and the past was yet present. Here may have been the brook where Isaac dug his wells; the plain where Joshua bade the sun stand still; the spring where Gideon selected his men and prepared them for battle; the vale where young David humbled the haughty Goliath; the barren hillside where Saul and Jonathan fell to the Philistines; the well of Gibeon, where Abner's young soldiers dueled with Joab's; Mount Carmel, where Elijah smote the false prophets, and the desert mountain on which he sensed the presence of the Almighty in the sound of thin silence.

Those early experiences were the seminal cultural and spiritual influences of my youth, and they inspired the course of my career as an environmental scientist. A fascination with the stories of the Bible and with the actual living settings in which they had taken place grew over the decades to become both a vocation and an avocation, a career in science and a labor of love.

Years later, after earning academic degrees in the natural sciences from universities in the United States, I returned to the newly established state of Israel to take part in mapping the country's land, vegetation, and water resources. Now armed with surveying instruments and aerial photographs, I still carried Grandpa's old Bible, the one in which he had inscribed the ancient adage *Vehagita bo yomam valaila*: "You should meditate upon it by day and by night" (Joshua 1:8).

Especially intriguing to me were the deserts of the Negev and Sinai, where—as described in the Book of Exodus—the unity of God was proclaimed from a mountaintop and a motley assemblage of fugitives from slavery was given the Law and forged into a nation. And so it was that, in 1951, I joined a group of equally naïve dreamers-of-glory to establish the frontier settlement of Sdeh Boqer in the very heart of the Negev highlands. Here, the depth of an entire year's rainfall seldom exceeds the hollow of a man's cupped hands. In this land of primeval grandeur and pervasive silence, the loudest noise ever heard was the incongruous rumble of a rare torrent suddenly hurtling down a ravine in the wake of a chance thunderstorm over distant mountains.

While living in the Negev, I spent some time with a tribe of Bedouin, then eking out an austere subsistence by grazing emaciated goats and camels on the scraggly shrubs that grew on the rock-strewn slopes and on the ephemeral grasses that sprung up in the winding wadis. Their mode of life resembled that of the biblical Patriarchs and the Hebrew tribes before their conquest of Canaan. To me, developing agriculture in that region seemed like a reenactment of the process by which those ancient people transformed themselves from roaming shepherds into village-dwelling cultivators.

One fateful day in mid-1953, we had an unexpected visitor. Taken for a tour of the Negev by officers of the Israeli army, David Ben-Gurion, the first prime minister of Israel, stopped at our encampment and asked what we were doing. After showing him our efforts at reclaiming land and reseeding pastures, we took him to the edge of the plain, where the land falls abruptly into the spectacular chasm known as the Valley of Zin. Ben-Gurion stood silently on the dizzying rim for a long while, as though transfixed. Gazing into the spectacular boulder-strewn ravine and across its barren moonscape toward the distant haze-shrouded mountains of Edom, he murmured to himself: "Oh to find a wayfarer's lodge in the wilderness." The words probably were not intended to be heard, but I happened to be standing next to him. Whimsically, I asked an impertinent question: "Do you mean the rest of the passage as well?"[2] He looked at me intently for a moment, and then said with a hint of a smile: "I see you know your Bible." That exchange was the beginning of a personal relationship between a very wise old man and a brash young one, a relationship that further inspired the course of my life and work.

On that chance visit, Ben-Gurion made his sudden decision, astonishing for a political leader at the peak of his power, to resign from the government and join our little outpost in order to take part personally in our pioneering enterprise.[3] Although he later returned to the government temporarily, he re-

mained a member of Sdeh Boqer. He eventually lived out his life in that settlement and is buried there. And it was at Sdeh Boqer that Ben-Gurion, like a
reincarnated biblical prophet, enunciated his vision: "The energy contained in
nature—in the earth and its waters, in the atom, in sunshine—will not avail us
if we fail to activate the most precious vital energy, the moral-spiritual energy
inherent in the inner recesses of our being; in the mysterious, uncompromis-
ing, unfathomable, and divinely inspired human soul."[4]

OVER THE succeeding decades, I became a more serious student of the Bible,
even while I extended the range of my professional activity as an environmental scientist beyond the Negev and Israel. As I worked throughout the Middle
East, I sought to gain an ever-deeper understanding of the region's ecology and
cultural evolution and of the way it is reflected in the records of other ancient
societies as well as in the Bible.[5] To me, the most meaningful attribute of the
Bible, apart from its inspiring ideals and stirring poetry, is its humanity. So
many of its characters are immediately and believably human, flesh and blood
and guts, with all the tribulations and foibles that condition entails, that I find
myself completely captivated by them, whether in sympathy or antipathy or a
mixture or both. Not only are the characters familiar, but—especially to one
aware of the ecological context—so are their essential situations. Although the
Bible can best be understood in terms of its time and place, its human por-
trayals also make it relevant to all times and places.

The Bible conveys different meanings to different people. Some regard it
primarily as a theological tract, a set of religiously mandated laws of behavior
and ritual. Others consider it to be a record of a nation's history. Still others
view it as a collection of myths or as literary fiction. In fact, the Bible is a composite of all these elements and much more. It is a heterogeneous assemblage
of writings composed over many centuries that encapsulates the experiences,
lore, and spiritual quest of numerous generations of Israelites, who were living
and struggling in and around the country first called Canaan (and later called
Israel, Palestine, or the Holy Land). As such, it is replete with information that
is of interest to ecologists and cultural historians, inasmuch as it reveals the
dynamic interplay between a society and the variable circumstances in which
its culture evolved.

Visitors to the country the Bible calls the Land of Israel are typically surprised, first of all, by its small size, being not much larger than a single county
in one of the western states of America. The hallowed Jordan, the most storied of all rivers,[6] is but a rivulet that would scarcely qualify as a tributary of a
tributary of the Mississippi. The second surprise is that, despite its smallness,

the country is amazingly diverse, with dramatic contrasts among its districts. There are densely forested mountains and fertile intermontane valleys in the north, sandy and marshy stretches of coast in the west, a deeply incised rift valley in the east, and a craggy desert in the south—all in a country that is scarcely more than 250 miles (400 km) long and 50 miles (80 km) wide.

How could such momentous history have taken place in such a sliver of habitable land, wedged so precariously between sea and desert? Here, at the juncture of three continents, disparate peoples and cultures have met and alternately clashed and mingled and merged since the beginning of history. One of these peoples, the Israelites, left a unique written record of their experiences in the region. How did they adapt to the austere conditions here? How did they, being a small nation in a particularly vulnerable locale, manage to survive the vicissitudes of time while evolving their own character and culture through trial and tribulation? And how did they compose the great literary work that eventually influenced the course of Western civilization so profoundly? Was it by chance alone that the initial realization of nature's unity, leading to monotheism—and, eventually, to ethical monotheism—was conceived in this of all regions, and by this of all the region's peoples? How, living in the midst of turmoil and violence, at the margin rather than in the centers of the region's great civilizations, did the Hebrew prophets and sages envision and enunciate the principles of universal justice and offer hope as well as instruction for achieving a better world? The intriguing answers may be sought, in part, in the special relationship between the people and the environment, embedded—explicitly or implicitly—in their collective diary known as the Bible.

THE WORLD as a whole is now poised at the cusp of an environmental crisis of humanity's own making. Some critics have ascribed our civilization's abusive treatment of the environment to a fundamental tenet of the Bible's first creation account: the purported appointment of humans as unrestrained masters of all forms of life on Earth, permitted to use all its resources for their own benefit. This opinion was derived from a selective extraction of passages, lumped together and interpreted tendentiously. Others, equally partisan, have invoked alternative passages in an effort to defend the Bible from its accusers. Neither of these simplistic views considers the complexity of the biblical account; the currents and countercurrents of perception, action, and reaction that accompanied the struggles of the biblical protagonists with the forces of nature and the neighbors arrayed against them; and the efforts of the different biblical writers who portrayed them to make ideological sense of those struggles. In actuality, the Bible is an intricate tapestry of many interwoven strands—facts, fantasies, ideas, and beliefs—practically impossible to encapsulate in a single, neat category.

During the 1990s, I was invited to several conferences on religion and the environment, at which religious leaders—rabbis, ministers, and priests—convened with scientists in a laudable effort to find common ground and to mobilize in defense of our threatened ecosystem. As I listened to the deliberations, however, I often had the uneasy feeling that some of the biblical scholars drew their wisdom from a tradition of thought that had for too long been detached from nature. Their arguments, cogent in themselves, were largely doctrinal, moralistic, philosophical, or philological. Their view of the Bible tended to be idealized, sanitized, divorced from the actual circumstances that had governed the lives of the biblical characters, and oblivious to the ecology within which the events of the Bible had occurred. Although each passage in the Bible is thoroughly rooted in the place and time of its composition, the Bible has too often been interpreted in ways that disregard its primal setting. The deliberate unification of the Bible by its compilers and editors has served to obscure the heterogeneity of the original sources, and the translation of the entire assemblage into uniform language (Greek, Latin, and, eventually, almost every major language in the world) has further tended to give it a false appearance of homogeneity. More than mere language and metaphor thus may have been compromised;[7] so may have been some of the Bible's important historical and moral lessons.

Our efforts to achieve a true understanding of the Scriptures may be further thwarted by the voluminous exegesis that has been attached to the Bible over the many centuries that followed its completion and compilation in the form made available to us. Although intended to clarify, some of the commentaries (being subjective and opinionated) may actually achieve the opposite result, obscuring or even perverting the original meaning and intent. The unquestioning believer in the sanctity, literal truth, and absolute consistency of the Bible may try to resolve any and all of the many apparent contradictions in the text by cleverly contrived reasoning. Over the years and centuries, some of these (often homiletic) explanations have themselves acquired the aura of sanctity and are regularly taught in some religious seminaries as though they were part and parcel of the Scriptures themselves.

The task of reconnecting the Bible to its real milieu is inherently difficult, as the environment of the region has changed in the interim and is now changing ever more rapidly. The difficulty is compounded by the fact that our own vantage point is receding farther away from the biblical ambience, especially as the society in which most of us live becomes increasingly urbanized and deracinated.

A difficult task can turn into a personal challenge. And so it happened that I have undertaken to write this book. The premise is that the events, characters, and ideas in the Bible evolved within a particular combination of environmental circumstances, and that the former can be properly understood only in relation to the latter. This book is, therefore, an invitation to read the Bible afresh, insofar as possible (and it may not be completely possible), without the extraneous loads of partisan, far-fetched exegesis and of either religious or agnostic preconceptions.

There is, to be sure, a wide spectrum of opinions regarding the Bible, and many of them are held with strong conviction and intense emotion.[8] On the hot "infrared" end of the spectrum stand the literalists, who believe in the total veracity of every word and passage in the Bible. On the opposite side, the cool "ultraviolet" end of the spectrum, stand the strict rationalists, who hold that the biblical stories are entirely mythical fantasies (like Homer's *Iliad* and *Odyssey*) and lack any basis in reality. Between those extremes lie all the various shades of visible light. My own position is somewhere in the middle, in the green range of the spectrum as it were, which is appropriate for an ecologist. However, one cannot stake out an intermediate position in this highly contentious range without incurring the risk of cross fire from both sides. So it is not without some trepidation that I state my case and prepare to bear the criticism that is sure to follow.

What I wish to do, above all, is to invite my readers to reexplore a realm that is considered by many to be all too familiar, but that, in fact, still contains undiscovered or unappreciated riches of content and insight; to read the old Scriptures again as though for the first time; to strip away the eye-misting onion layers of sectarian doctrine and pious religiosity; to ignore excessive midrashic extensions;[9] and to look anew at aspects too long ignored or misunderstood. Let us read the Bible as it was meant to be read, not the way it is presented to us in the solemn enclosure of a church or synagogue or in the sanctimonious preachings of a glib televangelist, but with an open mind and an active imagination, projecting ourselves into the time and place of the original stories and pronouncements and experiencing them afresh (albeit vicariously) with an empathetic attitude.

Peering into the core and crux of the Bible, we sense the powerful tension between the earthly and the spiritual, as we glimpse the unembellished and unidealized lives of real humans in actual settings: not angelic saints but lustful, greedy, often befuddled men and women, who may yet be generous and even heroic. We ponder how the biblical figures—so similar to us in flesh and spirit, fear and desire, despair and hope—wrested their livelihood from a stingy and grudging environment, how they thought and spoke and acted, and how they created so much stirring and meaningful history in such austere circumstances. And as we consider their lives and quests, we remember that they matter to us because they were our cultural progenitors.

The main theme of this book, the figurative "crimson chord" (an expression taken from Joshua 2:28 and Song of Songs 4:3) that should carry the message from beginning to end, is an environmental reading of the Hebrew Scriptures. The aim is to reveal the role of ecology in shaping the life and lore of the Israelites and, thereby, to actualize the ancient Writ and evoke a more authentic understanding of it in a truer context.

My examination of the topic, however objective its intent, is frankly tempered by a personal sense of humility toward the abiding beauty, majesty, inspiration, and enchantment of the original Book of Books. The Bible is, in fact, a series of texts written by scores of unknown authors in different styles, a mix of history and mythology, psychology and theology, familiar but still enigmatic; yet it is the first definitive text of ethical monotheism. The entire string of captivating stories—Adam and Eve and the serpent, Cain and Abel, Noah and the Flood, the Tower of Babel, Abraham and the Covenant, Ishmael and Isaac, Esau and Jacob, Judah and Tamar, Joseph and his brothers, the sojourn in Egypt, Moses and the Exodus, Mount Sinai and the Ten Commandments, Joshua and the settlement in Canaan, war with the Philistines, Saul and David, the rebellion of Absalom, Solomon and the queen of Sheba, the split between the northern and southern kingdoms, Elijah and Ahab, the destruction of the kingdom of Israel, the temporary survival of the kingdom of Judah, the religious reformations of Hezekiah and Josiah, the exile in Babylonia and the return, and, through it all, the morally compelling preachings of the prophets—seems ageless and ever fresh.[10] While some passages in the Bible display a primitive and intolerant tribalism, others rise to a compassionate humanism and a reverence for nature expressed in sublime poetic cadences full of startling insights and universal wisdom. As one of the Jewish sages advised some sixteen centuries ago: "Turn it and turn it, for all is contained within it."[11]

But there is something more to be admitted here, in all due candor and humility. Any present-day description of the biblical saga inevitably elicits feelings and thoughts, fantasies and beliefs, regarding the epic past. Fable, faith, myth, and mystery seem to intertwine and to obscure a reality that is inherently ambiguous. Complete clarity of motive and meaning eludes us. Yet an exploration of this archaic realm is, in some sense, a journey home, a return to a place and time, long abandoned yet vaguely remembered, from which our spiritual and cultural quest began before seeming to go astray. It is a mission impelled not only by rational motive or scientific curiosity, but also by a silent ancient call, an inner longing for the soul, once revealed here, from which we have been distanced but never completely severed. Can a reconsideration of the environmental origins of our civilization reconnect us to our true cultural roots? Can it reawaken, as well as inform, our long-lost reverence for nature and our inner desire to live in harmony within it? And can we explore the physical environment that gave birth to the Bible without dispelling the spiritual environment that imbued it with such profound meaning?

1

ENVIRONMENT AND CULTURE

A Premise and Its Implications

When I behold Your heavens . . .
the moon and stars that You set in place,
what is man that You are mindful of him . . .
that You have made him little less than divine . . .
You have made him master over Your handiwork.

PSALM 8:4–7

THE DEVELOPMENT of human culture, wherever it takes place, is shaped by the environment that prevailed at its inception. That environment encompasses all the features of a region's physical geography (location and geologic structure, topography, climate, and soils), biotic community of plants and animals, and cultural geography (the character of the human population, past and present). As such, the environment is not merely a passive and static stage on which cultural evolution takes place, but, indeed, a set of dynamic processes inducing that evolution. At the outset, the environment conditions the material life of a society. Reciprocally, a society's responses to the opportunities, challenges, constraints, and hazards presented by the environment tend to modify the environment. Thus a society's interaction with the environment inevitably affects its values and attitudes—indeed, its entire worldview. (The reciprocal interaction thus defined comes under the general category of human ecology.)[1]

Once stated, this premise may seem self-evident, yet it is too often ignored, especially as it applies to understanding the Hebrew Bible.

Accordingly, the environmental dimension should be traceable in the Hebrew Bible, which is, in essence, a subjective record of the formative experiences, memories, perceptions, and evolving faith of numerous generations of the people called the Hebrews or the Israelites in the various locations where they lived. The Bible spans an entire millennium, during which the Israelites successively sojourned in all the domains and interacted with all the major cultures of the ancient Near East. Out of that comprehensive, cumulative, and encompassing experience, they gradually synthesized their own culture, in which elements absorbed from the region's older cultures were not merely juxtaposed, but recomposed and given entirely new meanings.

The uniqueness of the Hebrew Bible inheres not necessarily in the original-ity of each of its various cultural elements (some of which were borrowed or inherited from older cultures), but in the process by which the disparate ele-ments were transmuted and eventually fused to create a coherent (even if not absolutely seamless) worldview, national ethos, religious faith, ritual practice, and code of morality and law.

In the course of that process, disparate pagan gods—either zoomorphic or anthropomorphic—were first made subservient to a chief god and then grad-ually merged into the notion of a single God, who was ultimately sublimated beyond physical representation to become an entirely abstract, ethereal, and all-pervading presence that motivates the entire universe and provides hu-mans with guidance and inspiration toward proper behavior. That perception did not occur as a sudden intuitive leap, but as a progressive outgrowth of and response to the Israelites' actual living experiences. Instead of deifying and worshiping the separate forces of nature, as did their neighbors, the Israelites came to consider the entirety of nature and all its manifestations to be God's creation and instrument. As God is the creator of all nature, he directs its evo-lution—including the evolution of humanity—toward a higher purpose. The ultimate culmination of the Israelites' ideological evolution was not merely an amalgam of earlier notions, but, indeed, a radically different intellectual and spiritual insight.

It might seem, therefore, that the Israelites devalued nature per se, subju-gating it to a superior force. Some modern critics have drawn the conclusion that the Bible fosters disrespect for nature, thereby contributing to the human presumption to emulate God, as it were, by dominating and subjugating the natural environment. The simplistic view of the Bible is that it fosters a single idea of a dominant God, of subservient nature, and of the destiny of humans as God's appointed surrogates. In response, some defenders of the Bible have attacked the misrepresentation of biblical passages by its critics, but few of them have attempted to present a systematic interpretation of the complex role of the environment in the evolution of Israelite culture and religion, and of the Israelites' treatment of their environment.[2]

The Hebrew Bible has been intensively analyzed and interpreted from widely differing viewpoints. Theologians have investigated the origins, tenets, and practices of the Israelite religion, based as it was on the radical percep-tion of a single, universal, invisible, omnipotent, and morally demanding God. Ethicists have pondered the moral foundations of the biblical strictures and edicts. Jurists have analyzed the biblical system of laws and injunctions, and their relationship to antecedent, contemporary, and subsequent legal systems.

Historians have examined the veracity of the events described in the Bible. Archaeologists have sought physical remains on the ground and considered whether they confirm or refute the biblical accounts. Linguists and literary critics have studied the vocabulary, etymology, and style of the Scriptures. Folklorists, mythologists, and psychologists, in turn, have investigated the Bible from their separate or complementary points of view.

In contrast, there have been relatively few studies of the Bible by modern ecologists capable of discerning the influence of the complex environments in which the Israelites developed their nationhood and religion and in which they composed the various scrolls that were later conjoined to constitute their principal written legacy.[3]

This lacuna cries out to be redressed, for all the events described in the Bible took place in the distinctive ecological domains of the region that includes the Fertile Crescent and is known to be the birthplace of Western civilization. Unlike most of us, now living in artificial urban settings shielded from exposure to the elements, our cultural forebears lived in close intimacy with nature and experienced its phenomena directly. They felt the unmitigated lash of hailstorms and gale-force winds, bore the brunt of droughts, and suffered the fury of floods. They depended utterly on the changing moods of an unpredictable nature, which could either lavish its bounty generously or withhold it grudgingly. Their perceptions of and responses to the heterogeneous natural environment of the region not only dictated their material modes of subsistence, but also influenced their notions of creation and the creator, of humanity's role in the scheme of life on Earth, of their own national destiny, and of proper collective and individual behavior—indeed, their entire attitude toward the world in which they and their neighbors lived.

THE ANCIENT Near East is the region where humans first made the fateful transition (only ten to twelve millennia ago) from a nomadic lifestyle based on hunting, gathering, and scavenging to a more or less settled lifestyle based on domesticating and cultivating plants and animals. Being the part of the world that has been subjected to the longest continuous human exploitation, it can teach us much about the relationship between civilization and the environment. That relationship is reciprocal. As many are now well aware, human societies can either improve or degrade an environment. Fewer are aware, however, of how profoundly the environment must have conditioned the development of each society from its very beginning.

The region that was the biblical arena is an ecologically varied continuum, exhibiting a gradation of topography (from mountains to plains to valleys), cli-

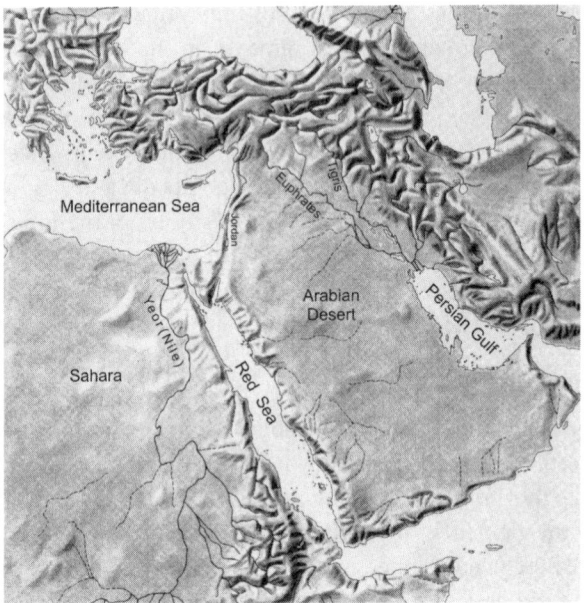

General relief map of the Near East.

mate (from relatively humid to hyperarid), flora (from dense forests to sparse xerophytes), and fauna. Superimposed on the inherent heterogeneity of the region is the variable impact of human habitation and exploitation over the millennia. Accordingly, the region is a palimpsest on which different societies in successive periods have left their imprint.

In a general way, we can recognize five principal ecological domains, existing in close geographic proximity to one another, but differing significantly in human habitability and potential resources: the *rainfed* (relatively humid) domain, the *pastoral* (semiarid) domain, the *riverine* domain, the *maritime* (coastal) domain, and the *desert* domain. The boundaries between adjacent domains are seldom sharp. Adjacent domains often exhibit transition zones in which ecological conditions vary gradually. Moreover, these boundaries tend to shift over time in response to fluctuations in climatic conditions and to the nature and intensity of human exploitation. In addition to experiencing the five natural domains, the Hebrews eventually experienced life in what can be called two cultural domains: the *urban* and the *exile* domains, both of which were instrumental in catalyzing the eventual composition of the Bible in the canonized form that has been bequeathed to us.[4] Historically, each of the region's domains presented opportunities, constraints, and challenges to human habitation, and consequently each gave rise to a more or less characteristic society, with its own modes of material subsistence and forms of cultural expression. The relations between the societies of adjacent domains were marked by both complementarity and competition. These relations included mutually advantageous trade in natural and produced goods, as well as hostile clashes over the vital resources of land, grazing or cultivation rights, and water.

Along with material adaptation, the society in every one of the domains evolved a characteristic culture, with its own folklore and locally meaningful religious beliefs and rituals. In each society, the people looked to the forces of nature that appeared to be most powerful and most fateful to their lives. They tended to personify and deify those forces and to worship them. Because those forces were bafflingly unpredictable (as were, for example, the irregular occurrences of catastrophic floods and droughts), people imagined the natural world to be a stage on which capricious gods vied with one another for supremacy. And since those gods could be either benevolent or malevolent—depending, so it seemed, on whether they were pleased or angered—shamans or priests devised elaborate rituals to curry their favor and avoid their wrath. Thus there developed the pagan, or polytheistic, religions of the ancient Near East.

Because the various deities were believed to resemble humans, they required proper abodes and savory food. Hence elaborate temples were built, at which animals were slaughtered and burned so that the aroma might please the gods and win their favor. These practices were promoted by a priesthood that had a vested interest in inculcating and perpetuating them.

The ruling gods differed among the domains. The nature gods of any one domain were locally, not necessarily universally, relevant, each holding sway over functions or phenomena perceived to be important in the particular environment. However, the gods of one domain were not entirely exclusive to

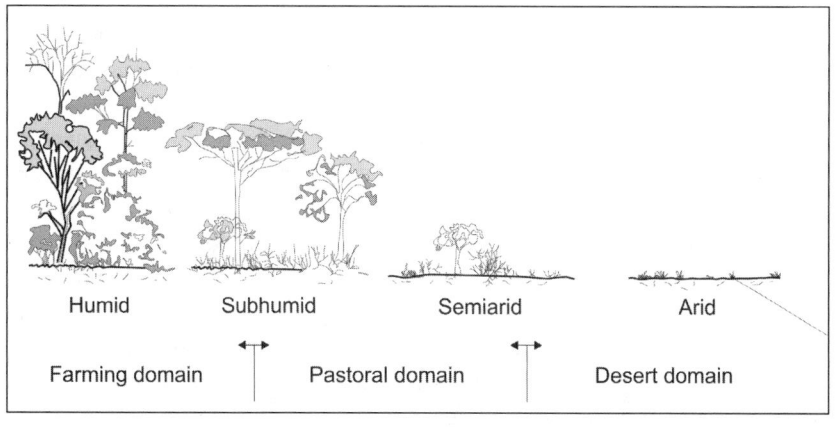

Succession of habitats (from humid to arid) and the natural climax vegetation (from forest to savanna to desert).

it. As societies in neighboring domains came into contact with one another through trade or war, they often adopted one another's gods into their own pantheons as minor deities alongside their local ruling gods. And when individuals shifted residence from one location to another, they commonly accepted the supremacy of their new abode's ruling gods for the duration of their stay.[5]

One small tribe, the Israelites, evidently deviated from the established pattern of the region's main societies. Early in their tumultuous history, the Israelites migrated from domain to domain before and even after their effort to settle permanently in Canaan. Having traversed and sojourned in all the region's disparate ecological habitats, they absorbed selective elements of diverse cultures and, by so doing, synthesized a new culture and a new faith. The multiple variants of polytheism, each of which was believed to be applicable in its particular domain, no longer provided a plausible explanation for the larger reality that the Israelites had observed in the different domains they had experienced. Consequently, they could begin to perceive the overarching unity of all creation. Their inclusive ecological experience thus conditioned them, mentally and spiritually, to combine the separate deified "forces of nature" into an overall "Force of Nature" and thus to coalesce the multiple gods of polytheism into a single God. That signal, stunning departure from the long-entrenched perceptions and practices of polytheism did not occur in one step. Rather, it apparently progressed through several tentative stages, leading from polytheism (belief in many contending gods, each with its own realm or function) through henotheism (belief in a principal god, with other gods playing minor or subservient roles) and monolatry (exclusive worship of one god among the various other gods) to tribal monotheism (belief in one God, associated with one nation). Only after much time and tribulation did the Israelites' faith attain its highest expression in the principles of ethical monotheism, identifying God with the abstract concepts of universal morality and justice.

Thus the peripatetic mode of life of the Israelites (or Hebrews)—first a clan, then a tribe, then a loose assemblage of tribes, and ultimately a coherent nation—during the formative stages of their cultural development, as they roamed from domain to domain in search of a land of their own and as they interacted with the indigenous cultures of each domain, constituted the seminal factor that led them to perceive the interconnectedness of all the phenomena of nature and hence the existence of a unifying supreme deity, as the first step in a process that led gradually toward pure monotheism.[6] Initially, the supreme god was likened to a father ruling over his (occasionally wayward) brood of lesser gods. In time, however, that supreme god became the Israelites' only God, not confined to one domain or function, but present everywhere as the creator and ruler of the entire universe and the supervisor of all human affairs. And as God's original entourage of subordinate deities faded away into irrelevance, it was replaced by the Israelites themselves, who viewed their nation as God's chosen servant and as exemplar and messenger to all other nations. Thus was born the notion of particular monotheism, a God initially associated with one special nation, chosen to serve as the harbinger to all humanity. In the future, their prophets predicted, all the other nations of the world would come to acknowledge and worship the same one-and-only true God.

The ecological influences on the life and lore of the Hebrews did not end when their early period of wandering culminated in their settlement in Canaan. Even there, security and stability eluded them. They had to cope with the adversity of the rugged hills and erodible soils of Canaan, the barrenness of the deserts of the Negev and Judea, the prolonged droughts alternating with capricious flash floods that occasionally inundated the valleys and lowlands, the violent westerly rainstorms that lashed the land in winter and the searing easterly winds that desiccated the land in summer, the occasional earthquakes that emanated from the numerous geologic faults, and the proximity of the storm-prone Mediterranean Sea. Theirs was a strip of land that, notwithstanding its small size, exhibited great contrasts of terrain, climate, and vegetation. The climatic instability of that semiarid country, as well as its extraordinarily sensitive geographic situation as a narrow corridor between Mesopotamia and Egypt (the two diametrically opposite centers of ancient civilization) made the Israelites especially vulnerable to repeated episodes of either famine or invasion, or both.

A natural human tendency is to seek a reason for every phenomenon. In recent centuries, we have come to rely on science to define the causes and consequences of physical and biological processes.[7] The ancients, lacking our scientific knowledge (which is ever incomplete, to be sure), sought etiological explanations in the realm of theology. They attributed quite a different significance to observed phenomena. If beneficial, they were taken to signify God's gift, expressing his approval. Otherwise, they were taken to signify God's punishment, expressing his disapproval.

The purely abstract notion of Yahweh as a formless spirit pervading the entire universe, which established itself ultimately as the distilled essence of divinity, in fact rested on a primal substratum of earlier inchoate perceptions, sensing the presence of supernatural powers in nature. Early Israelite religion, as reflected in many passages of the Bible, retained a primitive belief in numinous forces that imbued sacred stones (Genesis 28:11–19), sacred animals such as snakes (Numbers 21:9; 2 Kings 18:4), sacred trees or groves (Genesis 2:17, 3:3–7, 22, 12:6–7, 21:33, 35:4; Exodus 3:2–5), sacred springs and wells (Numbers 21:17–18), sacred caves (1 Kings 19:9–13), and mountains (Sinai and Horeb, Carmel, Ebal and Grizim, Moriah and Zion), as well as in lurking demons (Exodus 4:24–26).

Being a small nation in a precarious location, the Israelites had no one to whom they could appeal in times of dire need but their mysterious God. Their vulnerability and insecurity became ingrained in their collective perception and drove them constantly to seek ways to "find favor in the eyes of" their single, all-powerful, all-knowing, and just God, who alone could save them from ever-threatening destruction. Their special relationship to God, as they came to believe, was evidently conditional. It demanded that they strive ever more diligently to understand and obey his commandments, for otherwise he might turn against them and withhold the life-giving rains, afflict them with disease, blight their crops, or send one or another of their many potential enemies to subjugate and scourge them.

The belief in a single God seems to have offered something of a practical advantage. It promised freedom from the burdensome requirement to placate many gods of uncertain power and efficacy. It gave assurance that the acts of worshiping and praying to the one Almighty God were correctly addressed. And it offered hope, if not certainty, that he might respond, provided only that the believers truly adhere to his wishes and commandments. Psychologically, at least, the belief in the one just God provided a measure of confidence in an otherwise chaotic world.

But how should they please their demanding God? The most obvious way, following the traditional practice of polytheism, was to proffer precious gifts, and the common belief was that the more precious the gifts, the more likely they were to propitiate an angry or indifferent deity. The most precious gift any family could possibly provide was the life of its firstborn son. Indeed, the practice of child sacrifice was quite prevalent in the ancient Near East.

Alas, child sacrifice did not seem to bring security to the ever-vulnerable Israelites, given their exposed and unstable situation. And so the practice was abandoned (as exemplified in the symbolic story of Abraham's stayed hand when he was prepared to sacrifice his son Isaac [Genesis 22:1–19]), while the standard ritual of animal sacrifice continued. When it became evident that even animal sacrifice did not achieve its intended purpose, and neither did the offerings of any other material sacrifice, there began a process of soul-searching to discover what most displeased God and therefore had to be redressed and what action or attitude might, alternatively, please the Israelites' enigmatic God. Eventually, inspired religious leaders known as the prophets (rather than the establishment priests, who had a vested interest in continuing the mindless ritual of animal sacrifice and the acceptance of substantial offerings) came to focus attention on individual human behavior and collective social norms. God, the prophets proclaimed, expects not material gifts but righteous behavior. Their advocacy of justice and mercy as criteria by which to please an ethereal and spiritual God eventually led to the advent of an entirely new concept and code of ethics.[8] Although they believed their God to be universal, the Israelites sought an earthly location on which to focus their faith and ritual. As long as they were nomadic, their God resided in a transportable tent. When they settled on the land, they needed a permanent spot to which they could direct their faith and in which they could carry out their rituals. Hence the idea of the Temple as the abode of God. After they were exiled, their God was no longer focused on a single place but came to be regarded as present everywhere and always.

What sort of character did the Israelites ascribe to their God? In a kind of reversal of the Genesis story, in which God creates humans in his own likeness (an idealized status that they could never quite attain), the Israelites created or projected for themselves a God to help them overcome their frailty, a God who might fulfill their needs for assurance, protection, guidance, and inspiration. Having settled in the funnel-like land bridge between aggressive empires, and being ever vulnerable to attack, the Israelites needed a powerful, masculine figure to fight for them. Out of necessity, they conjured up a warrior God, fierce in battle against his enemies, a God of hosts (multitudes or armies) able to smite all who would dare to threaten his people. That he occasionally refrained from protecting his people could only be due to their own failure to fulfill his commandments. They therefore had to search their souls constantly, individually and communally, to discover their faults and to seek ways to rectify them. That endless introspection impelled their religious leaders to record the course of events in order to examine and reexamine their people's past and present actions in light of the demands presumed to have been made by God. The tendency to self-analysis was expressed most powerfully by the exhortations and castigations of the Hebrew prophets, and it was formalized in the ritual fasting on a solemn annual holiday: Yom Kippur, or the Day of Atonement.

The tribal nomadic origin, followed by settlement in Canaan and the attempt at consolidation in a kingdom that later split apart, followed by exile and dispersion during the reigns of the Assyrians and Babylonians, followed by the return of some of the Israelites under the protection of the Persians, created a dichotomy in the national soul between the ancestral wanderlust and a powerful attachment to the homeland. Having once lost their land, the Israelites continued to live in constant fear that they might lose it again. Having once regained their land, they also lived in constant faith that even if they were to lose it again, they might still regain it. Both that fear and that faith were ingrained in their national ethos and have persisted ever since.

The complex of heterogeneous formative influences to which the Israelites were subjected created other dichotomies as well. They perceived of God (who initially had several names, but eventually was referred to mainly as Yahweh) both as their private national Deity, specifically attached and attentive to them above all others, and as the universal God of all peoples and all living beings. And they viewed themselves at one and the same time as God's messengers to humanity, appointed to enlighten all nations, and as a nation destined to live apart from other nations as perpetual outsiders, an internalized vestige of their nomadic past. These various contradictions also became infused in their collective perception of themselves and of the world.

Experiencing the hazards of life in an inherently unstable region and being buffeted repeatedly by misfortune induced in the Israelites a constant yearning for what eluded them most—assurance and permanence. To compensate for what they could not attain in reality, they sought solace in an imaginary destiny crowned with glory. The harsher the reality, the more enticing and sublime became the fantasy. By what some may consider a sort of magical state of mind, the Israelites dreamed of an idyllic, messianic age, when they would be redeemed and all nations would live in a state of tranquility and harmony. Their prophets reinforced that dream by assuring the people again and again that God had not abandoned them and that they could indeed hasten the redemption by adhering ever more strictly to his commandments. Those who were disillusioned and chose to assimilate among other nations—when possible—left the fold and disappeared from the subsequent course of their people's history. Only those who clung to their faith ever more tenaciously retained their national and religious identity. That strong inner cohesion is at the root of the Jewish people's astounding survival over the many centuries in the face of all vicissitudes.

Ecological and geopolitical circumstances thus combined to influence the national character of the Israelites, as well as the character they ascribed to their God and hence their religiously sanctioned code of behavior. Ultimately, their self-enclosed, sectarian version of monotheism evolved into a universal philosophy and a prescriptive set of laws governing individual and communal conduct. And as the Israelite God was stripped of all the material forms that characterized the pagan gods, his spiritual essence came to imbue the religious conscience and consciousness of the faithful.

The idea that a single God created the entire universe and continues to direct everything in it implies that all nature is continuous, integrated, and consistent. The monotheistic premise that all phenomena in nature are essentially interconnected and operate under a single guiding principle is also the fundamental tenet—indeed, the precondition—of modern science.[9] Moreover, since the universe was created at a particular moment, and given that God created the world for an ultimate purpose toward which it is progressing, the concept of history also acquired a special meaning. Physical time had a beginning, and it was given direction, just as an arrow is thrust from a bow and given a trajectory. Human history therefore has a clear aim: to move toward an eventual age of universal redemption for all nations. This concept, again, was in stark contrast with the pagan notion of a static, cyclic, directionless, endlessly repeating history.[10]

SEARCHING AND RE-SEARCHING the Bible for ecological clues is particularly intriguing because latent within the terse Hebrew text are implicit messages, expressed in allusions, associations, and connotations. Those messages may have been left inexplicit simply because the writer of the account was so immersed in its context, and so confident that his contemporaries were likewise, as to be unaware that anything further need be explained. Across a time gap and culture gap of millennia, however, readers in a very different era, unaware of the setting and the nuances of the ancient language, may lose much of the meaning originally immanent in the text. The Hebrew language itself has metamorphosed over time, so that many words no longer carry the same nuances as they did long ago, nor do the metaphors convey the same associations. As many of the metaphors relate to what were then commonly experienced natural phenomena, as well as to the life and work of the ancient farmers and pastoralists, it now requires a special effort on our part to resurrect the original setting in order to better understand the biblical narrative.

The early books of the Bible were written in a dialect (including style and vocabulary) that is decidedly different from that of the later books. An example is the contrast between the Book of Kings (presumed to have been written before the Babylonian exile) and the Book of Chronicles (evidently written after the exile), both of which describe substantially the same history. The Hebrew language changed very significantly following that exile of 586 B.C.E., during the Persian period that followed, and again during the Hellenistic era in the fourth century B.C.E. Especially significant was the influence of the Aramaic language, whose script replaced the ancient Hebrew–Canaanite script. Hebrew gradually ceased being the spoken language of the ordinary people, although it remained the language of religious ritual. Indeed, the fact that the books of the Bible had been written originally in Hebrew helped preserve that ancestral language against the pervasive spread of Aramaic and, later, Greek.

Achieving a better understanding of the biblical narrative is even more difficult when it is translated into a language unrelated to the original, in uniform style throughout (thus masking the heterogeneity and obscuring the nuances of the original). A translation is necessarily a reconstructive paraphrasing.

Thus success in recovering the meaning and intent of the original Writ as it pertains to the environment depends on the ability to perceive the situation that prevailed where and when an event occurred or the text was written.

In the several millennia since the various books of the Bible were composed, the environment of the Middle East itself has changed. Some of the change has been caused by inexorable natural processes, such as climatic variation, sea-level rise, or tectonic shifts. More rapid and dramatic, however, is the change wrought by human activity. The relationship between humans and the environment has always been reciprocal, as every society, while adapting to local conditions, modifies those conditions as a result of its own adjustment to them.

Although the general attributes of the landscape and the macroclimate of the Middle East have remained essentially the same, some features of the ecosystems have been strongly influenced by human interference. Forests have been cleared; hillsides denuded and, in some places, terraced; rangelands overgrazed; rivers diverted onto plains; upland soils eroded; and valleys and estuaries choked with sediment. As a result, the diverse indigenous communities of flora and fauna initially prevalent in the region have been decimated and, in some cases, eliminated. (Some of the animals mentioned in the Bible, for example, are now exceedingly rare, and some seem to be extinct.)

Particularly where human exploitation of the land began earliest in history, some of the once-thriving areas have been reduced to near desolation. Civi-

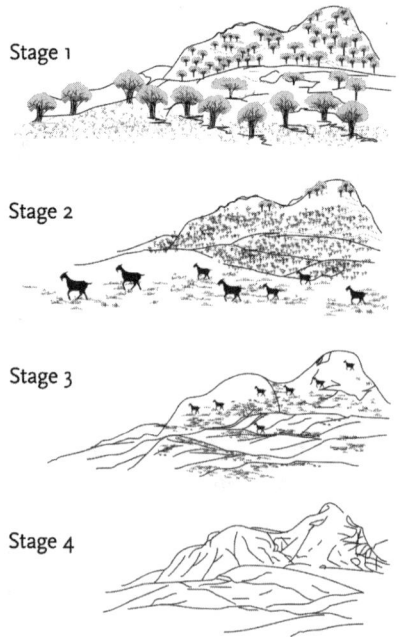

Stage 1

Stage 2

Stage 3

Stage 4

Land degradation (sometimes called desertification), a consequence of the long-term excessive exploitation of the land in the Middle East (as elsewhere): *stage 1*, the original natural vegetation consists of trees, shrubs, and grasses in a semiarid region; *stage 2*, the trees are decimated or cleared, and the land is cultivated and grazed; *stage 3*, erosion removes topsoil from the bared ground, resulting in loss of productivity; *stage 4*, ultimately, the carrying capacity of the land is diminished drastically, and the environment comes to resemble a desert.

lizations that succeeded all too well at first can be seen, in hindsight, to have inadvertently set the stage for their own eventual demise. In riverine southern Mesopotamia, irrigated farming without adequate drainage caused water-logging and salination of the soil. Equally devastating has been the impact of farming and grazing in the rainfed and pastoral domains. Where cultivation and grazing were practiced for many centuries on sloping terrain, without effective soil conservation, the land was bared and subjected to erosion. Hence the original mantle of fertile soil, perhaps 3 feet (1 m) deep, has been raked off by the rains (exposing the jagged bedrock) and swept down the valleys toward the sea. Especially vulnerable to degradation are the ecosystems in semiarid areas, which are naturally fragile. Land degradation may have been one reason why the Mesopotamians, Hittites, Aramaeans, Phoenicians, and others, each in turn, were tempted or compelled at some point in their history to venture beyond their homelands and invade those of their near and far neighbors.

The one factor that does not seem to have changed over time is human nature, motivated by a mixture of desires, perceived needs, and emotions. That is why many of the characters so realistically depicted in the Bible still seem vividly familiar in their original situations.

MODELS OF ENVIRONMENTAL ADAPTATION

A necessary step in research is to obtain the facts relevant to the subject of interest. After the salient facts are collected, a further effort is generally made to correlate the facts and thereby establish a pattern or set of principles that characterizes the system as a whole. The integration of facts into a coordinated system is known as a model. The formulation of models to make sense of a collection of disparate facts is a basic function of the human mind. A model remains an unproved conjecture, however, until tested rigorously to establish not only how well it accords with "old" facts, but also how well it predicts new facts. Such rigorous testing is routine in modern science, but was largely impossible in prescientific times. Then, people conceived models that seemed to make intuitive sense within a given set of circumstances and a given cultural context. Lacking the tools to test their models systematically and (more or less) objectively, they tended at times to embrace them as dogmas.

One possible model to explain the cultural evolution of a nation can be called environmental determinism. It postulates that attributes of a geographic milieu dictate the culture of those who live in it over a period of time. Absolute environmental determinism allows for no choice, disregards the character of the particular people (collectively and individually), and ignores any elements of chance. This model assumes that any and all groups of people living in the same environment necessarily acquire the same culture in automatic response to the opportunities offered by and the constraints imposed by the environment. This is a rather simplistic theory, which disregards the complexities in both the natural environment and the patterns of human perception and behavior.

A diametrically opposite model holds that cultural development is unique to each society, inasmuch as it is affected by the background and specific character of the people, by

continued

the particular individuals who assume leadership at crucial junctures, by random events, and by complex and unpredictable combinations of external and internal stimuli to which societies may respond in various ways.

While it is true that each society must necessarily adapt to its environment, the question is to what extent that adjustment dictates a particular mode of cultural development; that is, whether there is latitude for a range of possible patterns of response that may give rise to alternative cultures.

I propose a hypothesis of "environmental conditioning," by which culture is *conditioned*, but not necessarily or exclusively *determined*, by attributes of the natural environment. Humans, individuals as well as groups, are not automatons. Rather, they have innate personalities as well as antecedent experiences that lead to certain patterns of thought and behavior in preference to alternative patterns that may be as well (or, at times, even better) adapted to given circumstances, in the short run or in the longer run. This is especially true for nomadic groups that move from one environmental domain to another. They tend to bring to their new locations memories and habits and attitudes that formed in their earlier locations or in the course of their migration from one domain to another. The longer and more varied and intense the antecedent experience of a social group, the more likely it is to affect behavior in the new location. Moreover, in both its old and its new location, such a society must come into contact with neighboring groups that have their own cultures. Mergers and conflicts between groups may have profound and lasting effects on the coevolution or counter-evolution of interacting societies. So may the appearance of charismatic leaders who rise at crucial times and stamp their own personalities on the course of their nations' histories.

A further caveat is in order. Ascribing the course of a people's history to any single cause is, at best, simplistic. A case in point is "climatic determinism," a hypothesis that was in vogue in the early twentieth century and that overstated the importance of climatic factors in determining the character and fate of civilizations.[11] The histories of nations and regions are generally far too complex to be attributed entirely or primarily to a single factor, even a factor so encompassing as the environment. Any attempt to trace the way in which exposure to a variety of environments contributed to the evolution of a culture must not discount the importance of many social, economic, and political variables, as well as the characteristics of the particular people in interaction with those factors. Nor should it disregard the unpredictable and decisive role of uniquely inspired individuals (such as Moses, Paul, and Mohammad) who may appear at crucial historical moments and affect the subsequent course of events by the sheer force of their moral conviction and charismatic leadership.

To posit and explore the influence of the environments of the ancient Near East on the evolution of the culture of the Israelites is not tantamount to advocating pure and simple environmental determinism. There is no linear or automatic mechanism by which societies respond to a specifiable environmental stimulus. Societies are not machines. The process by which a culture evolves defies any attempt to encapsulate it in a direct cause-and-effect formula. A culture develops as a particular society seeks and chooses ways to respond to the opportunities offered by and the constraints imposed by a given environment. The more varied the environment, the more elaborate and less predictable are likely to be the possible modes of adaptive responses. A society exposed to a multiplicity of labile and risk-filled environments necessarily undergoes a highly complex process of evolution, affecting its economy and social structure, its attitudes and mores—indeed, its overall outlook and view of itself and the world. In other words, not all is determined by the environment, but much is influenced by it.

2

THE ECOLOGICAL CONTEXT

A Region of Disparate Domains

Elohim said, "Let the earth sprout vegetation." ... Elohim said,
"Let the waters swarm with living creatures, and birds that fly ... across
the expanse of the sky." Elohim created the great sea monsters, and all
the living creatures that creep. ... And Elohim blessed them all,
saying, "Be fruitful and multiply."

GENESIS 1:11, 20–22

THE TERM "ecology" derives from the Greek word *oikos*, which means "home" or "habitat" (abode), the place or environment within which a species or a community of species lives. Accordingly, ecology (from *oiko-logos*) is the study or science of the relationship between living beings and their habitats. Human ecology, by extension, is the study of how human societies interact with the Earth, including its soil, landforms, underlying mineral resources, overlying atmosphere, water (quantity, quality, and spatial and temporal distribution), climate, and entire panoply of organisms that share habitats. The historical study of human ecology considers three aspects of the reciprocal relationship between human civilization and the natural environment: first, how the environment shaped or conditioned the material and cultural development of a civilization; second, how a civilization viewed (understood or interpreted) the environment; and third, how a civilization affected (sustained, modified, or degraded) the environment over time.[1]

The region within which the Bible evolved, the ancient Near East (now generally referred to as the Middle East), encompasses parts of northeastern Africa and southwestern Asia. Ecologically, it constitutes an intermediate zone between the humid or subhumid environments of southeastern Europe and the hyperarid environments of the great desert belt that extends from the Sahara in the west through the Arabian Peninsula to the Thar Desert in the east. It was in parts of this intermediate zone that in the wake of the last ice age, which ended some ten to twelve millennia ago,[2] bands of modern humans first made the fateful change from a lifestyle based on nomadic hunting, gathering, and scavenging to one based on the sedentary cultivation of crops and transhumant herding of livestock. In the ancient Near East, they first domesticated plants and animals,[3] built villages and eventually cities, organized societies and states, invented ceramics and metallurgy, engaged in trade and periodic warfare, and developed systems of writing with which they recorded their myths, poems, religious tenets, historical events, commercial transactions, and legal codes.

The ancient Near East is thus our earliest cultural home, the birthplace of Western civilization.[4]

The cultural history of the ancient Near East is not a simple story of uniform steady development. Rather, it is a complex saga of fitful progress, beset by repeated crises and conflicts, during which various groups coevolved to form societies with distinctive cultures. The complexity of that process can be related, in part, to the inherently unstable environment within which it took place.

The environment of this region is extremely varied. Factors that contribute to its ecological heterogeneity are climatic zonation, latitude, proximity to the sea, topography, geology, soil types, as well as past and present modes and intensities of land use. The great variability makes it difficult to delineate the region's ecological domains in precise detail. Numerous habitat types can be discerned in each subregion, district, and subdistrict, depending on the spatial scale and the degree of specificity chosen to define them.

On an overall regional scale, however, we recognize five natural ecological domains, each of which spans a range of conditions. The irregular geographic configuration of these domains is such that they abut one another or interpenetrate in a patchwork pattern, so their boundaries are blurred. As each ecological domain gave rise to a characteristic culture, the closeness of the domains to one another forced the various societies to come into contact and interact, both materially and culturally. The interactions could be synergistic (marked by complementarity and mutual dependence) or antagonistic (defined by rivalry and conflict) or could alternate between the two.

The five ecological domains were (and still are) the humid highlands, the semiarid steppes, the river valleys, the seacoasts, and the deserts.[5] In addition to these natural domains, the Israelites experienced what can be regarded as two synthetic cultural domains: the urban and the exile domains.

The first natural ecological domain, the *humid highlands*, consists of the arc of mountain ranges that girds the so-called Fertile Crescent to the northwest, north, and northeast (including the Galilee, Lebanon, Taurus, and Zagros Mountains). These ranges face into the path of the cloud-bearing winds that sweep in from the west and northwest during the winter, and therefore they receive a comparative abundance of rain annually (from more than 40 inches [1000 mm] down to 16 inches [400 mm] or so). That rainfall gives rise to a relatively profuse cover of herbaceous and woody plants, among which are native species of grain-bearing grasses and legumes, as well as of trees and shrubs yielding edible fruits, that lent themselves to domestication. These plants include the progenitors of the crops that were domesticated by the region's early farmers, starting some ten to eleven millennia ago during the period called by archaeologists the Early Neolithic Age,[6] and thus it was in the humid highlands that regular farming first began.[7] In recognition of their climate and role in the history of civilization, the humid highlands are referred to henceforth as the *rainfed farming domain*.

The second domain, the *semiarid steppes*, consists of the undulating plains and low hills that lie in the rain shadow of the humid highlands. Here rainfall is of lesser amount (varying from 16 inches [400 mm] down to about 4 inches [100 mm]) and regularity, so rainfed farming is marginal, often too risky an enterprise to be practiced dependably year after year. However, the semiarid steppes do provide extensive vegetative resources that—although more sparse than in the humid highlands—can be utilized by grazing animals. Indeed, this is the subregion in which sheep, goats, and other species of livestock were domesticated and herded by tribes of seminomadic pastoralists at about the same time as plants began to be cultivated in the rainfed domain. Hence we refer to the semiarid steppes as the *pastoral domain*.

The third domain is the *river valleys* located in the semiarid and arid zones. These valleys receive the excess water flowing from the various catchments in the humid highlands. Because these runoffs originate from a zone outside the major river valleys, they are called exotic rivers. They tend to be seasonally variable, with the flows being most copious during the latter part of the rainy season (or, in some cases, during the season of snowmelt). We refer to these valleys as the *riverine domain*.

Some centuries after the domestication of crop plants in the rainfed domain, farmers discovered that they could import those plants into the riverine domain and grow them successfully along riverbanks and in floodplains following the recession of the annual floods. In time, farmers who settled along the rivers also learned to exercise greater control over the supply of water to their crops by means of artificial dikes, channels, and ditches designed to divert water from the rivers. Thus there arose the so-called hydraulic civilizations of the ancient Near East, which depended on irrigated farming for their subsistence.

The fourth ecological domain of the Middle East, the *seacoasts*, consists of the strips of land and water along the shores of the Mediterranean Sea, the Red Sea, and the Persian Gulf. Here the terrestrial and marine environments meet and interplay to form a unique ecological zone, with a characteristic community of interlinked aquatic and land plants and animals. Humans who lived along these coasts tended to become fishermen, seafarers, and land–sea traders. Eventually, some of them developed such characteristic pursuits as making glass from coastal sand, extracting dyes from sea snails, and building ships from the cedars, oaks, and cypresses that grew in the nearby highlands. Some settled permanently at advantageous sites—such as islands, coves, and estuaries—where they established seaports and maritime-trading centers. Others developed a form of maritime nomadism characteristic of the Mediterranean Sea Peoples, who combined coastal trading and raiding as means of subsistence.[8] The seacoasts compose the *littoral domain* or the *maritime domain.*

The fifth ecological domain is the *arid and hyperarid* zone, encompassing the vast dry lands that occupy the southern tier of the region. Here the relatively sparse population continued for a long while to maintain an austere

---

LIFE IN AN ARID ENVIRONMENT

Arid regions present a dual dilemma to human civilization: a restricted supply of water and a heightened requirement for water. The former is related to the paucity of rain; the latter, to the high evaporative demand imposed on plants and animals by the intense solar radiation and the warm and dry air. Arid areas are commonly defined as those where the potential evaporation (the average total seasonal evaporation that would occur if the surface were covered with free water) exceeds the average total seasonal precipitation (rainfall). A quantitative index of aridity is the ratio of potential evaporation to rainfall.

A definition based on average seasonal values, however, fails to account for another important feature of arid regions: the instability of climate. In general, the lower the rainfall in an area, the greater its variability. Annual rainfall in an arid region typically fluctuates between half and twice the long-term mean. A few seasons, occurring irregularly, may experience freak rainstorms (or cloudbursts), which tend to skew the average seasonal rainfall, making it higher than the most probable (mode) seasonal rainfall. Therefore, characterizing a region on the basis of its average rainfall may be misleading, as most seasons experience less than the average rainfall. Especially problematic is the incidence of drought. In an arid region, the occurrence of drought is a certainty. Only its timing, duration, and severity are in doubt.

Dealing with the variability of interannual rainfall is difficult. It is the tendency of humans to take full advantage of favorable years and to hope—or assume—that they will continue, thus ignoring the prospect that unfavorable years will come sooner or later. During seasons of abundant rainfall, pastures are lush and the flocks of pastoralists multiply. At the same time, the fields of farmers yield abundantly, so farmers tend to expand and intensify cultivation. During seasons of insufficient rainfall, pastures shrivel and fields lie bare. If the drought persists, overgrazing and overcultivating cause denudation of the land as a result of destruction of the native vegetation and erosion of the bare and pulverized soil from the action of wind and rain. These processes of degradation, now called *desertification,* can eventually reduce the productivity of an initially fertile area and leave the human population bereft of its basic livelihood.

continued

The problem is not merely material, but conceptual and spiritual as well. Nowadays, we think we understand, or at least accept as natural, the vagaries of climate. In ancient times, the fundamental scientific knowledge of climatic processes was lacking. People could not correlate the amount and distribution of rainfall with the movement of air masses, use synoptic maps to forecast weather, or monitor teleconnections with such distant phenomena as El Niño–La Niña cycles (to cite just a few examples of contemporary climate science).[1]

The objective reasons for the occurrence of favorable or unfavorable weather were a mystery to the ancients, so they could only conjecture what those reasons might be. What forces might be involved? The gods or the one God, of course. Favorable weather must be an indication of divine approval; unfavorable weather, of divine disapproval. Such notions seemed particularly attractive because they offered their adherents the possibility of affecting events. If they could find what pleased the gods (or God) and act accordingly (by means of sacrifice, prayer, or any other ritual), they might elicit better weather. Contrariwise, in the event of unfavorable conditions (diseases and pests, as well as droughts or floods), they would ask themselves what they had done to cause the disapproval. In this manner, concern about weather and other manifestations of nature was integrated into the ancients' worldview and religious faith.

These tendencies were especially pronounced in and crucial to the people—like the Israelites—who lived in the ecologically marginal frontier between the desert domain and the rainfed and riverine domains, where the climatic zonation shifted most capriciously.

---

1. Teleconnections are correlations between climatic phenomena that occur at locations some distance apart from each other. For example, during so called El Niño years, unusually wet conditions on the west coast of South America appear to be contemporaneous with unusually dry conditions in Indonesia.

existence as nomadic hunters and gatherers, availing themselves of the meager resources of the desert. Some also engaged in localized farming in isolated oases, and others herded ungulates in the thinly vegetated rangelands of the semidesert, especially in the vicinities of oases. Whenever drought became severe, some of the desert people banded together to conduct incursions into the adjacent pastoral and rainfed domains and consequently were greatly feared by the pastoralists and farmers there. In time, the denizens of the desert domesticated the camel, that hardy animal often called the ship of the desert. Thenceforth, some of them became caravaneers, conveying prized goods (including medicinal, aromatic, and spice plants, as well as minerals, gems, gums, and silk) from such distant locales as southern Arabia,[9] eastern Africa, India, and Central Asia to the centers of populations along the Mediterranean. These arid and hyperarid areas constitute the *desert domain*.

Each of the human societies that resided in its respective domain developed a distinctive mode of subsistence, social structure, and culture. An important feature in the culture of every one of the domains was the recognition and worship of the dominant natural forces that, in effect, controlled the course of human life and the welfare of society. The prevalent nature-based religions ex-

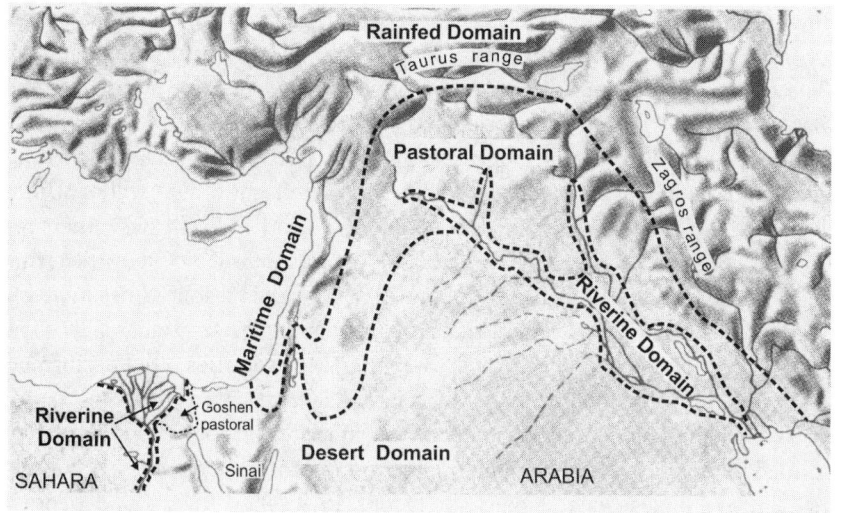

Principal ecological domains of the Fertile Crescent. Discernible are the arc of
the rainfed domain of Canaan and Lebanon in the west, the Taurus Mountains in
the north, and the Zagros Mountains in the east; the riverine domains of the Tigris–
Euphrates in the east and the Nile in the west; the semiarid pastoral domain in the
rain shadow of the rainfed highlands; the desert domain, which includes the Sahara
and the Sinai and Arabian Peninsulas; and the maritime domain of the eastern
coast of the Mediterranean Sea.

pressed themselves in characteristic beliefs and rituals. And because the landscape of the region is so full of contrasts, and the climate so variable, the inhabitants of the ancient Near East tended to view the forces of nature as being in
continuous conflict. Rainstorms alternated with searing desert winds; drench-
ing floods with droughts; periods of plenty with famines; good fortune with
misfortunes of every kind. It seemed as though these various phenomena were
set in motion by the actions of competing gods, some benevolent and some
malevolent. Hence the early polytheistic religions were directed toward propi-
tiating the provident forces and warding off the evil ones, in a constant quest
for attaining a measure of assurance in an environment that was inherently unstable, baffling, and insecure. Each domain had its chief god, imagined to be
attended by subordinate gods who were like members of his family and acted
as his servants and messengers. In time, every profession or class assumed its
own deity, such as the protecting god of sailors or miners or brick makers. Such
minor gods proliferated and eventually numbered in the hundreds, although
many of them overlapped in presumed function, if not in name.

In the rainfed domain, agricultural societies coalesced from separate clans
into villages, then into associated clusters of villages in particular districts, and

finally into city-states in which a nuclear settlement came to serve as a center of trade, artisanship, and worship for a hinterland of farming communities. The chief deities of these societies were generally a rainstorm god, such as Ba'al, and an earth goddess, variously named Ashera, Astarte ('Ashtoret), or 'Anat. In Canaanite agrarian societies, Ba'al was portrayed as a sky god riding the clouds, spewing forth or withholding his beneficent rain at will or whim. Ba'al's gift of rain was cast on the recumbent Mother Earth, who then responded with her fecundity. The process of life was thus depicted by the farming communities in the rainfed domain—in symbolic and sometimes in literal terms—as the primal mating of the sky god with the earth goddess. To elicit that process and promote the Earth's fertility, these societies often performed elaborate religious rituals that included sacrificial and sexual rites.

The pastoral societies differed from the agricultural societies in that they tended to emphasize and worship the brute and procreative prowess of dominant male animals, such as bulls and rams. However, as they, too, depended on seasonal rainfall, the pastoralists often worshiped a pantheon of animal gods and rain gods. And because pastoral pursuits were often integrated with

Ba'al, the storm god, wielding a club. This carved stela from the site of the city of Ugarit, in present-day Syria, dates to the Bronze Age (second millennium B.C.E.).

Ashera, the Canaanite fertility goddess. This carved figure dates to the Bronze Age or Early Iron Age.

An Egyptian version of the bull god, Apis,
after which the Israelites may have fashioned
the golden calf (Exodus 32).

rainfed farming, the agrarian communities reciprocally tended to worship animal gods as well.

The hydraulic societies of the riverine domain appealed to a panoply of deities, among whom were river gods, believed to control the annual floods. In rainless Egypt, where the connection between the annual spate of the Nile and its sources in faraway Ethiopia was obscure, one of the many deities was the popular river god Hapi. As the provider of the regular inundations, Hapi was portrayed in murals and sculptures as a rotund hermaphrodite with feminine breasts and masculine genitals, thus presumably endowed with the power of self-generation. In Mesopotamia, where the threat of soil salination by rising groundwater was felt most acutely, the good river deity was perceived to be countered by the lurking Tiamat, an evil goddess of the briny subterranean waters, who threatened to poison the Earth. Since animal husbandry was integrated with irrigated farming in the riverine domain, animal gods were worshiped by the hydraulic societies along with the river gods.

The coastal and island peoples who took to the sea worshiped their own version of the storm god, Ba'al, as well as the imaginary monsters of the deep (for example, Yamm, Tanin, and Leviathan), who seemed to have the power to lift and lower the tides, roil the waters, whip up awesome waves, and fan gale-force winds. When mollified, these awesome gods could provide calm seas, gentle winds, and safe passage. When enraged, they could sweep the ships of seafarers off course and lead them astray, smash them against rocky shoals, or swallow them up entirely. The people of the maritime domain, who depended on the sea for their livelihood, tended to establish cultic sites on high promontories or other conspicuous landmarks from which the sea could be observed and its mood predicted. Such landmarks, in turn, were visible from the sea and thus could serve to aid navigation.

Hapi, the hermaphroditic god of the Nile's inundations, offering a tray
of abundant foods. This carving is on a sandstone pillar at the temple
of Rameses II at Abydos, in Upper Egypt.

The desert people naturally worshiped their own deities—among them the
all-seeing god of the sun and the god or goddess of the moon,[10] the twain assumed to preside over the clear skies by day and by night, respectively. Desert
dwellers also honored the mysterious mountain gods who lived in the bowels
of the Earth and whose spirits lurked in caves. (These capricious gods, with
their pent-up powers, were thought to lie dormant for long periods of time,
only to awaken suddenly and cause the Earth to tremble, cliffs and boulders to
shatter, volcanoes to erupt, swirling winds to kick up dust that would blot out
daylight, and dry streams to gush forth with torrents of frothing water.) They
also tended to ascribe magical powers to the desert's venomous creatures,
which they revered as fetishes.

The societies in the various domains did not live in isolation from one another. Rather, they were aware of and influenced by the cultures of their neighbors.[11] The people of each domain often accepted the gods of other domains
as minor deities, worshiping them alongside their own major gods. In none
of the various domains, however, did monotheism arise spontaneously. This
view deviates from a prevalent notion that monotheism originated in the desert, influenced, as it were, by the seeming monotony, immutability, and sense
of eternity that pervades that landscape.

A necessary condition for the advent of monotheism was a realization that all natural phenomena—hence all the gods presumed to control them—were, in reality, manifestations of the overarching unity of nature and therefore of nature's creator. That insight evidently arose among the Israelites, very probably as a consequence of their exposure to the various environments and cultures and religions of all the disparate domains of the ancient Near East. Inklings of that realization had occurred before the Israelites enunciated it, but only with them did it reach its apotheosis in the form of a full-fledged doctrine of universal monotheism.

LINGUISTIC ASSOCIATIONS

To gain insight into the original meaning of the vocabulary of the Bible, we must step back in time and imagine ourselves in the world in which the peoples of the ancient Near East lived and developed their expressive languages.

Of particular relevance to understanding the Bible are the Semitic languages, which evidently had a common origin, but diverged over three or so millennia B.C.E. into several distinct variants. The languages of the Akkadians, Assyrians, and Babylonians are known as East Semitic (although they borrowed their systems of writing, the hieroglyphics, from the non-Semitic Sumerians). In contrast, the group of languages that includes Aramaic, Canaanite, Hebrew, Moabite, Phoenician, and Ugaritic is generally known as West Semitic. Arabic is considered to be a South Semitic language. These distinctions are somewhat arbitrary, however, as the related languages evolved not in isolation but in continuous interaction with one another, as well as with the non-Semitic languages of the region, including Egyptian, Hittite, Persian, and Sumerian.

The two main Semitic languages that have survived the vicissitudes of time and that are very much alive today are Arabic and Hebrew. Of the many words that are common to these languages, a number have acquired, over the centuries, a telling difference in nuance. Both languages originated with tribes whose mode of life was primarily pastoral. In time, the ancient Hebrew language, undoubtedly influenced by the language and culture of the Canaanites among whom the Israelites settled, adapted to serve the needs of a farming and urban community. Arabic, however, continued for many centuries to serve a desert or pastoral people. Although it eventually gained universality and became a highly sophisticated language of commerce, poetry, science, and philosophy, Arabic has retained elements of its original vocabulary and structure.

In both Arabic and Hebrew, the word *lahem* or *lehem* represents, in a generic sense, the essential food of the people. The staple food of hunters and pastoralists is meat, whereas that of farmers is grain raised in the field and its derivatives—flour and baked bread. Hence the word *lahem* in Arabic refers to meat, whereas in Hebrew *lehem* means "baked bread." (The place named Beth-lehem may have been a granary originally.) The words *lohem* (fighter) and *milhama* (war) in Hebrew may derive from *lehem*, food having been a frequent cause of conflict. The Hebrew word *herev* (sword), incidentally, is related to the Arabic word *harb* (war), the sword having served for many centuries as the main weapon of war.

Another interesting difference in nuance is associated with the word *ahal*, which in Arabic means "family." In Hebrew, the related word *ohel* means "tent." The association seems clear: in a pastoral society, a family normally shares a tent. The Hebrew term *shevet*

continued

(tribe) is related to the word for a shepherd's staff, which helps keep the flock together. The Hebrew word *malmad* signifies "cattle prod," which was a hooked stick used in lieu of a whip to urge on the oxen when plowing or riding in an ox-drawn cart. The letter *lamed* derives its shape from that implement. The terms *melamed* (teacher) and *lomed* (learner) are of the same root, implying that teaching is akin to prodding.

The Hebrew word *ḥoq* (law, rule, statute) is related to *ḥaqoq* (to carve, to inscribe), suggesting that laws were literally carved in stone (as were the Ten Commandments) or at least impressed into clay tablets that were later hardened. The related Arabic word is *ḥaq* (justice, truth).

The Hebrew word *adam* (man) is related to *adamah* (earth), *adom* (red), and *dam* (blood), as well as to *adim* (fertile component of the soil) in Arabic. The typical soil in Canaan, known in Latin as *terra rossa*, has a deep red color. It is called *ḥamra* (red) in Arabic, which is related to *ḥomer* or *ḥemar* (earthen material, clay) in Hebrew.

A very significant word in Hebrew is *geshem* (body, substance). It is related to the Arabic term *gism*, which carries the same essential meaning. However, in the Bible and in subsequent Hebrew usage, the word is used in lieu of *matar*, which in both languages means "rain." Why should "substance" imply "rain"? The answer is that the ancients did not fully understand the process that causes invisible vapor to condense into clouds and then to fall as rain. Nor did they see how rain-bearing clouds could form spontaneously. In the absence of physics, they turned to faith. Rain, they believed, was the substantiation of God's grace, a manifestation of his generosity. So a euphemism became identified with the most important natural gift, for which the people prayed constantly. The Bedouin, incidentally, sometimes refer to rain as Raḥmat Allah (God's mercy). Biblical Hebrew includes several words for rains of different intensities: *matar* is a substantial downpour; *revivim* is a gentle shower, as are *se'irim*, *delef*, and *zarzif*; *yoreh* (cast down) is the early rain, whereas *malqosh* (late) is the end-of-season rain; and, finally, *tal* is the dew that freshens the plants in the morning.

AS TOLD in the Bible, the Hebrews originated as a single clan in the riverine domain of Mesopotamia, from where they embarked on a venture of serial migrations. The chronology of their origin and early wanderings is uncertain, but those who give credence to the biblical account estimate that it may have taken place during the early or middle part of the second millennium B.C.E.[12] The biblical progenitors of the Israelites, the Patriarchs and Matriarchs (described in Genesis), moved to the pastoral domain in eastern and southern Canaan, where they herded flocks of goats and sheep. Then, impelled by a severe and prolonged drought, they migrated to the western riverine domain of Egypt, where—according to the dramatic story that was indelibly imprinted in their ancestral memory—they sojourned (and were enslaved) for several generations.

As their story continues, the Israelites left the riverine domain of Egypt at the time of the Exodus (variously estimated to have occurred between 1300 and 1200 B.C.E.) to wander in the desert domain of Sinai and the Negev, where they purportedly lingered for some forty years.

After the Exodus from Egypt and the wanderings in Sinai, the Israelites entered Canaan from the east and strove to settle in its pastoral and rain-fed domains. As they spread throughout the hill country and approached the coastal plain, they inevitably encountered another invading group: the Philistines.[13] The latter were an offshoot of the Sea Peoples of the maritime domain, who had raided the coasts of the eastern Mediterranean and settled along the southern coast of Canaan at about the same time as the Israelites had migrated into central Canaan, apparently during the twelfth century B.C.E. As the Israelites moved westward toward the coast while the Philistines moved eastward from the coast, the two nations inevitably met and clashed. At first, the Philistines appear to have had the advantage, thanks to their technological superiority, having entered the Iron Age before the Israelites acquired the resources and skills to do so. Within a few generations, however, the Israelites apparently were able to catch up with and even surpass the Philistines and overpower them.

Having wandered repeatedly from one part of the Near East to another, and having experienced all the major domains in the region, the Israelites absorbed elements of each domain's culture and religion while retaining their distinctiveness. Rather than adopt the religion of any of them, they integrated all those influences and gave them a new form and meaning. Their originality was manifested in the coalescence of observations and ideas and in the interpretation given them.

The concept of one omnipresent and all-powerful force, a great common denominator unifying the entire realm of nature, was a radical departure from the prevailing polytheistic view. Its genesis could not have been a sudden, unanimous, and irreversible revelation. Instead, it must have been a painstaking process, the monotheistic idea first dimly perceived by a few and resisted by the many, and then accepted only gradually and reluctantly, after repeated advocacy by religious and political leaders in successive generations. There is abundant evidence of that process in the exhortations of the Hebrew prophets against those who were all too prone to revert to the worship of the locally popular nature gods. Eventually, however, the unified vision of God and nature took hold until it became a major ideological force that, over time, transformed the cultural foundations of not merely the region, but indeed the entire Western world.

The wanderings of the early Israelites throughout the domains of the ancient Near East and their encounters with practically all its cultures were necessary preconditions for them to conceive the notion of monotheism. A necessary condition, however, is not always a sufficient condition. The readiness to accept a new paradigm of the deity would not by itself have brought about the establishment of a new religion, were it not for the historical convergence of the people's mental preconditioning with the imperative of their society to meet the challenges imposed on it by external conditions.

The culminating experiences that gave sharp focus to a vague ideology and concrete form to a new religion apparently did not take place in any of the five natural domains, but subsequently in two additional cultural domains: the *urban* and the *exile* domains. As with a chemical reaction, the presence of the right elements may not precipitate a new compound unless the reaction is catalyzed by a generative charge—a spark, as it were—that impels those elements to conjoin. Judaic monotheism began to coalesce into a formally organized, coherent religion in the city of Jerusalem, the capital of the Davidic dynasty. The consummation of monotheism was prompted by the urgent needs of a fledgling, beleaguered monarchy, for which a unifying principle was an essential instrument of state policy aimed at rallying a loose collection of bickering tribes and forging it into a cohesive nation motivated and mobilized to fend off its enemies. At a crucial moment in history, a monarchy needing a centralizing appeal and a nation willing to rally to it converged. That convergence incorporated the inchoate memories of the wanderings and deeds of the nation's ancestors, the ideological exhortations of visionary prophets, and the ritual practices formalized by the hereditary priesthood to constitute a unique national religion.

Still, the process was not complete. The political and ecclesiastical imperative of unity did not eliminate all vestiges of indigenous polytheism. We know this from the repeated and vehement pronouncements of the Yahwistic prophets and from the historical explanations given by the monotheistic writers of the Bible for the nation's continuing insecurity and recurrent misfortunes.

The Israelite tribes remained in Canaan for some centuries, but never united completely. As a result of internecine divisions, their nation split into the northern kingdom of Israel, with its capital in Samaria, and the southern kingdom of Judah, with its capital of Jerusalem. Consequently, each half-nation was so weakened as to be unable to withstand the onslaught of powerful invaders. The leadership of the northern kingdom was expelled from Samaria by the Assyrians in 720 B.C.E., and the kingdom of Israel thereafter ceased to exist. The leadership of the southern kingdom eventually met a similar fate, meted out by the Babylonians (who had succeeded the Assyrians) in 586 B.C.E. They destroyed Solomon's Temple, deposed the Davidic monarchy, and exiled the people of Jerusalem and its environs.

That disaster could have spelled the end of the southern nation of Judah, as it had the northern nation of Israel. Instead, by a most improbable anomaly, the Judean nation survived. The greater the disaster, it seems, the greater the need to cling to faith. The destruction of the Temple took the practice of the religion (based, as it was, on ceremonial sacrifices) out of the exclusive control of the priestly caste and forced a reexamination and redefinition of the ethical and spiritual basis of the Yahwistic creed. To replace the elaborate and gory Temple-centered rituals, the nation's spiritual leaders created a portable "virtual Temple" that would continue to draw the loyalty and summon the faith of the traumatized nation.[14]

The necessary function was served by the composition of a Holy Writ, embodying the history and destiny of the people, believed to be ordained by and dedicated to the service of Yahweh. Since the reading of the Writ was, in principle, accessible to all literate members of the community everywhere (not merely to a hereditary priesthood in a certified location), it began to have a democratizing and universalizing effect. And so it was that the very misfortune of exile prompted the nation's spiritual renewal—leading not only to its physical and political restitution, but also to its cultural outreach far beyond its own ethnic or geographic confines.

The process of compiling the Bible required a heroic intellectual effort—indeed, a labor of love and of faith—aimed at the collection, correlation, and recomposition of the disparate, fragmentary oral memories and written scrolls that had been preserved from earlier centuries in the different domains.

The faith in Yahweh was vindicated and reinforced when the Babylonian Empire itself was vanquished, scarcely two generations after the exile from Jerusalem, by the rising power of Persia in the east. The Persians then allowed a substantial number of the exiles from Judah to return to their homeland and to their city of Jerusalem, there to revive their national and religious life. Although the Temple was reconstructed after the return from exile and the priests restored the traditional rituals of animal sacrifice during the Second Temple period, they could not regain their monopoly over the practice of the religion. Instead, the learned rabbis, drawing from the spiritual and ethical legacy of the prophets of old, played an increasingly creative role in the subsequent religious life of the Jewish people.

The effort to complete the Bible took place over the next few centuries. All the material was redacted so as to constitute a unified book, which was eventually formalized and sanctified.[15] The product of that effort, the words inscribed on soft parchment, became the core of the nation's faith and culture, more solid and indestructible than any temple of stone could ever be.

3

THE FIRST RIVERINE DOMAIN

Influence of Mesopotamia

A river went out of Eden to water the garden.
GENESIS 2:10

A STRONG INFLUENCE on the cultural development of the Israelites emanated from Mesopotamia,[1] the reported birthplace of the first Hebrew ancestors. The very word "Hebrew" ('Ivri) appears to derive from the verb *'avor*, which means "come across." Accordingly, the Hebrews ('Ivrim) were so called because they were said to have crossed the great river—the Euphrates[2]—at the outset of their long and tortuous trek across the various domains of the ancient Near East to where they eventually settled: the land of Canaan, between the Jordan River and the Mediterranean Sea. Thus the influence of Mesopotamia may have had its origin in the earliest ancestral period.

That initial impact was reinforced as a consequence of the conquest and exiling of the Samaria-based kingdom of Israel by the Assyrians in 720 B.C.E. and, even more so, the conquest and exiling of the Jerusalem-based kingdom of Judah by the Babylonians in 586 B.C.E. Their period of exile brought the Israelites once again into direct and intimate contact with Mesopotamian culture, and the riverine domain in which it evolved, significant elements of which were absorbed by them and are reflected in several books of the Bible.

IN THE second half of the fifth millennium B.C.E., after the art of farming had been established in the rainfed domain of the Near East, a group of uncertain origin colonized the lower courses of the Tigris and Euphrates Rivers. Their neighbors to the north called them Sumerians; they referred to themselves as

A canal-watered plantation in ancient Assyria, being an idealized depiction of a "Garden of Delight" (as in the biblical *gan eden* [Genesis 2:8–10]).

"the dark-headed" and to their country simply as "the land." Sumer was a flat plain of brown alluvium, dusty when dry and miry when wet, swept by desiccating desert winds[3] and deluged periodically by sudden torrential overflows of the twin rivers.

The first settlements were clumps of huts in marshes along the watercourses of the lower Euphrates, much like the reed huts of present-day marsh dwellers who inhabit some parts of the area. With ingenuity and diligence, the Sumerians gradually transformed their land from a seasonally parched plain with interspersed swamps into a garden of extensive grain and forage fields and date-palm plantations. A Sumerian myth refers to the introduction of cereals from the distant highlands. Another alludes to farmers as "men of dikes and canals." The canals served as waterways and fishponds; they conveyed water from the rivers to groves and grasslands that furnished feed for sheep and cattle, as well as to grain fields of legendary fertility.[4]

The surplus production of their farmers enabled the Sumerians to develop the world's first urban society. Their earliest cities were small mud settlements. Later, houses were built of mud bricks joined by bitumen. Later still, the Sumerians used baked bricks (heated and fused to stone-like hardness) to construct their houses and temples. The latter were towers built on multistoried terraces. In time, the cities developed into city-states, ruled by dynasties. Among the Sumerian cities were Eridu, Uruk, Nippur, Lagash, Kish, and Ur.[5]

At this early stage, during the second half of the fourth millennium B.C.E., the Sumerians made their greatest contribution to the advance of civilization: the invention of writing, to facilitate record keeping for land management and administration. Using clay tablets, the Sumerians wrote with a reed stylus, the tip of which was pressed into the clay to make wedge-shaped marks—a script now called cuneiform, after the Latin word *cuneus* (wedge). The patterns formed by those marks were schematic representations of objects used to signify syllables, objects, or ideas. When baked in a kiln, the clay tablets hardened

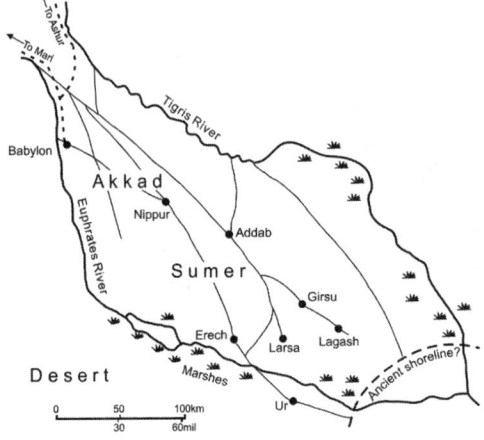

The courses of the Tigris and Euphrates Rivers in southern Mesopotamia—the land of Sumer and Akkad—showing the presumed locations of the ancient coast of the Persian Gulf and of the channels of diversion canals.

and became practically permanent (although brittle) records. Thousands of them, in various states of preservation, have been discovered and deciphered by archaeologists. Some of the records pertain to agriculture, whereas others are literary works, mythological epics, and mundane transactions.

Western civilization owes much to the Sumerians. They developed sailboats, wheeled vehicles, the potter's wheel (the first industrial machine with continuous rotary motion), yokes for harnessing animals and animal-drawn plows, machines with weighted levers for lifting water, accounting procedures, literature (including epics and love songs), and law codes. Notwithstanding its achievements, the Sumerian civilization as such did not last. As the Sumerians began to decline, during the third millennium B.C.E., the Akkadians, who spoke a Semitic language, gradually learned the ways of the Sumerians and eventually superseded them.

Toward the middle of the twenty-fourth century B.C.E., Sargon the Great, an Akkadian minister of the Sumerian king of Kish, gained control over Sumer

A seeder plow, an important practical invention, depicted on an ancient Mesopotamian seal. The seeds were dropped through a funnel-shaped tube into the bottom of a freshly opened furrow and then covered by loose soil to ensure germination.

Levered devices used to draw water
from canals to irrigate cropland
in the riverine domain of ancient
Mesopotamia.

and established the first known empire. He shifted the center of his power
northward, up the valley of the twin rivers. After a time, Akkad, too, began to
decline. The end of the third millennium witnessed the rise of the Amorites,
who, like the Akkadians, were a Semitic people. Around 1700 B.C.E., Hammurabi consolidated his power and made the city of Babylon the hub of an empire
that extended from the Persian Gulf northwestward almost to the Mediterranean Sea. But the hegemony of the Babylonian Empire was also short-lived.
It was replaced by a Cassite dynasty from the Zagros Mountains that gave way
to the rising power of the kingdom of Ashur, centered on the Tigris River in
northern Mesopotamia. For a time, the seat of power shifted back to Babylon,
until the late sixth century B.C.E., when the riverine kingdoms of Mesopotamia were supplanted by the rising power of Persia.

PRIMEVAL NATURE is depicted in Mesopotamian mythology as a monstrous
chaos that was to be overcome by the constant labor of people, supported
by patron gods who encouraged the works that would establish order and
regularity. Thus the mythical feat of the hero-god Enlil, or Marduk, in slaying the goddess of chaos, Tiamat, and creating the world from her sundered
body reflects the work of the Mesopotamians in reclaiming the swamps, raising their cities above the floodplain, and subduing the capricious Tigris and
Euphrates Rivers by means of levees and canals. Myths about the gods offered
explanations of baffling phenomena and a measure of assurance in the face of
nature's threatening vagaries.

The epic poem *War of the Gods* (*Enuma elish*), a Babylonian myth of creation, tells of the beginning of the world and of the building of the great city

COMPARISON OF MESOPOTAMIAN AND MOSAIC LAWS

Much has been written about the possible influence of Mesopotamian legal codes on Mosaic Law. The Code of Hammurabi has featured prominently in this regard. An Amorite (of pastoral Semitic origin), Hammurabi reigned as king of Babylonia in the eighteenth century B.C.E. His legal code was carved in the Akkadian language on a diorite stele, topped by a figure of the king being appointed to the task by the sun god Shamash. (The name Shamash is embedded in the word for "sun" in the Semitic languages, including *shemesh* in Hebrew and *shams* in Arabic.) The stele was discovered in 1901 by French archaeologists in the Elamite capital, Susa (the city of Shushan mentioned in the Books of Esther and Daniel),[1] to where it apparently had been carried as a trophy of war. Other copies of the code were later discovered in several locations.

King Hammurabi of Babylonia (1792–1750 B.C.E.) receiving the laws from the seated sun god, Shamash.

Many of the laws were clearly grounded in the specific circumstances that characterized riverine southern Mesopotamia in the early second millennium B.C.E., while others were less narrowly focused on riverine conditions and more universally applicable. Some of the laws in that category were remarkably similar to laws in the Mosaic code. The following are but a few examples:

> If a seignior accused another seignior and brought a charge of murder against him, but has not proved it, his accuser shall be put to death. (Hammurabi)

> If a man appears against another to testify maliciously and gives false testimony against him, the two parties to the dispute shall appear before Yahweh, before the priests or magistrates in authority at the time, and the magistrates shall make a thorough investigation. If the man who testified is a false witness, if he has testified falsely against his fellow, you shall do to him as he schemed to do to his fellow. (Deuteronomy 19:16–19).

> If a seignior stole either an ox or a sheep or an ass or a pig or a boat . . . if it belonged to a private citizen, he shall make good tenfold. If the thief does not have sufficient to make restitution, he shall be put to death. (Hammurabi)

> If you see your fellow's ox or sheep gone astray, do not ignore it; take it back to your fellow. If your fellow does not live near you or you do not know who he is, you shall bring it home and it shall remain with you until your fellow claims it, then you shall give it back to him . . . you must not remain indifferent. (Deuteronomy 22:1–3)

> If a seignior has stolen the young son of another seignior, he shall be put to death. (Hammurabi)

> He who kidnaps a man—whether he has sold him or is still holding him—shall be put to death. (Exodus 21:16)

If a son has struck his father, they shall cut off his hand. (Hammurabi)

He who strikes his father or his mother shall be put to death. (Exodus 21:15)

If a seignior has destroyed the eye of a member of the aristocracy, they shall destroy his eye. If he has broken another seignior's bone, they shall break his bone. If he has destroyed the eye of a commoner or broken the bone of a commoner, he shall pay one *mina* of silver. If he has destroyed the eye of a seignior's slave or broken the bone of a seignior's slave, he shall pay one half his value. If a seignior has knocked out a tooth of a seignior of his own rank, they shall knock out his tooth. If he has knocked out the tooth of a commoner, he shall pay one-third *mina* of silver. If a seignior has struck the cheek of a seignior superior to him, he shall be beaten sixty times with an oxtail whip. (Hammurabi)

If anyone maims his fellow, as he has done so shall it be done to him: fracture for fracture, eye for eye, tooth for tooth. The injury he inflicted on another shall be inflicted on him. (Leviticus 24:19–20)

The similarities may well be due to the influence of antecedent Mesopotamian culture on the development of Israelite culture and laws. The differences, however, are at least as significant as the similarities. In some instances, the Code of Hammurabi is less humane than Mosaic Law, as it mandates vengeful punishment, a principle called *lex talionis*:

If a man has struck the daughter of a free man and caused her to cast that which was in her womb, he shall pay ten shekels of silver. If the woman died as a result, they shall kill his daughter.

The edicts of Hammurabi are not grounded in a universal moral philosophy, ensuring the rights of all classes.[2] Although the early parts of the Bible recognize slavery, they specify the rights of slaves to humane treatment and—even more important—to freedom at the end of seven years. Moreover, the Bible repeatedly reminds the Israelites that they themselves had once been enslaved in Egypt and must therefore treat the disadvantaged (widows, orphans, and strangers) with justice and compassion.

---

1. The 8-foot (2.5 m) black diorite stele was originally placed in the temple of Esagila in Babylon, Hammurabi's capital. In the twelfth century B.C.E., it was taken to Susa by the Elamite king Shutruk-Nah. It was transported by its French discoverers to Paris nearly a hundred years ago and has been exhibited at the Louvre ever since. Written in cuneiform Akkadian, the Code of Hammurabi consists of 282 legal paragraphs, about 40 of which were erased (although some of them have been restored from other tablets found elsewhere). The edicts were preceded and followed by lofty statements of purpose (to promote justice and public welfare), along with a series of blessings for the faithful and curses against the malevolent. The stele also included effusive self-praise by the king.

2. Other law codes have been discovered in the area of Mesopotamia, some of which even predate the Code of Hammurabi. Among them are the Codes of Lipit-Ishtar, Ur-Nammu, and Eshnunna. Still another set of laws was found in the early twentieth century in Ashur, the capital of Assyria, dated to about 1100 B.C.E. A noteworthy feature of these Codes is the lowly status of women, the supremacy of men in marriage, and the harshness of punishments for women's presumed transgressions: "A man may flog his wife, he may pull out her hair, he may damage and split her ears. There is nothing wrong in this." And in cases of embezzlement by a wife in collusion with a slave: "If either a slave or a slave-girl has received anything from the hand of a man's wife, they shall cut off the nose and ears of the slave or slave-girl. . . . The man shall cut off his wife's ears."

Nothwithstanding its great achievements, the civilization that began in southern Mesopotamia did not last. Gradually, the center of the culture in that riverine domain tended to shift northward. What caused that shift? It seems to have been due largely to the decline of irrigated agriculture in the very area where it originated.

The diversion of river water onto the valley land led to a gradual process of soil degradation. The first aspect of the process was the accumulation of silt. Early in history, the upland watersheds of the Tigris and Euphrates Rivers were deforested and overgrazed. Erosion resulting from the seasonal rains in the uplands proceeded to strip off the denuded soil and pour it into the streams, which, in turn, carried it as suspended sediment hundreds of miles southeastward. As the silt-laden floodwaters wound their way toward the lower reaches of the valley, more and more of the sediment settled along the bottoms and sides of the rivers, thus raising their beds and their banks above the adjacent plain. Rivers that are elevated above their floodplains are inherently unstable: during periodic floods, they tend to overflow their banks, inundate large tracts of land, and from time to time change course abruptly. The silt also tends to settle in channels and fields, and thus to clog irrigation works.

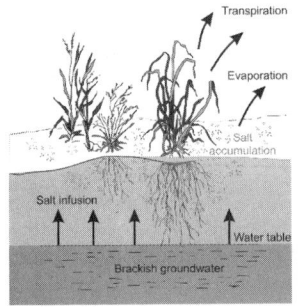

The processes of waterlogging and soil salination. The rising water table in poorly drained land saturates the soil, thus restricting aeration of the root zone. As soil moisture evaporates at the surface, it gradually infuses the topsoil with increasing concentrations of salts that eventually impede the growth of crop plants.

The second and even more severe component of the degradation process was soil salination. All water applied to the land contains some salt, and as the water seeps downward, it tends to dissolve additional salts originally in the soil and subsoil of arid regions. The flood irrigation of low-lying land causes large amounts of water to percolate into the soil from the canals and irrigated fields. In arid regions, rainfall is insufficient to leach out the salts from the soil. Moreover, the lateral underground drainage in a low-lying valley is typically too slow to dispose of the excess seepage caused by flood irrigation. The practically inevitable result is that the water table rises inexorably until it comes close to the surface. When that happens, the soil becomes waterlogged, the growth of crop roots is restricted, and evaporation tends to concentrate salts at the soil surface until the soil is rendered sterile. Initially, and for quite some years or even generations, the process of groundwater rise is invisible and may go unnoticed. But it is relentless, and its insidious effect is to doom an entire thriving district to eventual failure.[1]

The irrigation-based societies of ancient Mesopotamia had to contend with the inherently unstable hydraulic regimens of the twin rivers. The civilization of Mesopotamia thus depended on the complex dynamic interplay among the unseen rain and snow falling on the distant upper watersheds, the resulting flooding of the rivers (especially in spring), and the equally mysterious upsurge of the briny groundwater in the lower valley.

The Mesopotamians' perception of nature, expressed in their mythology and cosmology, naturally reflected the concerns of their societies about the capricious hydrological phenomena that determined their fate.[2]

1. The modern remedy for the twin scourges of waterlogging and salination of irrigated lands in river valleys is artificial drainage by means of ditches or perforated pipes placed below the water table. The ancient Mesopotamians lacked the knowledge and the means necessary to drain their land. Their only recourse was the practice of fallowing their fields periodically, which may have helped delay salination, but could not by itself prevent it in the long run. See M. Artzy and D. Hillel, "A Defense of the Theory of Progressive Salinization in Ancient Mosopotamia," *Geoarchaeology* 3 (1988): 235–238.

2. R. McC. Adams, *Heartland of Cities: Surveys of Ancient Settlements and Land Use on the Central Floodplain of the Euphrates* (Chicago: University of Chicago Press, 1981).

of Babylon under the protection of its god, Marduk. It begins with the words: "When the skies above were not yet named, nor earth below pronounced by name."[6] Initially, there were just two gods: Apsu, who represented the primordial waters under the Earth, and Tiamat, who personified the briny waters of the sea. They begot four generations of gods who fought with one another until Marduk came into being, superlative in every way. Tiamat was jealous and plotted to kill him. A terrible struggle ensued, from which Marduk emerged victorious. He trampled the lower part of Tiamat's body and then sliced her in half "like a fish for drying." From one half of her, he made a roof for the sky; from the other half, he made the Earth, which confined the subterranean waters below. Finally, Marduk proceeded to organize the rest of the universe, the months of the year, the moon, the clouds, and—of course—the Tigris and the Euphrates. Much of this is echoed in the two biblical creation stories, found in Genesis 1 and 2.

Part of the Babylonian myth of creation *War of the Gods* (*Enuma elish*), inscribed in cuneiform on a clay tablet from ancient Mesopotamia (ca. 700 B.C.E.).

The Bible begins with the words "When Elohim began to create heaven and earth—the earth being unformed and void, with darkness over the surface of the deep and a wind from Elohim sweeping over the water—Elohim said, 'Let there be light'" (Genesis 1:1).

Among the obvious parallels with the Mesopotamian myth is the preexistence of primordial waters filling the abyss, which in Hebrew is called *tehom*—a word similar to the name of the goddess of chaos, Tiamat. Of the thirty-six mentions of the word *tehom* or *tehomot* (the plural form) in the Bible, nearly all pertain to the "waters of the deep." Another strikingly similar theme is the separation of the sky from the ground, in the course of which the waters were divided into two distinct bodies: the waters on or below the Earth and the waters above the firmament: "Elohim said: 'Let there be an expanse in the midst of the water, that it may separate water from water.' Elohim made the expanse, and divided the water which was below the expanse from the water which was above the expanse. And it was so. And Elohim called the expanse sky" (Genesis 1:6–8).

The common translation of the Hebrew word *shamayim* (Genesis 1:1) is "sky," which misses the verbal association of the original: *sham-mayim* suggests, literally, "there is water" or "source of water." The Hebrew word translated into "expanse" (or, in alternative translations, "firmament") is *raqi'a*, which connotes "something hammered out." The verbal image suggests that God fashioned a curved canopy (as though by hammering or stamping a sheet of malleable metal, such as copper) and placed it at some height above the Earth. This imagined canopy was presumed to hold the reservoir of water stored above it, from which God could release rain at will by opening "the floodgates of the sky" (Genesis 7:11). (The ancients evidently did not perceive that the water descending from the sky is recycled water from the Earth, transferred to the atmosphere in the form of invisible vapor that later condenses into clouds and precipitates as rain; rather, they apparently thought that evaporated water disappeared entirely.)[7]

The second creation story in the Bible begins with the words "Such is the story of heaven and earth when they were created. When Yahweh Elohim made earth and heaven—when no grasses of the field had yet sprouted, because Yahweh Elohim had not yet sent rain upon the earth and there was not a man to till the soil" (Genesis 2:4).

The need for a human to make the ground productive was a notion clearly derived from Mesopotamia, where the valley was mostly barren until the Sumerians built diversions from the Tigris and Euphrates Rivers to water the land. So Yahweh Elohim created a man and planted a garden. But the garden had to be watered. Again, following the Mesopotamian model, the creator provided river water: "A river issues from Eden to water the garden, and it then divides and becomes four branches" (Genesis 2:10). The names assigned to the four rivers are significant. The first two—Pishon, which "winds through the whole land of Havilah, where the gold is . . . and bdellium is there, and lapis lazuli," and Gihon, which "winds through the whole land of Cush"—may pertain to Africa, perhaps to the two major tributaries of the Nile.[8] The other two rivers are explicitly Mesopotamian: the Tigris and the Euphrates (Genesis 2:11–14).

Notwithstanding the many similarities between the Mesopotamian epic and the biblical creation stories, there are fundamental differences. One is, of course, the multiplicity of gods in the epic and the unity of God in the Bible. In Genesis, God—whether called Elohim alone (in the P version of Genesis 1) or Yahweh Elohim (in the J version of Genesis 2)[9]—was the singular, uncontested, almighty power from the very start. God's labor of creation was not a cataclysmic battle between opposing gods (one representing chaos and the other, the creative force). In the first biblical account of creation, chaos was merely a depersonalized initial condition of fluid disorder that God proceeded to organize in an orderly, methodical manner. In the second biblical account of creation, earthen material and water vapor were the preexisting resources, out of which God created the human and planted the Garden of Eden. There are other differences as well: in the second biblical account, the human was fashioned out of 'afar, the material of the soil, whereas in *War of the Gods*, Marduk formed humankind out of the blood of the rebellious god Kingu, killed as a traitor because he had taken the wrong side in the cosmological conflict between Marduk and Tiamat.

THE FLOOD PHENOMENON was distinctly riverine, an annual occurrence in both Mesopotamia and Egypt. Most of the time, a flood was controllable and beneficent. Occasionally, however, especially in Mesopotamia, a flood could be violent and devastating. The biblical description of the Great Flood suggests a dual phenomenon: simultaneously, water rose from below and fell as rain from above: "All the fountains of the great deep burst apart and the floodgates of the sky broke open" (Genesis 7:11).[10] The Flood ended when the two sources were stopped: "The fountains of the deep and the floodgates of the sky were stopped up, and the rain from the sky was held back" (Genesis 8:2). Evidently, the "fountains of the deep" (which hydrologists call groundwater) and the "floodgates of the sky" (which meteorologists call precipitation) were considered to be two unconnected bodies of water that—either separately or in combination—were subject to the command of God.

The story of a catastrophic flood had its clear antecedents in Mesopotamian mythology. The famous *Epic of Gilgamesh*, discovered in the ruins of Nineveh

Fragments of the *Epic of Gilgamesh*, parts of which presage the story of Noah and the Flood (Genesis 6–8), inscribed in cuneiform on clay tablets found at the site of the ancient city of Ashur, in present-day Iraq.

and elsewhere in Mesopotamia, has been dated to around 2000 B.C.E. and thus predates the Bible by many centuries. Its protagonist, Gilgamesh, king of Uruk, saddened and frightened by the death of his friend Enkidu, embarked on a quest to avoid his own mortality. Gilgamesh sought out his venerable ancestor Utnapishtim, who, along with his wife, was the only living witness to the Great Flood and subsequently had been granted immortality. (The agent of immortality was a magical plant, just as was the Tree of Life in Genesis 3.)

Utnapishtim then told Gilgamesh the story of the flood:

> When their heart led the great gods to produce the flood . . . their words he repeats to the reed hut. . . . Man of Shuruppak [a city on the Euphrates] . . . build a ship! . . . Aboard the ship take thou the seed of all living things. . . . Her dimensions shall be to measure . . . ten dozen cubits the height of each of her walls. . . . Thou shall ceil her. . . . Ten dozen cubits each edge of the square deck. . . . Six measures of bitumen I poured into the furnace . . . I poured inside.[11]

Then came the flood, which returned the whole world to a state of fluid chaos. Only Utnapishtim and his retinue of animals survived. When the flood began to abate, he sent out scouts, beginning with a dove:

> The dove went forth, but came back; since no resting place for it was visible, she turned round. Then I sent forth a swallow. The swallow went forth, but came back; since no resting place for it was visible. . . . Then I sent forth and set free a raven. The raven went forth and, seeing that the waters had diminished, he eats, circles, caws, and turns not round. Then I let out all to the four winds,

and offered a sacrifice. . . . The gods smelled the sweet savor, the gods crowded like flies about the sacrificer.[12]

The *Epic of Gilgamesh* extends over three thousand lines and is written on twelve tablets. These excerpts reveal some of the parallels to the biblical story of Noah. But there are also important differences, the most obvious pertaining to the purported reason for the flood. In the polytheistic Mesopotamian tradition, the event was the whim of the capricious gods. The biblical account, typically, introduces a moral dimension. The Flood was the divine punishment for the corruption of humanity:

> Noah was a righteous man. . . . Noah walked with the Elohim. . . . [But] the earth became corrupt before the Elohim; the earth was filled with lawlessness . . . for all flesh had corrupted their ways on earth. Elohim said to Noah: "I have decided to put an end to all flesh." (Genesis 5:9–13)

> Yahweh saw how great was man's wickedness on earth, and how every plan devised by his mind was nothing but evil all the time. . . . And Yahweh said, "I will blot out from the earth the man whom I created—men together with beasts. . . ." But Noah found favor with Yahweh. (Genesis 6:5–8)[13]

The similarities between the Mesopotamian and biblical stories, however, are very striking: "Make yourself an ark of gopher wood . . . with compartments . . . and cover it inside and out with pitch. . . . [14] The length of the ark shall be three hundred cubits, its width fifty cubits, and its height thirty cubits. . . . And of all that lives . . . ye shall take two of each into the ark, to keep alive with you; they shall be male and female" (Genesis 6:14–15, 19).

After the rain stopped and the floodwaters receded, "Noah opened the window of the ark . . . and he sent out the raven, and it went to and fro until the waters had dried up from the earth. Then he sent out the dove to see whether the waters had decreased from the surface of the ground . . . the dove came in to him toward evening, and there in its bill was a plucked-off olive leaf! Then Noah knew that the waters had decreased on the earth" (Genesis 8:8–11).[15]

When Noah and his family and all the animals on the ark emerged, "Noah built an altar to Yahweh, and took of every clean beast, and of every clean fowl, and offered a burnt-offering on the altar. And Yahweh smelled the pleasing odor,[16] and Yahweh said to Himself: 'Never again will I doom the earth because of man, since the devisings of man's mind are evil from his youth; nor will I ever again destroy every living being . . . so long as the earth endures, seedtime and harvest, cold and heat, summer and winter, day and night shall not cease'" (Genesis 8:18–22).

Following the sacrifices, Elohim entered into a covenant with Noah and his offspring, as well as with all the Earth's creatures, to refrain from unleashing another flood. As a sign of his promise, Elohim stated: "I have set my bow in the clouds,[17] and it shall serve as a sign of the covenant between Me and ... you and every living creature among all flesh, so that the waters shall never again become a flood to destroy all flesh" (Genesis 9:12–15).

But Elohim also issued a special moral injunction to humanity that is totally absent in the *Epic of Gilgamesh*: "I will require a reckoning for human life of every man for that of his fellow man! Whoever sheds the blood of man, by man shall his blood be shed; for in His image did *Elohim* make man" (Genesis 9:5–6).

Noah's descendants then proceeded to repopulate the Earth. An agricultural cycle of seeding and harvesting was established, in which the main products were grain and wine. It was not long, however, before humans once again incurred Yahweh's disfavor. And this story, too, was centered in the riverine domain of Mesopotamia:

> Everyone on earth had the same language and the same words. And as they migrated from the east, they came upon a valley in the land of Shinar [Sumer?] and settled there. They said one to another: "Come, let us make bricks and burn them hard." Bricks served them as stone, and bitumen served them as mortar. And they said: "Come, let us build a city and a tower with its top in the sky." ... So Yahweh came down to look at the city and the tower man had built, and Yahweh said: "If, as one people with one language for all, this is how they have begun to act, then nothing that they may propose to do will be out of their reach. Let us, then, go down and confound their speech there, so that they shall not understand one another's speech." Thus Yahweh scattered them from there over the face of the whole earth; and they stopped building the city. That is why it is called Babel. (Genesis 11:1–9)

This passage includes an example of Yahweh consulting with an invisible retinue, to whom he speaks in the plural, as though he were chief of a council of lesser gods (a vestige of henotheism). The Tower of Babel apparently referred to the ancient Mesopotamians' penchant for building ziggurats, high temples in the form of terraced pyramids with successively receding stories. The common building material on that stoneless plain was clay bricks, either sun dried or (preferably, for strength) fired in a furnace or kiln. The entire story is evidently meant to explain why the inhabitants of the world, supposedly having originated in a single family, came to speak a multitude of tongues. The name Babel (Bavel) is taken here to be a pun on the Hebrew verb *balol* or *balbel*, which means "to mix up" or "to confuse."

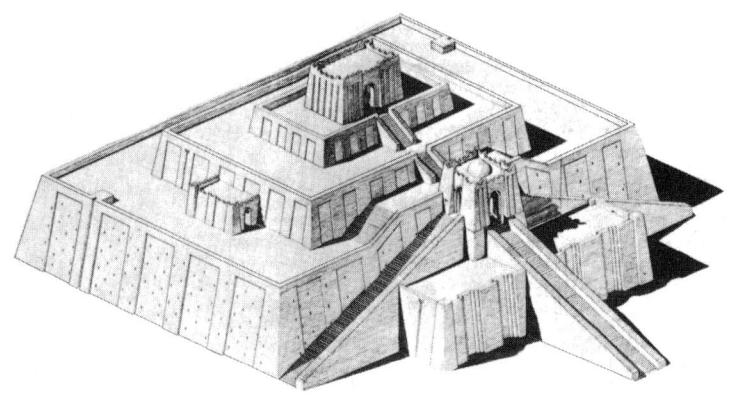

Schematic reconstruction of a multi-tiered ziggurat, referred to in the
story of the Tower of Babel (Genesis 11:1–9).

QUITE ANOTHER INFLUENCE from Mesopotamia on the Israelites was the
biblical calendar, according to which the year was assumed to begin in the
spring. In that river-watered land, the biological and agricultural year clearly
begins when, after a long period of dryness, the rivers are in spate (due to the
melting of the snows in the highland watersheds of the Tigris and Euphrates
Rivers) and the irrigation season can be initiated. In contrast, the flooding of
the Nile in Egypt typically occurs at the end of summer or the beginning of
autumn. Autumn is also when the rains generally arrived in rainfed Canaan.
That is very probably why, following the settlement of the Israelites in Canaan,
the start of their calendar year was shifted from spring (Nissan, roughly corresponding to April) to autumn (Tishrei, roughly corresponding to September).

The time the Israelites spent in riverine Mesopotamia before the completion and canonization of the Bible was rather limited. Yet the impact of Mesopotamian culture on them was considerable and can be discerned, explicitly
or inferentially, in many parts of the Bible. That influence found its expression
in the adoption and adaptation of myths, metaphors, linguistic terms, literary
forms, and legal codes. As was their wont in regard to other cultural influences, the Israelites absorbed many elements of Mesopotamian culture but
modified and fused them into their distinctive culture, in which the exclusive
worship of a single God and obedience to his moral commandments became
the major themes.

4

THE PASTORAL DOMAIN

Legacy of the Bedouin Patriarchs

Now Abram was very rich with livestock. . . . And he proceeded by stages
from the Negev. . . . Lot, who went with Abram, also had flocks and herds
and tents, so the land could not support them staying together.

GENESIS 13:2–6

THROUGHOUT ITS diverse parts and passages, the Hebrew Bible is replete with vivid depictions of the pastoral life. Even long after the majority of Israelites became sedentary farmers and city dwellers, the memory of their origin as nomadic or seminomadic shepherds and the continuing practice of sheep and goat husbandry by a segment of the society kept the lore of pastoralism alive. It expressed itself in language and legend, in numerous figures of speech and allusions, a persistent strand woven into the very fabric of Israelite culture.

One clear example is the familiar and beloved Psalm 23. Its opening verse, "Yahweh is my shepherd," is more than mere poetic metaphor; it assigns a role model for the relationship between the people of Israel and their God. He is their caring shepherd; they are his trusting and obedient flock.[1] The major leaders of Israel—Abraham, Jacob, Moses, and David—were shepherds as well. The image of green pastures and still waters expresses every shepherd's longing for soul-restoring security, just as the image of the valley of the shadow of death expresses the solitary shepherd's haunting fear of the dangers that may lurk in the narrow ravines of the semidesert.

THE STRETCHES of territory that lie within the semiarid steppe of the Middle East consist of a checkered landscape of hills and plains, typically with a savanna-like vegetation of sparse trees, shrubs, and a mixture of annual and

A scene in the pastoral domain: sheep grazing in a sparse grove of olive trees. The trees are very old, as evidenced by their hollowed-out trunks, but olive trees may continue to foliate and bear fruit for many centuries.

perennial grasses and herbs. There are pronounced contrasts among areas of disparate geologic formations (hard limestone, soft chalk, sandstone, granite, or basalt) and pedologic characteristics (deep or shallow soils, with varying contents of clay or sand). Differences in the composition and density of vegetation reflect variations in the local climate, especially as it determines the water regimen of the habitat. South-facing slopes, for instance, are subject to a greater intensity of solar radiation and hence tend to be warmer and drier than north-facing slopes. Similarly, slopes that shed a portion of the rain falling on them (runoff) tend to be drier than the bottomlands that receive and absorb that increment of water (run-on) from adjacent slopes.

On a larger scale, west-facing slopes of mountain ranges are more likely to intercept the rain clouds rolling in from the Mediterranean in winter than are east-facing slopes, which lie in the rain shadow of such clouds. An example of the rain-shadow effect is the abrupt boundary along the watershed divide between the subhumid western side of the Judean and Samarian ranges (facing the Mediterranean) and the extremely arid eastern side (facing the Jordan Valley and the Dead Sea).[2]

As a rule of thumb in this region, semiarid areas receive, on average, less than 16 inches (400 mm) of rain a year, as against a potential evaporation of 60 inches (1500 mm) or so. During the rainy season (October to April), which is the growing season of wheat and barley, the balance between potential evaporation and rainfall amount is much more favorable, however. Areas that receive between 12 and 16 inches (300 and 400 mm) of rain permit extensive grain production in most years, but with a risk of failure in, say, one year of three or four, because of the unstable nature of the climate and the periodic occurrence of drought. Such areas constitute the zone where semisedentary farming and seminomadic herding may be practiced conjunctively or, conversely, where separate communities of farmers and herders compete for limited resources of land, vegetation, and water.

Areas where the annual rainfall is, on average, less than 12 inches (300 mm) are generally submarginal for extensive farming because of the high risk of failure. They can be classified as semideserts and are in the realm of seminomadic pastoralism. Areas where the average annual rainfall is less than 8 inches (200 mm) are in the domain of transhumant, or nomadic, pastoralism. (In such areas, grain production, if it is practiced at all, usually is confined to depressions that receive runoff water from adjacent slopes.) Finally, areas with less than 4 inches (100 mm) rain a year are in the desert domain, where even grazing is marginal. (The exceptions are isolated oases, along the banks of rivers that draw their waters from a more humid zone, wadis that collect runoff naturally, or places where special measures are taken to collect runoff artificially in a system called water harvesting.)[3] Even in deserts, however, the grazing of small herds of hardy livestock, such as goats and camels, is still practiced, although the carrying capacity (the number of livestock that can be supported per unit of land area) is very low.

During the winter, flocks of sheep and goats often can subsist on the moisture contained in the (relatively) succulent vegetation available to them in that season, particularly in the depressions and wadi beds. During the hot and dry summer, however, an area must be within 6 miles (10 km) of a source of water (spring, well, or cistern) if it is to serve as pasture. That is the distance that a typical flock normally can traverse twice a day, usually in the morning and evening (on the way to and from pasture). Thus flocks are corralled overnight near a water source and are trekked to pastures within that radius. When the source of water and/or the pastures are depleted, shepherds must move to another area that has both resources. Choosing the areas and sequences of seasonal movement, or transhumance, in order to ensure the welfare of their flocks and to avoid violent conflict with competing clans or tribes always has been a crucial aspect of the lives of nomadic pastoralists.

The boundaries between the territories of neighboring tribes of nomadic pastoralists are seldom fixed for long. They depend on the nature of the intertribal relationship (whether friendly or hostile), which itself may change from time to time, as well as on the periodic stresses imposed by the vagaries of climate or the intervention of external powers. The extent of the area makes the demarcation of boundaries difficult in any case. Disputes over territory, grazing rights, and, especially, water resources may be resolved by negotiation and compromise or by force. Cisterns, which require maintenance (clearing of silt, plastering of cracks, and repairing of feeder channels) and contain a finite volume of water that quickly can be depleted, are "owned" by particular tribes and are guarded jealously.

BY THE very nature of their lives, nomadic pastoralists leave few written records or other types of durable material remains that can provide the sort of archaeological evidence often found in the sites of sedentary communities. Subsisting as small clans that utilize the sparse range of the semidesert on an intermittent basis, pastoralists of the arid zones do not tend to build solid houses of stone or adobe. Instead, they live in portable tents made of perishable cloth, supported by equally perishable wooden poles and pegs. They also use twigs to set up makeshift corrals for their flocks, and when the wood and twigs deteriorate, they burn them for fuel—leaving only ashes as flimsy remnants that sooner or later blow away in the wind or are washed off the surface by occasional rains.

Extensive research has been done in recent decades on the archaeological remains of ancient pastoral societies in Sinai and the Negev Desert.[4] Among those of interest to archaeologists are camping sites, vestiges of corrals, burial mounds, hunting traps (kites), rows or piles of stones (cairns) that may have served as memorials, ritual monuments, and rock drawings or carvings. Ar-

A Bedouin family tent in the Negev Desert.

chaeological remains may also include cult objects (figurines), ornaments (bracelets, beads, and the like), potsherds, and flints (blades, scrapers, axes, arrowheads, and other tools), as well as storage pits for grain (some locally produced, but most obtained by trade from the rainfed domain) and, of course, the bones of people and animals.

The remnants of ancient pastoral encampments found in the semiarid and arid domains indicate the way in which ecological conditions influenced the lives of herders in the semidesert. These sites usually are very small, consisting of just a few tent emplacements, alongside the remains of one or more enclosures (fenced with stones and twigs) to contain the flocks. Clearly, the pasture and water in the environs were too meager to support large concentrations of animals that would quickly overgraze an area and require constant, restless trekking. Instead, pastoralists necessarily adopted a more tenable regimen of intermittent movement from campsite to campsite, in each of which an extended family or small clan could typically remain for some weeks.

Only the possibility of remaining at a given site for a prolonged period and returning to it repeatedly could justify the construction of facilities for the storage of grain and seasonally used tools. Otherwise, a clan would be burdened with the need to carry all its possessions everywhere it moved. Storage areas usually were located in natural caves, behind mud walls abutting vertical cliffs, or in trenches dug into the earth and lined with stones and straw mats. In many cases, these caches were covered with brush or gravel to hide them from scavenging animals or competitive clans that might plunder the site in the absence of its "owners." That word is in quotes to signify the tenuous nature of all possessions and territorial rights in a domain where nothing could remain stable for long (not the climate, the availability of pasture and water, or the pattern of human occupation).

The archaeological evidence found in the arid rangelands of the Negev and Sinai suggests that pastoralism in these areas apparently began during the Pre-Pottery Neolithic Age—the ninth to eighth millennia B.C.E. One criterion of early pastoralism is the change in the composition of the skeletal remains of the animals that were consumed by residents of a given location. Whereas hunters tended to kill animals of various ages and both sexes more or less at random, pastoralists tended to be more selective. They slaughtered some of the young males of a flock for food, while rearing the young females for reproducing and milking, slaughtering them only when they were older and no longer productive. Hence it is sometimes possible, in certain sites, to infer the transition from hunting to animal husbandry.[5]

The notion that nomadic pastoralists were independent and "free as the wind" is a simplistic myth. In fact, their lives always were tenuous and greatly dependent on interaction (through trade and other modes of exchange) with neighboring sedentary societies. Nor was their domain—the semiarid steppe or savanna—free from external influences. On the contrary, it was a space often subject to incursion by and control of whatever power exercised hegemony over the region at any time. Evidence abounds that the Egyptians, for example, maintained outposts in Sinai for many centuries.

The story of nomads from the arid zone invading and conquering a settled agricultural land has repeated itself many times in the history of the ancient Near East. Such was the conquest of southern Mesopotamia by the Akkadians, a Semitic tribe who superseded the Sumerians there. Such, too, was the invasion of Egypt by the Hyksos, a Semitic people who then ruled the country for a century and a half, in the second millennium B.C.E., before being expelled. And such, as merely a link in a chain of analogous events, may well have been the entry into and eventual settlement of the land of Canaan by the Israelites.

AS TOLD in Genesis 11, Abraham's father, Terah, emigrated with his family from "Ur of the Chaldeans."[6] No details were given of the family's earlier life, nor were any reasons given for their migration. The family apparently proceeded along the arc of the Fertile Crescent, traveling first northwestward

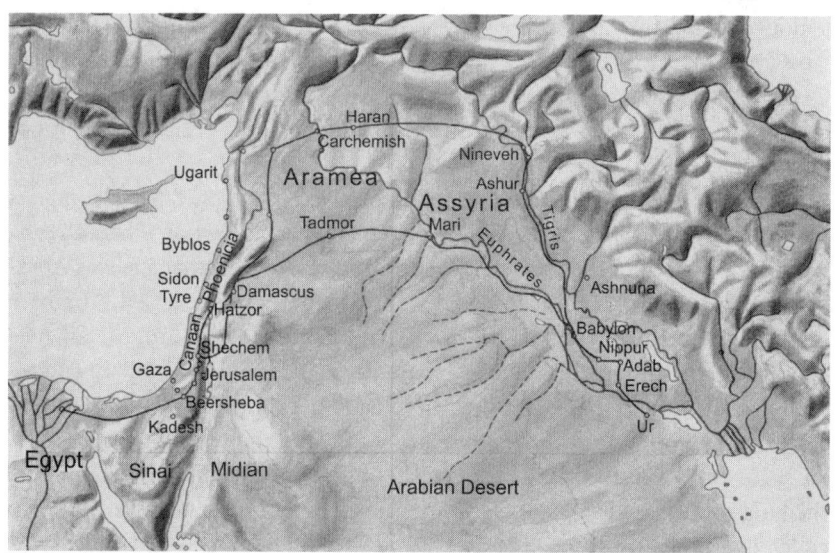

Ancient sites in the Fertile Crescent, either mentioned in the biblical accounts of the pastoral Patriarchs or known to have existed during their presumed period (second millennium B.C.E.).

along the Euphrates River. En route, they stopped and sojourned in the land of Haran, in the present-day borderland between northeastern Syria and southeastern Turkey. There the family's patriarch, Terah, died.

While other members of the clan remained in Haran, Abraham, reported to have been instructed directly by Yahweh (Genesis 12:1–6), left the land of his birth entirely and continued the journey to the distant land of Canaan, where, Yahweh promised, his progeny would become a great nation. At first, Abraham followed the curved path from Haran to central Canaan (Shechem), from where he gravitated southward to the plains along the fringes of the Negev Desert. There he subsisted by grazing his sheep and goats in the semi-arid pastoral domain. Thenceforth, Abraham and his family led a life that was, in some respects, similar to that of the Bedouin clans that have subsisted in the region during the subsequent several millennia.

Soon after Abraham's arrival in southern Canaan, the area was struck by a drought, and there was famine throughout the land. Abraham then reportedly did what countless fringe-land pastoralists had done before and would do after him: he trekked to the riverine domain of Egypt to sojourn there.[7] Egypt was generally unaffected by droughts that occurred in southwestern Asia from time to time, since its water regimen is dictated by a completely different climatic system. Egypt has always depended on the annual flooding of the Nile, a result of the summer monsoons rolling in from the Indian Ocean and causing rainfall over the highlands of Ethiopia. In contrast, the rains in southwestern Asia occur entirely in the winter. They come in mostly from the west (over the Mediterranean) and are very much less regular than the Indian Ocean monsoons, thus making the region more prone to droughts than the riverine Nile. (The two contrasting water regimens may even exhibit an opposite-phase fluctuation.)[8]

As the Bible tells it, Abraham was well received in Egypt, especially thanks to the beauty of his wife, Sarai (later to be renamed Sarah), whom Pharaoh took into his house under the deliberately conveyed impression that she was Abraham's sister rather than his wife.[9] When the deception was revealed, Abraham and his entourage were urged to leave Egypt (Genesis 12:10–20). Meanwhile, fortunately, the drought in Canaan apparently had ended, so Abraham and his entire clan returned there with their flocks and herds and tents.

This was not the only time Abraham was willing to trade his wife for favors. He employed the same stratagem to ingratiate himself with Abimelech, the king of the small town of Gerar in Canaan. When this ruse, too, was discovered, Abraham related that Sarah—her modified name—was indeed his half sister (being the daughter of his father but not of his mother, as stated in Genesis 20:2–16). The same story of trading a wife to gain the favor (or at least to assuage the hostility) of Abimelech is repeated with Isaac and Rebekah (Genesis 26:6–11). It illustrates the tenuousness and vulnerability of the life of seminomadic pastoral clans in the vicinity of settled areas and their perpetual need to curry favor with those who controlled the territory within which they sought subsistence. It also illustrates another feature of the pastoral societies of the time: the lowly status of women, who could be traded for favors.

Back in the pastoral domain of southern Canaan, Abraham soon found that the carrying capacity of the land was limited even when there was no drought. A conflict arose between his shepherds and those of his nephew Lot (for "the land could not support them staying together, for their possessions were so great" [Genesis 13:6]). Abraham then offered Lot the choice of land to be his grazing right, and Lot picked the well-watered floodplain of the Jordan. In other words, Lot chose Canaan's smaller-scale analogue of the riverine flood-plain of the Nile, whereas Abraham remained in the semiarid savanna range-land between Hebron and Beer-sheba.

Several events in the life of Abraham and his clan are especially indicative of human interactions with the environment in the semiarid pastoral domain.

Sarah had been a barren woman. The Bible ignores the possibility that her difficulty was due to her close, even incestuous, relationship with Abraham, which may have inhibited them from procreating. So Sarah gave her hand-maiden Hagar to Abraham to bear a child on her behalf.[10] (Polygamy was prevalent then, especially with men of substantial wealth.) Soon, however, jealousy grew between the two women, as Hagar—having become Abraham's concubine—began to treat her mistress disrespectfully. When Sarah castigat-ed her, Hagar escaped to the desert, where an angel reportedly rescued her

Aerial view of the lower course of the Jordan River, meandering on its way from the Sea of Galilee to the Dead Sea. Note the contrast between the dense vegetation in the river's floodplain and the barren hills on both sides. In the Bible, the Jordan Valley is described as being like "the garden of Yahweh, like the land of Egypt" (Genesis 13:10).

and returned her to the clan. The angel foresaw that the son she would bear, to be named Ishmael, would grow to become "a wild ass of a man, his hand against everyone, and everyone's hand against him" (Genesis 16:12). ("Wild ass" was not necessarily a pejorative, as it referred to a tough and swift animal that roamed the desert freely.)

Then Sarah herself became pregnant and bore Abraham his second son, whom they named Isaac. In contrast to his half brother, Isaac was a docile and obedient boy, much beloved by his father. Now the jealousy between the women intensified, until Sarah, the primary wife, prevailed on Abraham to banish Hagar and Ishmael to the desert,[11] where an angel rescued her once again. Ishmael then grew up in the wilderness and "became an archer" (Genesis 21:20). Merely to survive in the desert, a man would have had to become a skilled hunter and an aggressive fighter.

And so Abraham remained with his younger son only. Alas, Abraham's faith soon faced a supreme test, as he was commanded to offer Isaac as a sacrifice to The Elohim (Genesis 22).

Child sacrifice was very prevalent in the ancient Near East. The idea was to give the gods the most precious of all possessions, in order to elicit divine favor in return, especially in times of dire need.[12] The sacrifice was a sort of bargain between a person in distress and the divine, a plea for reciprocation: "Here I am making the ultimate sacrifice to express my utmost faith in you, my god; will you not help me in return?" Especially common was the custom of sacrificing the first portion of the crop or flock or the firstborn child. But Abraham could not quite bring himself to do that. A younger father may have been ready to sacrifice his firstborn son, on the assumption that he could thereby win favor and be granted several other sons. But Abraham, so the story goes, was one hundred years old when Isaac was born (Genesis 21:5) and could have had little hope for another son.[13] At the last minute, the angel of Yahweh stayed Abraham's hand and provided a ram as a substitute sacrifice.[14] This rejection of child sacrifice was a major step toward a more humane view of God and a greater concern for human life in general.[15]

Herdsmen in the pastoral domain have always suffered from uncertainty, impermanence, and potential conflict. Thus they have always been imbued with a yearning for security, permanence, and the very stability that their peripatetic mode of life does not usually provide. And so it was that, after Sarah's death, Abraham sought a permanent site (the cave of Machpelah) to serve as his clan's burial place. To obtain it, he had to purchase a plot from its owner.

Why was Abraham so eager to buy the cave of Machpelah? In sedentary farming society, power was based on land ownership. Because nomadic herders did not usually own land, they had no legal status. At best, they could gain grazing rights over a tract of rangeland or water rights over a well or cistern. Such rights were tenuous and ever contestable. Hence pastoralists were the

ABRAHAM: THE FATHER OF MANY NATIONS

According to the Bible and the Qur'an,[1] Abraham was the progenitor of both the Hebrews (Jews) and the Arabs, as well as the first man to believe in one God, Yahweh or Allah. To the Jews, he is Abraham Avinu (Abraham Our Father), whereas to the Muslims, he is Ibrahim El Khalil (Abraham the Friend [of Allah]). In the Bible, Abraham's original name was Abram (exalted father), and his wife's was Sarai (noble woman). Abram's name was changed to Abraham (father of many nations) (Genesis 17:5), while Sarai's was changed Sarah (Genesis 17:15). (The addition of the letter *h* may have meant to imply the blessing of Ya [Yahweh], the divine name that is often represented by the Hebrew letter *heh*.) The name Abram is mentioned 61 times in the Bible; the name Abraham, 175 times.

The Arabs are stated to be the descendants of Abraham's firstborn son, named by Hagar, his mother, Ishmael (El has heard, or El shall hear) (Genesis 16:11). It was with Ishmael, then thirteen years of age, that Abraham is said to have initiated the practice of circumcision, as a sign of his clan's covenant with God. The ninety-nine-year-old Abraham then performed the rite on himself and on all the males in his household (Genesis 1:23–27). According to the Qur'an, "Ibrahim and Isma'il raised the foundations of the House" (Sura 2:127), which is believed to be the Qa'abah of Mecca, the holiest shrine of Islam. Ibrahim is mentioned in 25 of the Qur'an's 114 chapters (*suras*).

The Hebrews are said to be the descendants of Abraham's second son, whom he begat at the age of one hundred. The boy was named by Sarah, his mother, Yitzhaq (shall laugh), in reference to her advanced age of ninety years (Genesis 18:12–15, 21:6). Yitzhaq (called Isaac in transliteration) was circumcised when he was eight days old (Genesis 21:4).

So faithful was Abraham that he was prepared to sacrifice his favorite son in response to what he perceived to be God's command. At this point, the Bible and the Qur'an differ. In the Bible, that son was Isaac, and the place was Mount Moriah (Genesis 22:1–2), later to be identified as the hill, north of ancient Jerusalem, on which Solomon built his Temple and dedicated it to the worship of Yahweh. In the Qur'an, neither the child nor the place of the intended sacrifice is stated explicitly (Sura 27:102), but Muslims believe that Ishmael was the child whom Abraham was prepared to sacrifice and that the place was the Qa'abah of Mecca. The two accounts are similar, yet the seemingly minor differences can be important to the respective believers.

One of the current points of contention is the Tomb of the Patriarchs, presumed to be the site of the cave of Machpelah in Hebron. The name of that town is related to the Hebrew word *haver*, which means "friend." The Arabic name, El Khalil, carries the same meaning, in reference (and deference) to Abraham, Allah's Friend, who purchased the plot and was buried there (having lived to the ripe age of 175) by both his sons.[2]

Archaeologists have found no evidence to confirm the existence of the biblical Abraham. The name, however, was not uncommon among the Western Semites of the second millennium B.C.E.,[3] and neither was the lifestyle ascribed to Abraham. Hence, despite the contradictions, anachronisms, and anomalies contained in Genesis,[4] it seems plausible that such a Patriarch may have existed at an early stage in the history of the Hebrews. It is also possible that the Abraham of the Bible is a composite of several ancient father figures, whose deeds were magnified and exalted as they were transmitted, at first orally and eventually in written form, from one generations to the next.

---

1. Qur'an means, literally, "recitation."

2. W. F. Albright, "Abraham the Hebrew: A New Archaeological Interpretation," *Bulletin of the American School of Oriental Research* 163 (1961): 36–54. See also T. L. Thompson, *The Historicity of the Patriarchal Narratives: The Quest for the Historical Abraham* (Berlin: de Gruyter, 1974), and J. van Seters, *Abraham in History and Tradition* (New Haven, Conn.: Yale University Press, 1975).

3. C. H. Gordon, "Biblical Customs and the Nuzi Tables," in E. F. Campbell and D. N. Freedman, eds., *The Biblical Archaeologist Reader* (Garden City, N.Y.: Doubleday, 1964), 2:20–30.

4. I. Finkelstein and N. A. Silberman, *The Bible Unearthed: Archaeology's New Vision of Ancient Israel and the Origin of Its Sacred Texts* (New York: Free Press, 2001).

eternal outsiders, looking in and yearning to possess a bit of permanence, formalized by publicly sanctioned ownership, not as a matter of favor or charity but by indisputable right. Owning a section of land, however small, was a symbolic way of achieving the status of residence. Burying successive generations in a family plot would further reinforce and perpetuate that status. All this is symbolized by Abraham's purchase of a cave in which to entomb and honor his family's deceased.

Trade between pastoralists and farmers was a well-established practice, with customary rituals and niceties. But the strategy was then, as it is now, for each side to gain the maximum advantage with the minimum expenditure of effort or substance—while demonstrating the all-important appearance of dignity, chivalry, and honor. Even today, anyone who has had any experience at bargaining in the Middle East (where bargaining is a customary part of every transaction) can recognize the ritual followed by Abraham and Ephron the Hittite, the owner of the cave: "Sarah died in Kiriath Arba—now Hebron—in the land of Canaan. And Abraham proceeded to mourn for Sarah, and to bewail her. Then he rose . . . and spoke to the Hittites, saying: 'I am a resident alien among you; sell me a burial site among you, that I may remove my dead for burial'" (Genesis 23:1–4).[16]

At first he was offered the requested burial site for free, but Abraham declined the overly generous offer, which he knew to be mere pretense. The requirement of honor was to demonstrate generosity in return. So Abraham replied in the presence of the neighbors of Ephron: "Let me pay the price of the land; accept it from me, that I may bury my dead here" (Genesis 23:13).

After those niceties, the real bargaining was to commence over the price: "Ephron replied to Abraham: 'My lord. . . . A piece of land worth four hundred shekels of silver—what is that between you and me? Go bury your dead'" (Genesis 23:15).[17] Judging from other transactions in the Bible, the price set for this property was exorbitant.[18] By custom, Abraham was expected to make a counteroffer, suggesting a smaller sum. Instead, he surprised the assembled locals by his immediate acceptance: "Abraham paid out to Ephron the money that he had named in the hearing of the Hittites—four hundred shekels of silver at the going merchant's rate." By this symbolic act, Abraham established his family in the land of Canaan for posterity.[19]

A precondition for herding animals in an arid region is the availability of fresh water. Because water is so scarce, it always has been a potential cause of contention. While Abraham sojourned in the territory near Gerar, a rivalry developed there between his men and the men of Abimelech, the local king, over the rights to a well:[20]

Abraham reproached Abimelech for the well of water, which Abimelech's servants had seized. But Abimelech said: "I do not know who did this. . . ." So

Abraham took sheep and oxen and gave them to Abimelech, and the two of them made a pact. Abraham set seven ewes of the flock by themselves . . . [and he said] "You are to accept these seven ewes from me as proof that I dug this well." Hence the place was called Beer-sheba, for there the two swore an oath.[21]

The rivalry over water intensified during the lifetime of Abraham's son Isaac, who lived in the same semiarid pastoral domain of the northern Negev: "After the death of Abraham, Elohim blessed his son Isaac, and Isaac settled near Beer lahai-roi" (Genesis 25:11).[22] Isaac then moved to the northwestern Negev, near the permanent settlement of Gerar, at the edge of the rainfed farming domain. There he engaged in farming as well as in grazing and was said to have "reaped a hundredfold the same year" (Genesis 26:12).[23]

Sowing grain is a risky venture in the pastoral domain; it can be highly successful in years of above-average rainfall that is optimally distributed during the season. The occurrence of such seasons is infrequent and unforeseeable. Occasionally, several such years may occur in succession, only to be followed by a year or a number of years with below-average rainfall.[24]

Isaac's engagement in farming as well as grazing expressed the natural desire of pastoralists to attain a more secure mode of subsistence or even to make the transition from nomadic herding to sedentary farming. Alas, that opportunistic attempt, although it may be successful in favorable years, is certain to encounter the resistance of the native agriculturists, who perceive it as an encroachment on their own fragile domain. And so it was with Isaac:

A Bedouin pastoralist tilling a stony field with a traditional plow drawn by a camel.

"The man grew richer and richer until he was very wealthy; he acquired flocks and herds, and a large household, so that the Philistines envied him" (Genesis 26:12–14).[25] Typically, the focus of the dispute was access to water. To graze his large flocks, Isaac needed several sources of water distributed throughout the area of his pasturage. But "the Philistines stopped up all the wells that his father's servants had dug ... , filling them with earth. And Abimelech said to Isaac: 'Go away from us, for you have become far too big for us'" (Genesis 26:15–16).

> So Isaac departed from there and encamped in the wadi of Gerar. ... And Isaac dug anew the wells which had been dug in the days of his father Abraham. ... But when Isaac's servants, digging in the wadi, found there a well of spring water, the herdsmen of Gerar quarreled with Isaac's herdsmen, saying, "the water is ours." So he called the name of the well Esek [contest] because they contended with him. And when they dug another well, they disputed over that one also. So he called the name of it Sitnah [hatred]. And he removed from there, and dug another well, and they did not quarrel over it; so he called it Rehoboth [spaciousness], saying: "Now at last Yahweh has granted us ample space to increase in the land." (Genesis 26:17–22)

Skirting the frontier between the sown land and the desert, Isaac shifted his pasturage southeastward to his father's old camping ground, Beer-sheba. There he built an altar, invoked Yahweh by name, and pitched his tent. Then Isaac's servants dug a well and proclaimed triumphantly: "We have found water!" (Genesis 26:32). Nothing could so gladden a pastoral clan in the desert fringe land as the discovery of a source of fresh water.

Isaac's interaction with Abimelech, the king of Gerar, was typical of the ambivalent relationship between pastoralists and sedentary societies.[26] It was marked by an alternation of conflict and accommodation. When the king came to Isaac in Beer-sheba, apparently for a friendly visit (but accompanied by Phicol, the captain of his troops!), Isaac said: "'Why have you come to me, seeing that you have been hostile to me and have driven me away from you?' And they said ... 'Let us make a pact with you that you will not do us harm, just as we have not molested you'" (Genesis 26:27–29). And, as was the custom, the agreement culminated in a festive meal: "Then he made for them a feast, and they ate and drank. Early in the morning, they exchanged oaths. Isaac bade them farewell, and they departed from him in peace" (Genesis 26:30–31). Here, again, is a reprise of an earlier story (Genesis 21:22–32), involving Isaac's father, Abraham.

In an arid region such as the northern Negev, where Isaac sojourned, the meager seasonal rainfall merely wets the upper layers of the soil (which constitute the rooting zone of the local vegetation). In most seasons, therefore, there is little, if any, excess moisture percolating into the deeper layers of the subsoil and recharging the groundwater. Hence the chances of finding groundwater at a reasonable depth by digging a well in any random location are practically nil. Groundwater may occur in specific and restricted locations, however, particularly along the intermittent watercourses (wadis) where occasional flash floods cause excess water to percolate into the permeable gravel deposits of the streambed. Where the gravel is underlain by impermeable clay or bedrock, the water may form a localized "perched" body of underground water—a resource that can be tapped by means of shallow wells. Evidently, that was what Isaac did to provide water for his flocks.

THE STORY of Esau and Jacob, the twin sons of Isaac and Rebekah, is but one in a string of archetypal biblical tales of sibling rivalry. That of Cain and Abel represents the age-old clash between farmers and pastoralists, and that of Ishmael and Isaac represents the equally sensitive division between desert hunters and pastoralists, reprised by the contrast between Esau and Jacob: "When the boys grew up, Esau became a skillful hunter, a man of the outdoors [open wilderness]; but Jacob was a mild man who stayed in tents" (Genesis 25:27).

So when the time came for old Isaac, enfeebled and blinded by age, to die, the two brothers vied for his final blessing. Significantly, a father's blessing had to be earned by the presentation of an appropriate food offering, just as the gods' favor had to be elicited by means of a sacrificial offering. As Esau was the firstborn son and his father's favorite, Isaac sent him to hunt and prepare venison for the occasion. Hunting was what Esau did best, his badge of manhood, which his father admired. But while he was away, his cunning brother Jacob (urged and helped by their mother, Rebekah, who preferred Jacob over Esau, perhaps because she had been born into the semisedentary pastoral tradition) preempted Esau and deceived their father. Jacob impersonated his brother (wearing Esau's clothes) and presented himself to Isaac, offering him the delectable lamb dish prepared by Rebekah to taste like venison:

His father Isaac said to him: "Come close and kiss me, my son." . . . And he smelled his clothes and he blessed him, saying:

"Ah, my son's smell
is like the smell of the fields that Yahweh has blessed.
May the Elohim give you of the dew of heaven
and the fat of the earth,

Springs and wells historically have been the focal points of social life, especially in nomadic and seminomadic societies. It was "by the well of water" near Haran that Eliezer, Abraham's emissary, met Rebekah, "the damsel very fair to look at." She then drew water from the fountain to serve him and his camels (Genesis 24), thereby proving to be kind as well as pretty. Consequently, she was invited, and agreed, to become the wife of Isaac, "and he loved her." That is, incidentally, the first mention in the Bible of a man's love for a woman.

Bedouin women in the Negev, drawing water from a shallow well to water their flocks. Containers for liquids were either ceramic jars or tanned and sewn animal skins. The ceramic jars were convenient for sedentary farmers, but were brittle. Hence nomadic pastoralists preferred to use animal skins to carry water, as did Hagar when she was banished to the desert (Genesis 21:14–15, 19).

Genesis 24 gives a particularly charming account of Rebekah's consent to leave her family and be taken by Eliezer to faraway Canaan, and of her first meeting with Isaac, at the end of her long journey from Haran to Canaan:

They [her mother and brother] called Rebekah and said to her: "Will you go with this man?" And she said: "I will." So they sent off their sister Rebekah and her nurse with Abraham's servant and his men. And they blessed Rebekah, and said to her:

"Our sister, may you grow
into thousands of myriads;
may your offspring seize
the gates of their foes."

Then Rebekah and her maids arose, mounted the camels, and followed the man. . . . Now Isaac . . . was settled in the region of the Negev. And Isaac went out walking in the field toward evening and, looking up, he saw camels approaching. Raising her eyes, Rebekah saw Isaac. She alighted from the camel and said to the servant, "Who is that man walking in the field toward us?" And the servant said, "That is my master." So she took her veil and covered herself. The servant told Isaac all the things that he had done. Isaac then brought her into the tent of his mother Sarah, and he took Rebekah as his wife. Isaac loved her, and thus found comfort after his mother's death. (Genesis 24:58–67)

Even more romantic is the description of how, a generation later, Jacob first met Rachel at the well where she came to water her father's flock (Genesis 29). Chivalrous Jacob single-handedly rolled away the heavy stone that covered the well, watered Rachel's sheep, and then kissed the maiden who was "shapely and beautiful." (Incidentally, only a family of shepherds would name its daughter Rachel, which means "ewe." The name Rebekah is also of pastoral origin, related to the word *marbek*, which means "fatling"—which may seem odd to us, but was prized by shepherds.)

The meeting of Jacob and Rachel in Haran was reenacted many generations later by Moses (Exodus 2). After escaping the wrath of Pharaoh, he wandered through the desert to Midian (presumed to have been in northwestern Arabia). The first place Moses sought out was, naturally, the communal well, where he happened to meet the daughters of the priest of Midian. After helping them water their flock by driving off hostile shepherds, the noble stranger was invited to the home of their father. The grateful priest then gave Moses his daughter Zipporah, to be his wife. (Zipporah, incidentally, is also a naturalist name, although not necessarily a pastoral one like Rebekah and Rachel; it means "female bird.") Moses thus repeated the pattern that Jacob had set, working as a shepherd in the household of his father-in-law. In so doing, he identified himself with the pastoral tradition of the Patriarchs and the antecedents of the nation that he was destined to lead.

abundance of new grain and wine.
May peoples serve you
and nations bow to you;
be master over your brothers
and let your mother's sons bow to you.
Cursed be those who curse you,
blessed be those who bless you." (Genesis 27:26–29)

After Jacob left and Esau returned, both he and his father discovered the deception. What followed was a poignant, heart-rending exchange:

Isaac was seized with very violent trembling. "Who was it then," he demanded, "that hunted game and brought it to me? I ate it before you came, and I blessed him; now he must remain blessed!" When Esau heard his father's words, he burst into a wild and bitter sobbing, and said to his father, "Bless me too, Father!" But he answered, "Your brother came with guile and took away your blessing." . . . And Esau said to his father, "Have you but one blessing, Father?" . . . And Esau wept aloud. And his father Isaac answered, saying to him:

"Your abode shall enjoy the fat of the earth
and the dew of heaven above.
Yet by your sword you shall live,
and you shall serve your brother;
But when you grow restive,
you shall break his yoke from your neck." (Genesis 27:33–40)

Both blessings were for fertile land and a guaranteed supply of water from heaven (symbolized by dew). However, Jacob the pastoralist was granted superior status as primary heir and future tribal chieftain, thus symbolizing the

INBREEDING OF THE EARLY HEBREWS

The early patriarchal family was endogamous. Inbreeding was not just prevalent but, indeed, the purposeful norm in the first generations of the Hebrew clan:

Abram married his half sister Sarai, daughter of his father but not his mother.

Abram's brother Nahor married Milcah, daughter of their other brother, Haran.

Bethuel, son of Nahor and Milcah, fathered Rebekah and Laban.

Isaac, son of Abraham and Sarah, married his cousin Rebekah, daughter of Bethuel.

Jacob, son of Isaac and Rebekah, wed his cousins Leah and Rachel, daughters of Laban.

Esau, son of Isaac and Rebekah, wishing to please his father, married his cousin Mahalath, daughter of Ishmael, son of Abraham and Hagar and half brother of Isaac.

The Bible relates many instances of endogamy:

Abraham was now old, advanced in years. . . . And Abraham said to the senior servant of his household, who had charge of all that he owned: "Put your hand under my thigh, and I will make you swear by Yahweh, the God of heaven and the God of the earth, that you will not take a wife for my son from the daughters of the Canaanites, among whom I dwell, but will go to the land of my birth[1] and get a wife for my son Isaac." (Genesis 24:2–4).

Rebekah said to Isaac, "I am disgusted with my life because of the Hittite women. If Jacob marries a . . . native woman, what good will my life be to me?" So Isaac sent for Jacob and blessed him. He instructed him, saying, "You shall not take a wife from among the Canaanite women. Go to Paddan Aram, to the house of Bethuel, your mother's father, and take a wife there from among the daughters of Laban, your mother's brother." (Genesis 27:46–28:1–2)

Esau took to wife Judith, daughter of Beeri the Hittite and Basemath, daughter of Elon the Hittite. And they were a source of bitterness to Isaac and to Rebekah. (Genesis 26:34–35)

Esau realized that the Canaanite women displeased his father Isaac. So Esau went to Ishmael, and took to wife . . . Mahalath the daughter of Ishmael, son of Abraham, sister of Nebaioth. (Genesis 28:8)

The tradition of within-family marriage continued even during the Israelites' sojourn in Egypt: "These are the families of the Levites according to their generations. Amram took him Jochebed his father's sister to wife; and she bore him Aaron and Moses" (Exodus 6:20). This type of inbreeding was later forbidden (Leviticus 18:12).

Thus it appears that the Hebrews were clannish and exclusive from the start. An alternative view is that the Bible's attribution of the origin of the Israelites to a single tightly knit family was contrived at a much later time as a device to cement the disparate tribes of the nation by inculcating a sense of familial bonding. In any case, the Israelites appear to have been strangers among the natives of every land in which they sojourned: in Haran, Egypt, and Sinai, as well as in the land that was to be their ultimate home, Canaan. By the time they were exiled from Canaan and dispersed throughout the ancient Near East, the sense of separateness, both ethnic and religious, had long been seared into their consciousness. The feeling of alienation was reciprocated by the neighboring peoples among whom the Israelites sojourned. As the Israelites were beset by enemies and felt constantly threatened, their attitude toward other nations was always marked by wariness. Their sense of being a nation apart may well have been an important factor in their reluctance to assimilate, their stubbornly maintained identity, and—ultimately—their survival as a nation.

---

1. Abraham refers to Haran as the land of his birth. That seems to contradict the statement in Genesis 11:31 that Abraham was born in Ur (in southern Mesopotamia) and later migrated to Haran (in the northern part of the Euphrates watershed).

ascendancy of the pastoral lifestyle over the hunting lifestyle. And, whatever the feelings of the old Patriarch Isaac, the prevailing code of honor required that a blessing once given, especially one invoking Yahweh Elohim, could not be withdrawn, even if obtained under false premises.[27]

Again at the instigation of his mother, Jacob traveled to the ancestral land of Haran. He did so not only to escape Esau's wrath, but also to comply with his mother's desire that he find a bride from among her kinfolk, not from the Canaanites. In Haran, he was accepted into the household of his uncle Laban (Rebekah's brother), who was a keeper of flocks of sheep and goats. His, however, was a fairly settled life, for he grazed the stubble fields and fringe lands of the riverine domain, rather than the savanna-like dry range of the semidesert. Jacob tended Laban's flocks for seven years to win the right to marry Laban's younger daughter, Rachel. But on the morning after his wedding, Jacob (the deceiver of his father and brother) learned that he himself had been deceived. The bride he was given was not the beautiful Rachel, whom he loved, but her older sister Leah, the one with the "weak eyes."[28] So he contended with Laban, who struck a hard bargain: Jacob would get Rachel, in addition to Leah, one week later, but only if he would commit himself to work for Laban for an additional seven years.

Although Rachel was Jacob's favorite, she had difficulty conceiving. Leah bore four sons but had difficulty maintaining Jacob's interest. Sibling rivalry naturally developed between the sisters over their husband's attention. It is expressed in the story concerning mandrake, an herbal medicament:

> Once, at the time of the wheat harvest, Reuben came upon some mandrakes in the field and brought them to his mother Leah. Rachel said to Leah, "Please give me some of your son's mandrakes. . . . I promise, he shall lie with you tonight, in return for your son's mandrakes." When Jacob came home from the field in the evening, Leah went out to meet him and said, "You are to sleep with me, for I have hired you with my son's mandrakes." And he lay with her that night. God heeded Leah, and she conceived and bore him a fifth son . . . [and later] a sixth son. . . . Now Elohim remembered Rachel; Elohim opened her womb. She conceived and bore a son and said: "Elohim has taken away my disgrace." So she named him Joseph [shall add], which is to say, "May Yahweh add another son for me." (Genesis 30:14–24)

Mandrake (*Mandragora automnalis*) is a perennial herb of the Solanacaea family.[29] It has no stalk, but thick roots and large leaves that grow out of the base of the crown. Although known to be poisonous, when taken in small amounts mandrake evidently was considered to be a sexual stimulant and fertility enhancer. As the story implies, it seems to have been efficacious for both sisters, but especially for Rachel, who had been barren until she partook of it.

When the time came, after the many years of servitude, for Jacob to leave the household of his father-in-law, Laban, and to become independent, there ensued a bargaining session over the severance pay due to Jacob that was, again, a cunning contest of wits and mutual deception:

> Jacob said to Laban: "Give me leave to go back to my own homeland. Give me my wives and my children, for whom I have served you." . . . He said: "What shall I pay you?" And Jacob said, ". . . Let me pass through the whole flock today, removing from there every speckled and spotted animal—every dark-colored sheep and every spotted and speckled goat. Such shall be my wages." . . . And Laban said, "Very well, let it be as you say." (Genesis 30:25–34)

Righteous protestations notwithstanding, when Jacob and Laban came to implement their agreement, each tried to outsmart the other. The first move was Laban's: "He removed that day the streaked and spotted he-goats—every one that had white on it, and all the dark colored sheep, and left them into the charge of his sons. And he put a distance of three days' journey between himself and Jacob, while Jacob was pasturing the rest of Laban's flock." Leaving behind only unspotted sheep and goats, however, did not guarantee that none of the lambs yet to be born would be spotted. Now came Jacob's turn to manipulate the flock to his own advantage. He took rods of fresh poplar, almond, and chestnut, and peeled off strips of bark, exposing the white interior. He set the rods in the watering troughs for the flocks to see, for he knew that they mated when they came to drink. Consequently, the flocks brought forth streaked, speckled, and spotted offspring. So Jacob "grew exceedingly prosperous, and came to own large flocks, maidservants and menservants, camels and asses" (Genesis 30:35–43).

The modern science of genetics denies this "Lamarckian" notion of influencing unborn lambs merely by exposing their mothers to the sight of striped rods. But that was not necessarily Jacob's only stratagem. In addition, he may have employed a system of selection and preferential breeding that did succeed in gradually increasing the proportion of spotted animals in the flock.[30]

Jacob then tried to escape from Laban while Laban was shearing his sheep elsewhere, but—what with all his flocks and wives and children and servants—the going must have been very slow. The description of that escape and its aftermath provides insight into the ventures and lore of pastoralists in the semiarid domain of the Fertile Crescent.

Jacob "fled with all he had. Soon he was across the river [Euphrates] and heading toward the hill country of Gilead." When Laban found out, "he took his kinsmen with him and pursued him a distance of seven days, catching up with him in the hill country of Gilead" (Genesis 31:21–23).[31]

Facing Laban, Jacob expressed his long-term grievance at having had to endure the lonely and arduous life of an indentured shepherd in the service of an ungrateful master: "What is my guilt that you should pursue me? . . . These twenty years have I spent in your service, your ewes and she-goats never miscarried, nor did I feast on rams from your flock. That which was torn by beasts . . . I myself made good the loss. . . . Often, scorching heat ravaged me by day and frost by night; and sleep fled from my eyes . . . and you changed my wages time and again" (Genesis 31:36–41).

Finally, although reluctantly, Laban accepted the separation, saying: "Now come, let us make a pact." To witness and commemorate the agreement, they erected a makeshift monument:

> Jacob took a stone, and set it up as a pillar, and Jacob said to his kinsmen: "Gather stones." So they took stones and made a mound, and they partook of a meal there by the mound. . . . And Laban said: "This mound is a witness between you and me this day." . . . And Jacob swore by the Fear of his father Isaac. Jacob then offered a sacrifice on the mount, and invited his kinsmen to partake of the meal . . . and they did eat bread, and tarried all night in the mountain. (Genesis 31:44–54)

These passages reveal much about the customs of pastoralists. Erecting commemorative stone mounds is one important custom. A pile of stones in a desolate, arid area can indeed be a monument to eternity, for no wind or rain or any other force of nature (except a very violent earthquake) would be likely to cause its collapse, and very few people would have any interest in dismantling it. Such cairns are still found here and there in the deserts and semideserts of the Middle East, bearing silent witness to those in the distant past who had their own reasons, unknown to us, for erecting them, presumably to commemorate an important event in their lives. Piles of stones were also used to protect new graves from being bared by erosion (caused by wind or rain) and from being dug up and desecrated by scavenging animals.[32]

A solemn agreement was to be sanctified by a religious act, a sacrifice, which culminated in a festive meal. Part of the sacrificed animal was burned completely as an offering to the gods (or God), while another part was eaten by the assembled participants. "Bread" in this context obviously referred to the broiled meat of the sacrificed animal.[33] Thus was resolved the rivalry within the family over property (livestock) and women (Laban's daughters, Jacob's wives). Letting his daughters go was a painful concession for Laban, for he had to relinquish a part of his lineage, descendants who thenceforth would live in a faraway land, probably never to be seen again. Before parting from his daughters and grandchildren, Laban swore by the several gods of the family patriarchs (Jacob's grandfather Abraham, Laban's grandfather Nahor, and their common great-grandfather, Terah), each of whom apparently had his own god or gods (Genesis 31:53).

After Laban's departure, Jacob, temporarily encamped in Gilead, had to cross the Jordan River to enter Canaan and to confront his estranged brother, Esau. That reunion was fraught with danger, for Esau might still bear resentment over what Jacob had done to him twenty years earlier. Before Jacob crossed the river, however, he was to experience two mysterious encounters: "Jacob went on his way, and angels of Elohim encountered him. When he saw them, Jacob said 'This is Elohim's camp.' So he named that place Mahanayim [encampments]" (Genesis 32:2–3). That enigmatic event, however, was only preliminary to the second encounter, which was so fateful as to cause a symbolic change of Jacob's name:

> That night he arose, and taking his two wives, his two maidservants, and his eleven children,[34] he crossed the ford of the Jabbok. After taking them across the stream . . . Jacob was left alone. And a man wrestled with him until the break of dawn. When he saw that he had not prevailed against him, he wrenched Jacob's hip at its socket. . . . Then he said, "Let me go, for dawn is breaking." But he answered, "I will not let you go, unless you bless me." Said the other, "What is your name?" He replied, "Jacob." Said he, "Your name shall no longer be Jacob, but Israel, for you have striven with beings divine and human, and have prevailed." Jacob asked, "Pray tell me your name?" But he said, "You must not ask my name." And he took leave of him there. So Jacob named the place Peniel, meaning "I have seen a divine being face to face, yet my life has been preserved." The sun rose upon him as he passed Penuel, limping on his hip. (Genesis 32:23–32)

Jacob's earlier encounters with the divine had occurred through the medium of dreams (Genesis 28:12–13, 31:11). Now he met several actual angels and then one in particular who engaged him in a wrestling match. Was he a person or a pagan river god or perhaps a manifestation (or messenger) of the one God, as Jacob believed?

To "have striven with beings divine and human, and have prevailed"—that was quite an accolade for an itinerant shepherd and fugitive who always had to live by his wits. At last, Jacob had achieved the status of a Patriarch and a tribal chieftain, a true heir to Abraham and Isaac, able to communicate directly with the divine. His new appellation, Israel, was thenceforth applied to the entire clan and to all future generations, collectively to be known as Bnei Yisrael

(Children of Israel). The entire land of Canaan was later given the eponymous designation Eretz Yisrael (Land of Israel).

Having dealt first with his father-in-law and then, reportedly, with the angel of God, Jacob was now ready to face his long-estranged brother, whom he had wronged long ago by stealing their father's blessing:

> Looking up, Jacob saw Esau coming, accompanied by four hundred men. He [hastily] divided the children among Leah, Rachel, and the two maids, putting the maids and their children first, Leah and her children next, and Rachel and Joseph last. He himself went on ahead and bowed low to the ground seven times until he was near his brother. Esau ran to greet him. He embraced him and, falling on his neck, he kissed him, and they wept. (Genesis 33:1–4)

Simple-minded and cordial a man was Esau, and he generously had forgiven Jacob for deceiving him twenty years earlier. Jacob nonetheless felt the need to placate Esau by offering him gifts of livestock. Esau declined at first, but took the gifts in the end. (In the time-honored mores of the Near East, taking a gift implies an obligation to reciprocate, but refusing a gift is construed as an insult.)

Esau then offered to lead Jacob to his own home territory in Seir (the province of Edom, southeast of Canaan), but Jacob declined, giving a shepherd's reasons: "My lord knows that the children are frail and that the flocks and herds, which are nursing, are a care to me; if they are driven hard a single day, all the flocks will die. Let my lord go on ahead of his servant, while I travel slowly, at the pace of the cattle before me and at the pace of the children, until I come to my lord in Seir" (Genesis 33:13–14). But Jacob had no intention of following his brother and depending on his generosity. Instead, he traveled toward his ancestral land of Canaan.

Another reason for Jacob's refusal to accept Esau's invitation to join his territory could have been the fear that the two clans might, sooner or later, clash over rights to pasturage and water. (As with Abraham and Lot [Genesis 13:6], two clans with separate herds of livestock could not dwell together where the resources were limited, especially in a time of drought.) In modern terms, we would say that the land's limited carrying capacity could be exceeded, so it was better for the two brothers to separate. This reason is confirmed in a subsequent account of their separation:

> Esau took his wives, and his sons, and his daughters, and all the members of his household, his cattle and all his livestock, and all the property he had acquired in the land of Canaan, and went to another land because of his brother Jacob. For their possessions were too many for them to dwell together and the land

where they sojourned could not support them because of their livestock. So Esau settled in the hill country of Seir [Edom]. (Genesis 36:6–8)

Jacob's clan chose to encamp near the city of Shechem and purchased a tract of land from the sons of Hamor, the king of the city (Genesis 33:19).[35] Wishing to take up residence there, Jacob built an altar on the site, dedicated to El Elohei Yisrael, Israel's God of gods.

Soon, however, the tranquility that Jacob had hoped for was shattered. His daughter Dinah ventured out to meet the Canaanite girls.[36] While alone, she was accosted by Shechem (the young man obviously named after the city, being the son of Hamor). He first raped her and then beseeched his father to ask Jacob for her hand in marriage. When Jacob's sons learned what had happened to their sister, they decided to avenge the egregious breach of the family's honor. (The concept of "family honor" was deeply ingrained and fiercely defended in the tribal culture of pastoralists in the semiarid steppe. The unwritten law of the domain had to be very harsh: rape was punishable by death, to be carried out by the honor-bound brothers of the victim.)

The scheme conceived and executed by Jacob's sons showed them to have been as cunning as their father and much crueler: "Hamor spoke with them, saying: 'My son Shechem longs for your daughter. Please give her to him in marriage. Intermarry with us: give your daughters to us, and take our daughters for yourselves; you will dwell among us, and the land will be open before you; settle, move about, and acquire holdings in it.' Then Shechem said to her father and brothers, 'Do me this favor, and I will pay whatever you tell me'" (Genesis 34:8–12).

The prospect of an alliance between the denizens of a permanent urban center (including its agricultural environs) and the pastoralists of its wider hinterland could perhaps have been a productive combination. But pastoralists, even while yearning for security, could not easily change their lifestyle. Moreover, the Israelite clan's exclusivity precluded such accommodation and assimilation. Jacob's sons only pretended to agree to Hamor's proposition, while plotting their revenge: "The sons of Jacob answered Shechem and his father Hamor—speaking with guile because he had defiled their sister Dinah—and said to them, 'We cannot . . . give our sister to a man who is uncircumcised. . . . Only on this condition will we agree with you; that you will become like us in that every male among you is circumcised.' . . . Their words pleased Hamor and his son Shechem. And the youth lost no time in doing the thing" (Genesis 34:13–15). Then Hamor and Shechem went to the gate of their city to persuade the men of the city to submit to circumcision. Their reasoning was unabashedly self-serving: "Only on this condition will the men agree with us to dwell among us and be as one kindred: that all our males become circumcised."

Then came the revenge, and it was most gory: "On the third day, when they [men of the city] were in pain, two of Jacob's sons, Simeon and Levi, brothers of Dinah, took each his sword, came upon the city unmolested, and slew all the males. They put Hamor and his son Shechem to the sword, took Dinah out of Shechem's house, and went away. The other sons of Jacob came upon the slain and plundered the town, because their sister had been defiled" (Genesis 34:22–27).[37]

But Jacob was horrified. He reprimanded his hot-blooded sons, saying: "'You have brought trouble on me, making me odious among the inhabitants of the land, the Canaanites and Perizzites; my men are few in number, so that if they unite against me and attack me, I and my house will be destroyed.' But they answered: 'Should our sister be treated like a whore?'" (Genesis 34:30–31).

Here, again, Simeon and Levi expressed the pastoral clan's obsessive preoccupation with honor and exclusivity. They acted out, and thereby reinforced, their sense of alienation, distrust, and hostility toward the people of the land in which they wished to live and resisted the prospect of assimilating with them. Seminomadic pastoralists and sedentary villagers were of different cultures. They could coexist and trade and occasionally clash, but would not readily merge. Typical of the pastoralists was the notion of collective punishment: if one member of a tribe, village, or town had transgressed, the transgressor's whole community was held culpable and was to be punished indiscriminately.

The violent incident at Shechem forced Jacob's clan to leave the vicinity of that city. It also cast a pall of hostility and fear over the clan's relationship with other population centers in the rainfed domain of Canaan. In distress, Jacob sought reassurance from on high:

So Jacob said to his household and to all who were with him, "Rid yourselves of the alien gods in your midst, purify yourselves, and change your clothes. Come, let us go to Beth-El [house of El], and I will build an altar there to the El who answered me when I was in distress and who has been with me wherever I have gone." They gave Jacob all the alien gods that they had, and the rings that were in their ears, and Jacob buried them under the terebinth that was near Shechem. As they set out, a terror from Elohim fell on the cities round about, so that they did not pursue the sons of Jacob. (Genesis 35:2–5)

Thus it was that the early Hebrews (henceforth called Israelites) continued the pattern of living on the fringes of the rainfed land as perpetual outsiders. Wherever they sojourned, Jacob built an altar and offered a sacrifice to Elohim. Then, as they continued moving southward and approached Ephrath, near the town of Bethlehem,[38] pregnant Rachel went into labor and died in childbirth (Genesis 35:19).[39] The surviving child was named Benjamin (son of

LORE AND MORES OF PASTORALISTS:
HOSPITALITY AND REVENGE

In contrast to the regular, repetitive pattern of life in the rainfed and, especially, riverine domains, life in the climatically labile pastoral domain was inherently irregular. Nothing there was steady or stable in the long run. The environment made impossible the establishment of a uniform, centrally governed social order. Instead, it favored small and highly flexible groups, ever willing and able to shift locations, face risks, and adjust to new possibilities (however temporary). Adjustment required a talent for spontaneous improvisation and resourcefulness.

Among the characteristic responses to the imperatives of pastoral life in an arid environment were perception of the larger territory rather than attachment to a specific locale, constant alertness and readiness to seize opportunities, fierce competitiveness, ability to change behavior and mode of life rather than remain fixated on a rigid norm, self-reliance, attachment and fierce loyalty to the clan rather than the larger society, yearning for unattainable permanence and stability, and fascination with traditions of storytelling and with deeds of individual derring-do.

All these were incorporated into a complex culture that, with all its seemingly contradictory and conflicting tendencies, was transmitted as tradition from ancient times to numerous subsequent generations, even long after the original environment and the lifestyle it engendered became anachronistic.

To illustrate the ecological conditioning of pastoral lore and mores, let us consider two traditional patterns of behavior: hospitality and revenge. Both are exemplified quite vividly in biblical stories.

A scene from the life of Abraham is described in Genesis 18 with characteristic brevity, yet with such evocative realism that we can visualize how it might have happened just yesterday, rather than some thirty-five centuries ago.

Abraham was encamped with his clan at a small grove of terebinth trees, somewhere in the savanna-like semiarid rangeland of southern Canaan. Sitting at the opening of his tent one warm afternoon and gazing at the surrounding expanse of shimmering wilderness, Abraham was startled at the sudden appearance of three strangers.

Not knowing their identity, Abraham, we might expect, should have been wary. The unannounced intruders may have been marauders, roaming across the lawless land in search of plunder. Did Abraham therefore draw his sword and summon his menservants to ward off the danger? No. Instead, Abraham—alone and apparently unarmed—ran to greet the strangers. He bowed to them, begged them to rest in the shade of the tree, and offered them water to wash their tired feet. Hurriedly, he then urged his wife to bake cakes. Next he ran to the herd to fetch a tender calf and had his servant dress it for the guests. Finally, he served them personally with the cakes, meat, milk, and curds.[1]

Only then did the guests reveal themselves to be angels, who were there to bring Abraham both good and bad tidings: Abraham's wife, Sarah, notwithstanding her age of ninety, would bear him a son in one year hence, but the city of Sodom, where Abraham's nephew Lot resided, was about to be destroyed. Initially, however, Abraham was completely unaware of the strangers' identity. Why, then, did he trouble himself just to accommodate anonymous passersby? The standard answer is that Abraham was a man of extraordinary generosity, a charitable and compassionate man, ever ready to assist the needy.

But Abraham's personality alone cannot be the entire explanation. Bedouin throughout the Middle East have always prided themselves on hospitality to strangers. It is their longstanding custom. So the question shifts from the specific case of Abraham to the traditional hospitality that is characteristic of nomads throughout the pastoral domain.

A custom so pervasive and lasting must serve a purpose for those who practice it. A common explanation is that it behooves each nomad to be hospitable to others because, on another occasion, the situation may be reversed and the host may need the help of a former guest. That is to say, hospitality in the present is a way of acquiring a sort of insurance for the future. But what if, at a time of dire need, the former host is unlucky enough to encounter a total stranger rather than the former guest (who, indeed, may not remain forever grateful and willing to reciprocate a long-past kindness)? A more plausible explanation for Bedouin hospitality must be postulated on the attainment of an immediate, concrete benefit, rather than on the vague expectation of a future benefit in some hypothetical circumstance. That is to say, altruism must somehow serve direct self-interest. But in what way?

Consider the situation of a typical pastoral family or clan eking out an existence on the dry range. Its stay in any one place is necessarily temporary. Within weeks, the pasture is likely to be overgrazed, the water in the cistern depleted. Where should the family go next for subsistence? North, south, east, or west? Where is there good pasture and a source of water? Where are the friendly clans, and where are the hostile ones? Where might this clan find a suitable bride for a grown son or a husband for a marriageable daughter? Is a foreign invasion taking place, or an infestation of locusts? If so, where? These are crucial, life-or-death questions.

What the clan needs is *information*. But the territory is too extensive, variable, and insecure to be explored independently by a small clan. In the absence of any organized mode of communication, the best sources of information are the reports of wayfarers. Therefore, the imperative is to be kind to travelers. Give them food and water, and sound them out. If they came from the west, they can describe what is there. And if they are pleased with the hospitality accorded them, they may stop again on their return and report on the situation in the east as well. And if a clan gains wide recognition for welcoming visitors, perhaps more wayfarers will come by and provide even more frequent and complete information.[2]

"Information is power" goes the cliché in our so-called Information Age. We tend to think of it as a modern idea. In fact, it is an age-old idea. Moreover, it has long been a necessity. Its expression in the pastoral domain of the Middle East is intrinsic precisely because it is rooted in the ecology of the region and in the mode of life that evolved there. A self-serving need for information thus induces altruistic behavior.

Bedouin hospitality, exemplified in the story of Abraham and the angels, is but one example of the way in which ecological circumstances that prevail in each domain condition individual and social behavior. To be sure, nothing is absolutely predetermined. Individual choice and character enter here, as elsewhere. Hospitality involves costs, not only benefits. The food, water, and lodging provided to strangers diminish the meager resources available to a clan that subsists on the thin edge of survival, especially in years of drought. An even greater risk is that some recipients of the generosity might abuse it. In the long run, however, it appears that the benefits of hospitality in the pastoral domain must have outweighed the costs in enough cases to make it worthwhile, for otherwise the custom could not have persisted.

The Bible attests that the pastoralist custom of hospitality to strangers was not practiced by town dwellers. When the angels arrived in Sodom, Lot received them most generously, in the same manner as Abraham. Although Lot had taken up residence in the town, he maintained the pastoral tradition. But the townspeople of Sodom acted quite differently (Genesis 19). They regarded strangers as intruders and treated them with hostility. Equally inhospitable and cruel to a wayfarer were the denizens of Gibeah (Judges 19).

continued

Hospitality to strangers is the benevolent aspect of pastoral life. There is, however, a malevolent aspect, and it is at least equally potent and prevalent. Its most violent expression is the so-called blood feud. At the tribal level, it can cost the lives of scores of innocent people over a period of several generations.[3] Underlying and impelling the practice is the assumption of collective responsibility and guilt. When a person from one clan commits an offense against a person from another clan, the offender's entire clan is held liable. All members of the victim's clan can exact revenge; indeed, they are duty-bound to see to it that revenge is carried out. The purpose of this custom is not only to seek justice for past transgressions and deter future ones, but also—in some cases primarily—to redeem the injured honor, status, and pride of the clan as a whole. (The expanded, nationalistic manifestation of the blood feud can result in international war and cost the lives of many thousands, at present no less than in the past.)

The ecological background to the blood feud is not difficult to perceive. In the semi-desert, where people are few and resources scarce, shepherds typically roam the range alone or in pairs, leading small flocks to sparse pastures along winding creek beds and narrow ravines. Clashes over access to meager supplies of water or grass can often break out between competing shepherds. Such clashes tend to turn violent and may result in murder. In the lonely fastness of the semidesert, however, few, if any, witnesses may be present, so an offender can all too easily escape detection. In the absence of any constituted authority capable of investigating crimes, ascertaining guilt, and ensuring due process of law, it remains for the clans themselves to make and enforce their own law.

It tends to be a cruel law. If the identity of the offender is suspected but his clan shields him, or if his identity is unknown, all members of his clan are held to be liable, without limitation of time. Many years may pass, yet the offense is indelibly etched in the tribal memory and the law must be applied eventually, if not by the first generation, then by the next and the ones to follow. The harsher the application of law, the greater its value as a deterrent to future crime—so goes the rationale. If innocents are victimized in the process, so be it, as long as the debt of blood and honor is redeemed. And then, of course, a counter-revenge is required by the most recently offended clan. As reciprocal retributions escalate into tribal warfare, the original offense may be forgotten entirely. The violence continues to feed on itself and to justify itself afresh with each new killing.[4]

There was, though, a countercurrent in the evolving culture of the Israelites that recognized the malevolent and destructive nature of the vengeful tradition and tried to moderate or even eliminate it. The quality of mercy is featured in the Bible even more prominently than its opposite. Witness, as one example among many, the characterization "Yahweh, Yahweh, El compassionate and gracious, slow to anger, abounding in kindness and faithfulness, extending kindness to the thousandth generation, forgiving iniquity, transgression, and sin" (Exodus 34:6). But then the very next sentence reverts to the negative: "Yet He does not remit all punishment, but visits the iniquity of parents upon children and children's children, upon the third and fourth generation."

The most famous and widely quoted statement of loving kindness appears first in Leviticus: "Love your fellow as yourself" (Leviticus 19:18). That injunction is applied to strangers as well: "The stranger who resides with you shall be to you as one of your citizens; you shall love him as yourself, for you were strangers in the land of Egypt: I Yahweh am your Elohim" (Leviticus 19:34).

The principle that each individual (not the clan collectively) is responsible for his or her own transgression is stated in Deuteronomy: "The father shall not be put to death for the children, neither shall the children be put to death for the father; every man shall be put to death for his own sin" (Deuteronomy 24:16).

After the settlement in Canaan, a practical means was to be devised for preventing the vendetta tradition from continuing unchecked. Six towns, strategically distributed in the country, were declared to be safe havens,[5] where a murderer could find asylum from the "avenger of blood" (Numbers 35:9–34). Guilt could also be expunged by intoning a prayer (Deuteronomy 21:1–9).

---

1. This menu contravenes the common interpretation of Jewish dietary laws and calls into question the entire basis for the ban on mixing dairy and meat products. Significantly, the guests ate the meal willingly and then revealed themselves to be messengers of God. The explanation that this event took place before the granting of the Torah in Sinai seems facile in light of the fact that David and his men, many centuries later, also ate meat with a dairy product (2 Samuel 17:29). An alternative explanation is discussed in the box "On Cooking a Kid in Its Mother's 'Milk,'" in chapter 9.

2. In addition to information, visitors can provide companionship, entertaining stories to relieve boredom, and tradable goods. But information gathering is, arguably, the most vital aspect of traditional Bedouin hospitality.

3. C. S. Jarvis, *Yesterday and Today in Sinai* (Edinburgh: Blackwood, 1933).

4. The word *naqam* (vengeance) in its various forms and declinations is mentioned no fewer than sixty-two times in the Bible, and its synonyms appear many more times. An example is the words of the prophet Nahum (whose name actually means "comforted"): "Yahweh is a jealous and avenging El, Yahweh is vengeful and fierce in wrath; Yahweh takes vengeance on His adversaries, and He rages against His enemies" (Nahum 1:2). Observers of the current scene in the Middle East can recognize the atavistic reversion to the ancient vendetta exemplified in the biblical story of the "eternal" enmity between the Israelites and the Amalekites: "Remember what Amalek did to you. . . . Therefore . . . you shall blot out the memory of Amalek from under heaven. Do not forget!" (Deuteronomy 25:17–19).

5. The Tent of Yahweh, the holiest site before the construction of the Temple by Solomon, was also considered a safe haven. At least once, however, the sanctity of that refuge was honored in the breach. It happened after Joab, King David's nephew and army commander, supported Adonijah rather than Solomon for the royal succession. After Solomon was crowned king, Joab escaped to the Tent of Yahweh, only to be slain there by Solomon's general, Benaiah (1 Kings 2:28–34).

my right hand). From Ephrath, the clan evidently moved farther from the rain-fed domain into the semiarid pastoral domain: "Israel journeyed on, pitched his tent beyond Migdal-eder" (Genesis 35:21). The name of the place means "tower of the flock," suggesting an outpost and observation tower, from which the grazing flocks could be observed and controlled.

Thereafter, Jacob-Israel faded gradually from the center of the biblical stage, which increasingly was taken over by his sons. His diminished status is highlighted by the blatant usurpation of paternal authority committed by the eldest son, Reuben: "While Israel stayed in that land, Reuben went and lay with Bilhah his father's concubine, and Israel found out" (Genesis 35:22). Bilhah had been the handmaid of Rachel, who had submitted her to Jacob, by whom Bilhah had borne two sons: Dan and Naphtali. Although Israel apparently did nothing about Reuben's indiscretion at the time, he certainly did not forget it. In his last message to his sons, he chided Reuben:

Reuben, you are my firstborn,
my might and first fruit of my vigor,
exceeding in rank and exceeding in honor.

An old shrine, traditionally believed to be the tomb of Rachel at Ephrath,
near Bethlehem (Genesis 35:20).

Unstable as water, you shall excel no longer;
for when you mounted your father's bed,
you brought disgrace—my couch he mounted! (Genesis 49:3–4)[40]

We learn much more about the mores and sexual practices of this pastoral
clan from the story of Judah, Jacob's fourth son, and his Canaanite daughter-
in-law, Tamar (Genesis 38). Judah took a Canaanite wife, Bath-Shua (daughter
of Shua), and sired three sons. When the eldest, named Er, came of age, Judah
chose a wife for him named Tamar. Er, however, "was displeasing to Yahweh"
and died before siring a son of his own. So Judah ordered his second son,
Onan, to perform the duty of a deceased husband's brother (a practice known
as levirate), to beget a child with the widow on behalf of his brother in order
to perpetuate the line: "But Onan, knowing that the seed would not count as
his, let it go to waste whenever he joined with his brother's wife. . . ."[41] What
he did was displeasing to Yahweh, and He took his life also. Then Judah said
to his daughter-in-law Tamar, 'Stay as a widow in your father's house until my
son Shelah grows up'—for he thought 'He too might die like his brothers'"
(Genesis 38:9–11).

Evidently fearing that Tamar might be carrying an innate curse, Judah
failed to fulfill his promise to her. So Tamar remained in her father's house and
languished there, until finally she made the brave decision to take matters into
her own hands:

A long time afterward, Shua's daughter, the wife of Judah, died. When his period of mourning was over, Judah went up to Timnah to his sheepshearers, together with his friend Hirah the Adullamite. And Tamar was told, "Your father-in-law is coming up to Timnah for the sheep shearing." So she took off her widow's garb, covered her face with a veil, and, wrapping herself up, sat down at the entrance to Enaim, which is on the road to Timnah; for she saw that Shelah was grown up, yet she had not been given to him as wife. When Judah saw her, he took her for a harlot for she had covered her face. So he turned aside to her by the road and said, "Here, let me come unto you." . . . "What," she asked, "will you pay for coming unto me?" He replied, "I will send a kid from my flock." But she said, "You must leave a pledge until you have sent it." (Genesis 38:12–17)

Judah then gave her the pledge she had asked for—his staff and his signet and cord—and he came into her, and she conceived by him. Then she returned to her home in Timnah and to her widow's weeds.[42] And when Judah sent his friend Hirah to deliver the animal that he had promised and to retrieve his belongings from the *kdesha* by the side of the road, she could not be found.

The term *kdesha* refers to a temple harlot, a class of women dedicated to the Canaanite worship of Ba'al and Ashera. The word is derived from *kdushah*, which means "sanctity"—that is, a temple harlot dedicated to the pagan fertility rite. A few temple harlots may have strayed on occasion, or been released, from their cultic duties to engage in common prostitution, or so at least thought Judah when he approached the disguised Tamar. Anyhow, their brief encounter had its consequences: "About three months later, Judah was told, 'Your daughter-in-law Tamar has played the harlot; in fact, she is with child by harlotry.' 'Bring her out,' said Judah, 'and let her be burned.' As she was being brought out, she sent this message to her father-in-law, 'I am with child by the man to whom these belong'" (Genesis 38:24–25). Then Judah admitted that he had wronged Tamar by denying her the customary right to be betrothed to her deceased husband's younger brother.

Tamar later gave birth to twin sons: Perez and Zerah. Thus ends the story of Tamar's remarkable courage and resourcefulness, as told in Genesis. What the biblical account reveals only much later (1 Chronicles 2:4–15) was the eventual consequence of Tamar's liaison with Judah: their tenth-generation grandson was none other than David, destined to be the king of Israel![43]

JACOB'S SONS continued the pastoral lifestyle, but they were consumed by jealousy toward their young brother Joseph, who, being the son of beloved Rachel, was their father's favorite: "At seventeen years of age, Joseph tended the flocks with his brothers . . . and he brought bad reports of them to their father. . . . Now Israel loved Joseph best of all his sons, for he was the child of his old age; and he had made him an ornamented tunic. And when his brothers saw that their father loved him more than any of his brothers, they hated him" (Genesis 37:1–4).

Because of the nomadic character of herding, the members of a pastoral clan occasionally lost contact with one another. Jacob, oblivious to the resentments borne by the older brothers toward Joseph, sent him to find out how his brothers were faring. After wandering in the field and losing his way, Joseph finally found his brothers in the vale of Dothan (in the hill district of central Canaan).

As Joseph's brothers saw him approaching, their resentment flared up and they schemed to rid themselves of him. After some deliberation, they threw him into an empty cistern and then decided to sell him to a passing caravan of Ishmaelite (Midianite) merchants—Semitic traders carrying aromatics, spices, and medicinal herbs from Gilead toward Egypt—for 20 shekels of silver. Then they dipped Joseph's garment in the blood of a goat and brought it to their father, to make him think that a wild beast had devoured Joseph.

THEREAFTER, the biblical story shifts to the riverine domain of Egypt, where Joseph was sold into slavery and cast into prison under a false accusation of attempted rape. Being remarkably resourceful, he managed not just to extricate himself, but to be appointed viceroy of Egypt. In that position, he saved the people of Egypt, as well as his own clan, from famine. (But that is a story to be told more fully in the next chapter.)

In times of drought in the rainfed and pastoral domains of southwestern Asia, nomadic or even seminomadic pastoralists tended to gravitate toward the lush wetlands on the eastern fringes of the Nile delta, both to seek pasture and to purchase grain:

> The famine spread over the whole world. So all the world came to Joseph in Egypt to procure rations. (Genesis 41:56–57)

> When Jacob saw that there were food rations to be had in Egypt, he said to his sons, ". . . Go down and procure rations for us there, that we may live and not die." (Genesis 42:1–3)

As the brothers carried out their father's order, they encountered Joseph in his capacity as viceroy of Egypt, in charge of all grain supplies. Although they did not recognize him, he recognized them.

Eventually, Joseph revealed himself to his brothers[44] and invited the entire clan to settle in Egypt in order to escape the continuing famine in Canaan. So they made the trek across northern Sinai: "Jacob set out from Beer-sheba. The sons of Israel put their father Jacob and their children and their wives in the wagons that Pharaoh had sent to transport him; and they took along their livestock and the wealth that they had amassed in the land of Canaan. . . . The total of Jacob's household who came to Egypt was seventy persons" (Genesis 46:5–6, 27).

When they arrived, Joseph introduced five of his brothers to Pharaoh:[45]

Pharaoh said unto his brothers, "What is your occupation?" They answered Pharaoh, "We your servants are shepherds, as were also our fathers. We have come . . . to sojourn in this land, for there is no pasture for your servants' flocks, the famine being severe in the land of Canaan. Pray, then, let your servants stay in the region of Goshen." Then Pharaoh said to Joseph, ". . . Let them stay in the region of Goshen. And if you know any capable men among them, put them in charge of my livestock." (Genesis 47:3–6)

The Israelites' migration to Egypt was a matter of expedience only. As nomadic pastoralists, they could not feel at home in the rigidly organized and regimented society of Egypt:

Jacob lived in the land of Egypt seventeen years . . . and the time drew near that Israel must die. . . . So he called his son Joseph, and said to him: ". . . Place your hand under my thigh[46] as a pledge of your steadfast loyalty: please do not bury me in Egypt. When I lie down with my fathers, take me up from Egypt and bury me in their burial place. . . . I am about to die, but Elohim be with you and bring you back to the land of your fathers." (Genesis 47:28–30, 48:21)

By "land of your fathers," Jacob was not referring to Mesopotamia or Haran, where the clan had reportedly originated, but to Canaan. The Patriarchs to whom Jacob alluded had wandered throughout the fringes of that land, never having possessed it. Abraham had left it to go to Egypt and had returned to it. Isaac had remained in southern Canaan, but had had to contend with his sedentary neighbors over grazing and water rights. Jacob and his offspring had left Canaan for only the duration of the drought there, but had tarried in Egypt for several generations. Yet they continued to yearn for Canaan. In actuality, it was a danger-filled and drought-prone land, already densely occupied by a populace that was generally unfriendly toward strangers. But from the day Abraham obeyed the call to go there, Canaan had become home to the Hebrews. Over their history, they would leave it repeatedly, whether voluntarily or forcibly, yet carry the idealized memory of it with them and strive to return. It became one of the defining themes of their national and religious culture.

THE ISRAELITES tarried in the pasture-rich land of Goshen (on the eastern fringes of the Nile delta) long after the drought that had caused them to leave Canaan had ended. They did not assimilate among the Egyptian people, however, but retained their ethnic separation and distinct lifestyle. And as they multiplied in numbers, and as each of the siblings' households became a tribe, they were resented and feared by the Egyptians. The Egyptians' suspicions of the Semitic aliens in their midst may well have been fueled by memories of Egypt's subjugation by the Hyksos, between the eighteenth and sixteenth centuries B.C.E. And so the Egyptians eventually enslaved the alien Israelites and forced them into hard labor.

Some generations later, there was born in Egypt a child named Moses, who was to become the greatest leader in the history of the Israelites, the man who, so the Bible tells us, led his people out of bondage, gave them a code of Law (the Torah), and forged them into a nation prepared to enter and possess their promised land. Significantly, Moses could assume his role as liberator and spiritual guide only after having had to leave the riverine domain of Egypt in order to experience life as a shepherd in the semiarid pastoral domain. That experience was an essential aspect of Moses's initiation and preparation for his mission, as it enabled him to understand and identify with the ancestral tradition and future destiny of the Israelite people.

According to the account of his life in the Book of Exodus, Moses was born into a family of the tribe of Levi, was saved miraculously from death as an infant, was adopted by Pharaoh's daughter, and was raised as a prince in the royal family of Egypt. One day, he ventured out of the palace and chanced to witness the cruelty of an Egyptian officer toward an Israelite slave. Moses intervened and smote the oppressor. Consequently, he had to flee from Egypt. His journey was the reverse of that of Jacob's clan, but was similar in a way to the journey of Abraham, who had left a riverine domain to take up shepherding in the pastoral domain. Moses went to Midian, presumably in the northeastern corner of Sinai or in northwestern Arabia. There he helped the daughters of Jethro (or Reuel), the priest of Midian, draw water for their flock. After marrying one of the daughters, Zipporah, he became a shepherd, tending the flock of his father-in-law. And it was while shepherding that flock alone in the wilderness that Moses encountered Yahweh Elohim in the burning bush and was appointed to be the shepherd of his people—the Children of Israel.

5

THE SECOND RIVERINE DOMAIN

Sojourn and Slavery in Egypt

*There was famine in all the lands, but throughout
the land of Egypt there was bread.*

GENESIS 42:54

THE RIVERINE DOMAINS of both Mesopotamia and Egypt depended on the ebb and flow of rivers that swelled annually and provided water for irrigation. Because an excess of water could be as damaging as its paucity, the societies in both domains depended on the effective regulation of river water. Although the hydraulic civilization of Egypt may have begun somewhat later than that of Mesopotamia, the two centers of culture were contemporaries for long periods of history and constituted diametrically opposed powers vying for hegemony over the entire Fertile Crescent. Their rivalry continued for many centuries and was most intense in the second and first millennia B.C.E., during which great armies (among the earliest massed armies in the history of warfare) marched back and forth to test which of the two centers would predominate. The nations unfortunate enough to be located along their paths were often trampled as a result of that antagonism. Both civilizations had their periods of greater and lesser ascendancy. At various times, each won and lost wars, dominated other nations, and was itself subjugated. Moreover, both suffered occasional outbreaks of pestilence, as well as episodes of floods and droughts that caused famine and periodically decimated their populations.

Through all these vicissitudes, the civilization of Egypt survived (albeit with shifts of power between Upper and Lower Egypt and their eventual unification), continuing in the same stable location. In contrast, the nations of Mesopotamia—Sumer, Akkad, Babylonia, and Assyria—rose and then declined, in turn, as the center of population and power shifted progressively from the southern to the central section of the Tigris–Euphrates valley and then alternated between the northern and the central section while the lower valley never recovered.

Egypt, in the famous words of Herodotus, was "the gift of the Nile."[1] There were actually two gifts: water and silt. These were the same gifts brought by the Tigris and the Euphrates to southern Mesopotamia, but there were important differences. Neither clogging by silt nor poisoning by salt were as significant in Egypt as in Mesopotamia, so the land of Egypt remained productive, while the land of lower Mesopotamia suffered degradation.

The productive land of Egypt is divided into two sections: Upper and Lower Egypt. The strip of floodplain that flanks the upstream section of the Nile in the south is called Upper Egypt. Over a south–north distance of hundreds of miles, the well-watered floodplain is only a few miles wide and is hemmed in by rugged deserts on both sides. Hence the traditional name for Egypt in Hebrew, Mitzrayim, means literally "a narrow place." In contrast, the broad delta in the north, where the river fans out into a series of distributaries, is called Lower Egypt, and it includes more than half of Egypt's potentially arable land. For many centuries, however, much of the delta consisted of undrained, reedy marshes with numerous branching streams that had to be regulated before the area could become truly productive, and that task was achieved only gradually.

The rich silt of the Nile valley comes mainly from the steep volcanic highlands of Ethiopia, lashed each summer by the monsoonal rains that roll in from the Indian Ocean. The downpours scour the slopes, scraping off their loose mantle of nutrient-rich dark soil and splashing it into the annual flood of the Blue Nile. Added to that mineral silt is the organic humus contributed by the White Nile from its source in the rain forests and swamps of equatorial central Africa.

The name Nile (Greek, Neilos; Latin, Nilus; Arabic, Nil) may derive from the Semitic root *nahal*, which means "vale" or "stream." Or Nile may have evolved from *nil*, a word of uncertain origin that means "blue" or "indigo"— perhaps related to the Sanskrit *nila* (dark blue). The ancient Egyptians named the river Ar or Aur (preserved in the Coptic Iaro, which resembles Yeor, the biblical name for the river), meaning "black," in reference to the color of the silt deposited by the river during its flood stage. That dark silt has given the land itself one of its ancient Egyptian names, Khami.[2]

When the gathering flood reached Egypt, it would overflow the riverbanks and deposit an annual increment of silt, estimated to have averaged about 0.04 inch (1 mm) thick, on the floodplain. This amount was not so excessive as to choke irrigation canals or cover young seedlings, but it was fertile enough to nourish crops. The Egyptians divided their year into three seasons. The first was the time of inundation, from when the river began to rise and flood the land until it started to recede sufficiently to permit sowing. The second season was the period of crop growth, between sowing and harvesting. The third was the interval of low water, between the harvest and the advent of the next inundation. Each of these seasons lasted for about four lunar months. The Egyptians considered the rise of the Nile to be the effective beginning of their year. But although remarkably regular, the river's regimen was not perfectly so. In some years, the Nile began to rise early, whereas in other years, it seemed to tarry and did not begin to flood until a month or so later. Hence the effective length of the Egyptian year might fluctuate from eleven to thirteen months. Eventually, the Egyptians devised a better way to determine the length of their year, by pegging their calendar to the rising of the brightest star, Sirius.

The early farmers of Egypt, around 5000 B.C.E., probably relied on natural irrigation by the unregulated floods to water the banks of the river. As soon as the floodwaters withdrew, they could cast their seeds in the mud. At times, however, the flood did not last long enough to wet the soil thoroughly, and then the crops would fail and famine would ensue. So the Egyptian farmers learned to build earthen dikes around their plots, thus creating basins in which a desired depth of water could be impounded until it soaked into the ground and wet the soil deeply enough to sustain the roots of crops throughout the growing season. The diked basins also retained the nutrient-rich silt and prevented it from running off with the receding floodwaters.

The earliest pictorial record of artificial irrigation is the mace head of the "Scorpion King," which depicts a ceremonial cutting of an irrigation channel. The Scorpion King is shown holding a hoe, with some laborers excavating the canal and others holding a basket and a broom, all standing along the channel. Rectangular irrigation basins can be seen in the background.

The Scorpion King (so called because his visage is always accompanied by a symbolic scorpion), inaugurating an irrigation canal. The carving is on the head of a mace (ca. 3100 B.C.E.), a personal weapon and symbol of authority in ancient Egypt.

The basis of Egypt's productivity was the nearly optimal combination of water and soil nutrients provided by an annual regimen that was more dependable and timely (although not perfectly so) than the capricious floods in Mesopotamia. It enabled the Egyptian farmers to produce a surplus that fed the artisans, scribes, priests, merchants, noblemen, soldiers, and—above them all—pharaohs, who used their coercive power to order the building of self-aggrandizing monuments. These monuments still stand, less in testimony to the vainglorious pharaohs who ordered them than to the diligence and organization of a society of farmers and builders, nurtured by the Nile and rooted in the land irrigated by it.

EGYPT: THE STABLE CIVILIZATION

In contrast to the fragility of the successive civilizations of southern Mesopotamia, the civilization of Egypt exhibited remarkable stability over several millennia, essentially in the same location. What explains this persistence of irrigated farming in Egypt in the face of its demise in southern Mesopotamia? The answer lies in the different soil and water regimes of the two riverine domains. Neither clogging by silt nor soil degradation by waterlogging and salination was as severe along the Nile as in the lower plain of the Tigris and Euphrates Rivers.

Whereas in Mesopotamia the inundation usually came in the spring, and evaporation in the summer tended to salinize the soil, in Egypt the Nile began to rise in mid-August and attained its maximum height in early October. Thus in Egypt, the flooding came at a much more favorable time for the planting of winter crops: well after the spring harvest and after the summer heat had killed the weeds and aerated the soil.

The narrow floodplain of the Nile (except in the delta), as well as the deep-cut riverbed and the rough topography of the adjacent desert, made it generally impractical to divert the water through long canals and spread it over large tracts, as was done in Mesopotamia. Hence there was no extensive raising of the underground water table. Instead, the water table was controlled by the stage of the river, which, over most of its length, lies below the level of the adjacent land. When the river crested and inundated the land, the seepage naturally raised the water table temporarily. Then, as the flood receded, the water level dropped and pulled the water table down after it. This all-important annual pulsation of the Nile and the associated fluctuation of the water table under a free-draining floodplain (with a permeable sandy subsoil) created an automatically repeating, self-flushing cycle by which the salts were leached from the irrigated land and carried away by the river itself.[1]

1. Unfortunately, the soil of Egypt, historically famous for its durability and productivity, is now threatened with degradation. The Aswan High Dam, inaugurated in 1970, has blocked the fertile silt that had been delivered by the Nile. The river itself, now running clear of silt, has increased its erosivity and has been scouring its own banks. And along the estuaries of the delta there is no more deposition, so the coast has been subject to progressive erosion and to intrusion of seawater. Finally, the artificial maintenance of a nearly constant water level in the river, necessary to allow year-round irrigation and continuous or successive cropping, has raised the water table. So Egypt is now subject to the maladies of waterlogging and salination, to which it had for so long seemed immune. See D. Hillel, *Out of the Earth: Civilization and the Life of the Soil* (Berkeley: University of California Press, 1992).

Aspects of riverine life and farming in ancient Egypt, showing the Nile and its distributary canals, fruit-bearing orchards, plowing and planting, and harvesting grain. This painting is on a wall of a tomb at Thebes.

Some scholars have conjectured that the arduous work of building the grand monuments, as well as the other public works of Egypt (including granaries, canals, levees, and fortifications), was done mainly by farmers turned wage earners during the slack periods between planting and harvesting.[3] Another possibility is that a great deal of the construction was performed by conscripted laborers forced to serve the authorities. Some of the work may also have been done by foreign slaves, more or less in the manner described in the Book of Exodus. The political and ecclesiastical structure of Egyptian society was strongly hierarchical. With the nearly absolute authority and proclaimed deity of the pharaohs, who were presumed to hold sway over the cosmic order and over famine and plenty, ancient Egypt conformed perhaps more closely

Agricultural activities in ancient Egypt—plowing, sowing, and hoeing—as depicted on a wall painting dated to the fourteenth century B.C.E.

than Mesopotamia to anthropologist Karl A. Wittfogel's concept of a despotic "hydraulic civilization."[4]

The affluence of the riverine lands has always attracted the hungry dwellers in the bordering desert and pastoral domains. A document from the nineteenth century B.C.E. mentions nomads begging to enter Egypt to keep themselves and their flocks alive.[5] Egyptian border garrisons generally warded off such would-be intruders. Occasionally, however, some were admitted. The Bible relates that Abraham, and later Jacob and his sons, sojourned in Egypt during times of famine in Canaan. After the mid-eighteenth century B.C.E., "wretched Asiatics" invaded and conquered Lower Egypt. These intruders, called Hyksos (rulers of foreign lands), were expelled about two centuries later.

For thousands of years, the people of rainless Egypt owed their very existence to a river that flowed mysteriously and inexplicably out of the greatest and most forbidding desert in the world. Had the Nile failed to rise and overflow, even for one season, all Egypt would have perished. Hence the rulers of ancient Egypt regularly sent emissaries upstream to observe and report the coming of the annual flood. And yet, clever though they were, the Egyptians remained totally ignorant of the ultimate source and cause of the Nile's life-giving waters, on which they depended absolutely. Not knowing whence the river came meant having no assurance that it would continue to nurture Egypt's population. Living in a state of uncertainty, the Egyptians could be sustained only by faith, as illustrated in a text dating to the reign of Djoser (ca. twenty-eighth century B.C.E.):

I asked . . . What is the birthplace of the Nile? Who is the god there? . . . He said to me: There is a city in the midst of the waters [from which] the Nile rises, named Elephantine. It is the beginning of the beginning. . . . It is the joining of the land, the primeval hillock of earth, the throne of Re, when he reckons to cast life beside everybody. . . . The Two Caverns is the name of the water; they are the two breasts that pour forth all good things. It is the couch of the Nile, in which he becomes young [again]. . . . He fecundates [the land] by mounting as a male, the bull, to the female; he renews [his] virility, assuaging his desire. He rushes twenty eight cubits [high]; he hastens at Diospolis seven cubits [high]. Khnum is there as a god. . . . As I slept in life and satisfaction, I discovered the god standing over against me. I propitiated him with praise; I prayed to him in his presence. He revealed himself to me, his face being fresh. His words were: "I am Khnum, thy fashioner. . . . I know the Nile. When he is introduced into the fields, his introduction gives life to every nostril. . . . The Nile will pour forth for thee, without a year of cessation or laxness for any land. Plants will grow, bowing down under the fruit. . . . The starvation year will have gone, and borrowing from granaries will have departed. Egypt will come into the fields, the banks will sparkle . . . and contentment will be in their hearts more than that which was formerly."[6]

Most of all, the Egyptians believed in the eternal cycle of life, death, and res-urrection—a never-ending sequence exemplified by the river itself. Indeed, they deified the river, praised its powers, and prayed for its benevolence. They associated the river with the god Hapi, depicted as a chubby hermaphrodite with large breasts, the genitals of a man, and a crown of papyrus reeds. And although from time to time the river turned capricious, alternately miserly or superabundant, it never withheld its waters completely. Each year, on a schedule seemingly as regular as the cycle of heavenly bodies, it surged and overflowed its banks and poured forth its dark brown waters over the land of Egypt. Ideally, that would happen at the end of summer, just in time for the planting season in autumn. Thus, wondrously, the Nile River created the most fertile of all lands.

The lore of the ancient Egyptians expressed their perceptions of the world, their vision of the regenerative journey made diurnally by the unblinking sun (traversing a cloudless sky) and annually by the Nile, as well as their hopes and fears regarding the human condition. They pinned their faith on a vast pan-theon of gods and goddesses, and recorded their notions in picturesque detail on the walls of tombs and temples and on scrolls of papyrus.

A brief but significant episode in Egyptian history was the reign of Amen-hotep IV (or Amenophis IV, ca. 1372–1355 B.C.E.), who was an iconoclast

A dual representation of the Nile god, Hapi, as unifier of Upper Egypt (symbolized by the lotus flowers on the left) and Lower Egypt (symbolized by the papyrus reeds on the right), found in Abu Simbel and dated to the thirteenth century B.C.E.

Since rainfall was unimportant in Egypt, the notion of a thundering storm god never took hold. The importance of river water led, in both Egypt and Mesopotamia, to the principle of waters as the beginning of all life. The same idea also entered the perception of the early Israelites and found expression in the two accounts of creation in the Book of Genesis.

In the long course of Egypt's cultural history, several cosmologies were formulated. In the tradition of Heliopolis (dating to the unification of Lower and Upper Egypt, around 3000 B.C.E.), the universe consisted initially of a limitless ocean of inert water that lay in silent darkness. (This concept is reminiscent of the Bible's account of the precreation world as being "unformed and void, with darkness over the surface of the deep, and a wind of Elohim hovering over the waters" [Genesis 1:2].) The Egyptians envisaged that primeval condition as a being called Nu or Nun. The vast expanse of lifeless water continued to exist even after creation and was imagined to surround the celestial firmament guarding the sun, moon, stars, and Earth as well as the boundaries of the underworld.[1] The Egyptians always feared that Nu might crash through the sky and drown the Earth, thus dooming humanity.

Atum, lord of the limits of the sky, arose out of Nu at the beginning of time to create the universe. Atum's semen produced the twin gods Shu and Tefnut, who fashioned Geb and Nut, who gave birth to Osiris, Seth, Isis, Nephthys, and—ultimately—the entire population of the land. The sky goddess, Nut, was depicted as arching over her consort, the earth god, Geb. (Egyptian mythology regarded the Earth as the male principle and the sky as the female, in contrast with Canaanite mythology, which portrayed the sky as a god who mated with the earth goddess.) After giving birth to four offspring, Nut was separated from Geb by Shu, who embodied the air permeated by the rays of the sun. In one depiction, the sun god journeys across the firmament on the underside of Nut's arched body. Reaching the western horizon at the end of the day, the sun god is swallowed by the sky goddess. He then travels through the inside of her body during the night and emerges reborn by Nut in the east amid a display of redness that is the blood of parturition.

Another version attributes the beginning of life to Ptah, the god of Egypt's ancient capital, Memphis. Ptah gave life to the other gods by the union of his heart and the speech of his tongue. By pronouncing their identities, he made all creatures come into being. (Creation by word of mouth is the mode described in Genesis 1 as well.)

During the New Kingdom (1550–1200 B.C.E.), the priests of Thebes attained the heights of eloquence in hymns extolling Amun as the supreme creator, who compassed the sky above and the underworld below. Amun was the burst of energy or the wind that stirred primal matter to begin the process of creation, in a sense similar to the "wind of Elohim hovering over the waters" and then commanding: "Let there be light!" (Genesis 1:2–3).

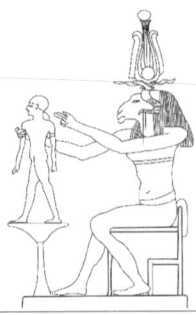

The god Khnum fashioning a man from clay on a potter's wheel, reminiscent of Yahweh shaping Adam from the soil of the Earth (Genesis 2:7).

In still another version of creation, Khnum, a ram-headed god of the cataract region of the Nile, molded the human form on a potter's wheel, presumably from clay (as in Genesis 2:7). Khnum's sacred ram symbolized procreativity in the natural world. He formed gods, animals, birds, fish, and reptiles as well as humans. Khnum also controlled the caverns of Hapi, the god of the inundation of the Nile.

Although the world had a finite beginning, the Egyptian concept was that thenceforth all events followed the regular pattern preordained by the creator of order in the universe, who also created the human race. Human beings are therefore forever subservient to the immutable plan set at the beginning of time, and their lives are predestined. The notion of a static world—marked by endlessly repeated cycles of day and night, winter and summer, life and death and resurrection without any discernible goal—imbued the Egyptians' view of the world, and it derived from the seemingly eternal regularity of their natural environment. The sun shone every day, the desert remained immutable, the river followed its predictable regimen, and there was no reason to suspect that all would do anything else than continue the same pattern forever.

The resignedly passive worldview of the Egyptians is also reflected in the Bible, most explicitly in Ecclesiastes:

> One generation goes, another comes,
> but the earth remains forever.
> The sun rises, and the sun sets—
> and glides back to where it rises.
> Southward blowing,
> turning northward,
> ever turning blows the wind;
> on its rounds the wind returns.
> All streams flow into the sea,
> yet the sea is never full;
> to the place [from] which they flow
> the streams flow back again. . . .
> Only that shall happen which has happened. . . .
> There is nothing new beneath the sun! (Ecclesiastes 1:4–9)

However, the notion of the Israelites that eventually gained predominance was contrary to the static view typical of the Egyptians. It was based on the assumption that the world, now beset by imperfection and transgression, is not meant by Yahweh to stay the same but to progress toward a state of greater order, harmony, and justice.

---

1. G. Hart, *Egyptian Myths* (London: British Museum Press, 1999), 11.

and a radical religious reformer. Soon after ascending to the throne, he broke with established norms and proclaimed his belief in the supremacy of the sun disk, called Aten. The young king renamed himself Akhenaten (also rendered Akhenaton or Ikhnaton), which means "Aten is satisfied" or "Aten's glory." Defying the vested interests of the priesthood and the inculcated beliefs of the people, he defaced every monument on which appeared the name of Amun, long considered Egypt's chief god. Instead, he advocated the sun disk as the supreme deity, with himself as Aten's intermediary to the people. His misshapen face and body were depicted alongside the beautiful face of his wife,

A stone carving depicting Akhenaten and his wife, Nefertiti, worshiping the sun disk, Aten, with its life-giving rays of light and warmth.

Nefertiti (who was his cousin), under the streaming rays of the sun disk. Alas, Akhenaten's reforms did not last.[7] His successor, Tutankhamun, reversed Akhenaten's policies and reestablished the authority of the traditional priesthood.

Much has been made of Akhenaten and his seventeen-year reign. Numerous writers have extolled him as one of history's most original reformers, a courageous individualist, and a harbinger of monotheism. Perhaps the most famous of them was Sigmund Freud, who—in an effort to explain (or explain away?) the biblical claim of originality in the revelation of monotheism to Moses—posited that Akhenaten was a mentor of the Hebrew lawgiver.[8] Akhenaten's philosophy was not humanistic, but absolutist, demanding total, unquestioning submission.[9] It lacked an ethical component. The idea of individual responsibility was foreign to him. Hence he could not have been the forerunner of the principles attributed in the Bible to Moses. In general, the hydraulic civilization of Egypt tended to produce an ordered society that was characterized by regimentation and uniformity. (In contrast, the nomadic pastoral civilization tended to produce a labile and mobile society, more conducive to individuality.)

A more balanced view of Akhenaten now seems to prevail. Certain aspects of his beliefs were indeed original. The perception of the sun god as a divine disk was divorced from anthropomorphic or zoomorphic idolatry. Still, the object of worship was not an abstract or a disembodied spirit, invisible yet present everywhere at all times (as the concept of Yahweh eventually became in the Israelites' religion), but remained a distinct, visible, and physical entity. Moreover, the imperious disk that reigned over the world was not vested with any feelings of love or compassion or concern for the plight of humans. Nor did Akhenaten's sun disk recognize that human beings have free will and can therefore be called on to choose good over evil, to seek justice, and to behave morally and mercifully. If anything, the deity seems to have been excessively dehumanized. Metaphorically, Akhenaten's unblinking sun offered no comforting shade.

In cloudless Egypt, the dominant presence of the sun had long been recognized and worshiped. The solar icon—the sphere (visible as a disk) that moves across the sky from dawn to dusk, that disappears below the western horizon, only to reappear in the east the next morning—had assumed the rank of a deity long before the time of Akhenaten. The cult of Amun had identified that god with Re, using the compound name Amun-Re to betoken a universal deity, a celestial king, who personified all solar attributes. And while Akhenaten and his family worshiped Aten, his courtiers were expected to worship Akhenaten himself. The majority of Egyptians, however, were either ignorant of or hostile to the exclusive nature of this king's faith.

Akhenaten was a gifted poet. His great hymn to the sun disk is a sublime paean to the beauty of creation:

> Thou appearest beautifully on the horizon of heaven,
> Thou living Aton, the beginning of life!
> When thou art risen on the eastern horizon,
> Thou hast filled every land with thy beauty.
> Thou art gracious, great, glistening, and high over every land;
> Thy rays encompass the land to the limit of all that thou hast made.[10]

A similarity of spirit and wording between Akhenaten's hymn and several Psalms (particularly Psalm 104) has been noted:

> How manifold it is, what thou hast made! . . .
> Thou didst create the world according to thy desire. . . .
> All men, cattle, and wild beasts, whatever is on earth, going upon its feet, and
>   what is on high, flying with its wings. . . .
> Everyone has his food, and his time of life is reckoned. . . .
> For thou hast set a Nile in heaven, that it may descend for them and make waves
>   upon the mountains. (Akhenaten)

> How many are the things You have made, O Yahweh!
> You made them all with wisdom.

The earth is full of Your creations. . . .
All of them look to You
to give them their food when it is due. . . .
He established the earth upon its foundations. . . .
You made the deep cover it as a garment.
The waters stood above the mountains. (Psalm 104:24, 27, 5–6)

The similarity may suggest that the Psalms were influenced by the Egyptian precedent or that both hymns followed a long-standing style and theme of religious poetry that had been prevalent throughout the ancient Near East for many centuries.

WE NOW RETURN to the story of Joseph and the consequent sojourn of the Israelites in Egypt. A caravan of Ishmaelite traders (Genesis 37:25), one of many bearing gum, spices, aromatics, and medicinal herbs to a nation that made extensive use of such products for treating maladies as well as embalming the dead, brought Joseph from Canaan to Egypt.[11]

In Egypt, Joseph was sold into slavery to Potiphar, an officer in Pharaoh's court. Joseph won the man's confidence initially, but then Potiphar's wife tried to seduce him. When Joseph rebuffed her advances, she accused him of having attempted to rape her and caused him to be thrown into prison.[12] Even in prison, however, Joseph managed to find favor with the prison master. He then befriended two fellow inmates who were former officers of the royal household, a butler and a baker, and interpreted their dreams. Here, as on many other occasions in the Bible, we encounter the notion of a dream as a divine message, portending the future: "They said to him: 'We had dreams, and there is no one to interpret them.' So Joseph said to them: 'Surely Elohim can interpret! Tell me [your dreams]'" (Genesis 40:8).

By implication, Joseph—endowed from early childhood with an audacious and self-confident personality—thus presented himself as a conduit to Elohim! It is doubtful that anyone in Egypt would have known about the Hebrew god Elohim, but the two imprisoned officers confided their respective dreams anyhow. The chief butler's dream of clusters of grapes on a vine, from which he pressed wine and served Pharaoh, fits into the context of Egyptian agriculture, specifically viticulture. (Similar scenes are depicted in numerous Egyptian wall paintings.) Joseph interpreted the dream as an omen that the butler would soon be cleared of charges and reinstated in the royal household. Then Joseph asked: "Think of me when all is well with you again, and do me the kindness of mentioning me to Pharaoh, so as to free me from this place. . . . I was kidnapped in the land of the Hebrews, nor have I done anything here that they should have put me in the dungeon" (Genesis 40:9–15).[13]

Joseph then interpreted the baker's dream correctly, too, but his dream, unlike the butler's, portended misfortune: three days later, Pharaoh had the baker hanged. But the butler, restored to his former position, did not immediately reciprocate Joseph's favor; he forgot Joseph—but not entirely.

Two years later, an occasion arose to remind him. Pharaoh himself had haunting and enigmatic dreams, which no one could interpret. The dreams turned out to be fateful ones, not only for Joseph but, indeed, for all of Egypt:

> Pharaoh dreamed that he was standing by the Nile, when out of the Nile there came up seven cows, handsome and sturdy, and they grazed in the reed grass. But presently, seven other cows came up from the Nile close behind them, ugly and gaunt, and stood beside the cows on the bank of the Nile; and the ugly gaunt cows ate up the seven handsome cows. And Pharaoh awoke. He fell asleep and dreamed a second time: Seven ears of grain, solid and healthy, grew on a single stalk. But close behind them sprouted seven ears, thin and scorched by the east wind. And the thin ears swallowed up the seven solid and full ears. Then Pharaoh awoke: it was a dream! (Genesis 41:1–7)

In the morning, Pharaoh was troubled, so he summoned the magicians and the wise men of Egypt and told them his dreams, but none could interpret them.[14] Then the butler remembered the young man who had interpreted his dream while they were in the dungeon, and he related the story to Pharaoh. So Pharaoh sent for Joseph and recited his dreams to him.[15] And Joseph provided this interpretation:

> Pharaoh's dreams are one and the same: the Elohim has told Pharaoh what He is about to do. The seven healthy cows are seven years, and the seven healthy ears are seven years. . . . The seven lean and ugly cows that followed are seven years, as are also the seven empty ears scorched by the east wind; they are seven years of famine. . . . Immediately ahead are seven years of great abundance in all the land of Egypt. After them will come seven years of famine, and all the abundance in the land of Egypt will be forgotten. . . . Accordingly, let Pharaoh find a man of discernment and wisdom, and set him over the land of Egypt. . . . Let all the food of these good years that are coming be gathered and let the grain be collected under Pharaoh's authority. . . . Let that food be a reserve for . . . the seven years of famine. (Genesis 41:25–36)

Whoever wrote this account obviously had a profound knowledge of Egypt and of its mostly regular but occasionally anomalous water supply, as well as

A master (*upper left*) oversees farm laborers as they harvest, thresh, winnow, and carry
grain for storage. The scene in this wall painting, from the tomb of Mena and dated to
the fourteenth century B.C.E., is reminiscent of the description of Joseph's supervision
of grain production and storage throughout Egypt (Genesis 41:47–49).

of its cultural traditions. What Joseph recommended was what nowadays is
called drought contingency planning. It is a measure that societies in arid regions must take to cope with extreme swings between periods of surplus and
of dire want. For, regardless of how regular the water supply appears to be
in most years, such are the vagaries of climate that a drought, or a series of
droughts, is certain to occur sooner or later. Nowadays we know that peren-
nially recurring shifts in weather patterns in various regions may be related
to such phenomena as El Niño–La Niña cycles.[16] A recent study has found
that those cycles may correlate with fluctuations in the annual flood stage of
the Nile, as measured by Nilometers.[17] So it was not at all unreasonable for
Joseph to assume that a succession of good years (with copious flooding of the
Nile) may well be followed by the opposite phase of the cycle—a succession of
several low-flow years. The number seven is, of course, a folkloric "magical"
number.[18] But in principle, Joseph's advice to store grain in years of abundance
for use in years of want was evidently so relevant to the situation in Egypt
that Pharaoh immediately accepted it and appointed Joseph to administer the
implementation of his plan:

> "Since Elohim has made this known to you, there is none so discerning and
> wise as you. You shall be in charge of my house. . . . I put you in charge of all

the land of Egypt." And removing his signet ring from his hand, Pharaoh put it on Joseph's hand; and he had him dressed in robes of fine linen, and put a gold chain about his neck. . . . Pharaoh then gave Joseph the name Zaphenath-paaneah [decipherer of hidden messages]; and he gave him for a wife Asenath the daughter of Poti-phera priest of On. . . . And Joseph was thirty years old when he entered the service of Pharaoh king of Egypt. (Genesis 41:39–46)

The expression "you shall be in charge of my house" was especially apt in Egypt, where the very word "pharaoh" meant, literally, "the big house" (that is, the palace), an obvious metaphor for a pharaoh-sanctioned position of power. But along with the privileges of Joseph's appointment as viceroy went an awesome responsibility:

Joseph traveled through all the land of Egypt. During the seven years of plenty, the land produced in abundance. And he gathered all the grain of the seven years . . . and stored the grain in the cities; he put in each city the grain of the fields around it. . . . Then the seven years of abundance that the land of Egypt enjoyed came to an end, and the seven years of famine set in, just as Joseph had foretold. There was famine in all the lands, but throughout the land of Egypt there was bread. . . . When the famine became severe in the land of Egypt, Joseph laid open all that was within and rationed out grain to the Egyptians. The famine, however, spread over the whole world. So all the world came to Joseph in Egypt to procure rations. (Genesis 41:46–48, 53–57)

Thus does the biblical account explain the establishment of centralization and authoritarian control in Egypt—indeed, the type of regime that Wittfogel called "oriental despotism." The absolute centralization of power in Egypt during a prolonged drought is described in the following passage:

Conveying grain to silos, such as those that may have been built by the Israelite slaves in Pithom and Raamses (Exodus 1:11), as portrayed in a wall painting in Upper Egypt.

Joseph gathered in all the money that was in the land of Egypt . . . for the rations that were being procured . . . into Pharaoh's palace; and Joseph brought the money into Pharaoh's house. And when money gave out in the land of Egypt . . . all the Egyptians came unto Joseph and said: "Give us bread, lest we die before your eyes." . . . And Joseph said: "Bring your livestock, and I will sell to you against your livestock if the money is gone." . . . And when that year was ended, they came unto him . . . and said . . . "nothing is left at my lord's disposal save our persons and our farmland. Take us and our land in exchange for bread, and we with our land will be serfs to Pharaoh; provide the seed, that we may live and not die, and that the land may not become waste." . . . Thus the land passed over to Pharaoh. (Genesis 47:14–20)[19]

THE STORY of the mission undertaken by Jacob's sons to purchase grain in Egypt emphasizes the precarious life of nomadic pastoralists. In the pastoral domain, the dearth of rain may cause the total loss of pasture and water. In such areas, there was little possibility of storing surplus supplies in years of plenty to sustain life in years of drought. In contrast, the irrigation-based riverine societies enjoyed relative food security. In Egypt, the perennial storage of grain was evidently a standard practice. Witness the account in the Book of Exodus of the enslaved Israelites being employed in the construction of 'arei miskenot (store-cities) for Pharaoh (Exodus 1:11).

A lively controversy has been brewing in recent years regarding the sojourn of the Israelites in Egypt. In the absence of proof of the Israelites' presence in Egypt, the Exodus, the wandering in Sinai, and the organized conquest of Canaan, some archaeologists have concluded that the entire saga may be nothing more than a myth.[20] Such a sweeping conclusion seems unwarranted. Apart from the scientific (and logical) principle that the absence of proof does not in itself constitute proof of absence, there is the biblical account itself. Along with the obviously mythical elements, many of the details in the narrative are so vivid and so realistic as to belie any notion that they were entirely contrived. Whoever wrote the story of the Israelites in Egypt must have known the country very well, either must have lived there or must have received the information from others who had. The background is believable, the names seem authentic, and the entire atmosphere and sense of place appear genuine.

That is not to deny that the writers of the Bible infused the stories with dramatic embellishments. When boiled down to its essentials, however, there appears to be a kernel of believability in the account of the Israelites' experience in Egypt. That experience may not have been shared by all the Israelites at the time, but perhaps by a segment of the loose collection of clans and tribes that eventually amalgamated to form the Israelite nation. Even if the stories are entirely unreal, what remains undeniably real is their influence on Israelite culture. Ultimately, the saga of the Israelites' sojourn in Egypt became so deeply imprinted in the collective consciousness of their descendants as to be considered a defining event in the birth and subsequent development of their nationhood and religion.

One theme that is entirely believable is that a severe drought in southwestern Asia would impel tribes of nomadic pastoralists to attempt to enter Egypt. This is known to have happened repeatedly in the history of the region. A pictorial illustration of it can be seen in a wall painting found inside a burial tomb in Beni Hassan, showing a convoy of Semites from Canaan arriving in Egypt with their families, goats, and laden donkeys. There is evidence, dating to the eighteenth century B.C.E., for the presence of numerous Asians serving in Egyptian households, probably as slaves.[21] They may have been brought to Egypt by Asian (Ishmaelite) traders, in the manner described for the transport and enslavement of Joseph (Genesis 37:28, 36, 39:1).

In Egypt, the Israelites did not become irrigation farmers, as were most Egyptians, and did not settle in the cultivated areas, but—at least initially— remained pastoralists and, as such, lived apart from the indigenous sedentary population. They did not graze the dry range on the fringes of the desert (as they had in southern Canaan), however, but the wetlands on the eastern fringes of the riverine domain of the Nile delta. The specific area in which they were to sojourn was the land of Goshen, which was probably the stretch of wetlands along the easternmost of the Nile's distributaries, perhaps including the valley now known as Wadi Tumeilat (an ancient arm of the Nile that has long since dried up). Joseph had directed them to Goshen in order to separate them from the domain of the Egyptians, along the main Nile and its distribu-

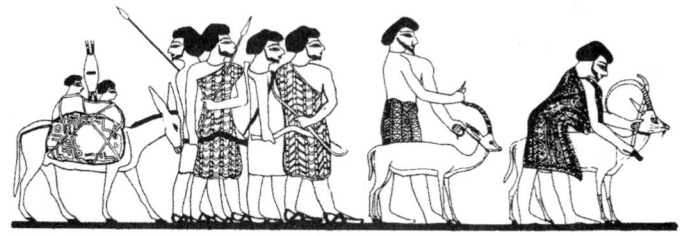

A procession of Semitic pastoralists entering Egypt, apparently
during a time of drought in southwestern Asia, reminiscent of the migration
from Canaan to Goshen by Jacob and his clan (Genesis 46:1–7).
This painting is on a wall in a tomb at Beni Hassan.

taries. He may have done so to avoid conflict between the nomadic Israelite pastoralists and the sedentary Egyptian farmers.

To the Egyptians, the nomadic pastoralists, who had arrived seemingly out of nowhere, were the quintessential "others"—elusive, fickle, unwashed, untamed, rough, and uncivilized. The Egyptians loathed the wild sons of the wilderness, who came to Egypt across the desert, all parched and ragged, with their smelly animals and strange customs and languages and faiths. And they feared the nomads' penchant for sudden, violent, marauding attacks against travelers and even against settled communities. Indeed, nomadic pastoralists from the west (Libya), south (Nubia), or east (Sinai) periodically tried to penetrate Egypt itself. So they were viewed as potential threats that must be warded off and subjugated, lest they disrupt the stability of Egypt's traditional, time-honored, and sanctified way of life. Thus it was an exceptional act of generosity and hospitality that Joseph wished Pharaoh to accord the Israelites, as the severe drought reportedly persisted throughout the entire region of northeastern Africa and southwestern Asia (Genesis 47:13).[22]

What we have here, all in all, does not seem like pure myth but like a plausible (if storied) account of how an Egyptian administration tried to cope with a severe and prolonged drought. The account touches on themes that were vital to the life of Egypt: centralized authority of the pharaoh, drought contingency planning, taxation, storage of grain, emergency food distribution, land tenure and ownership, private and state lands, sharecropping, independence of the priesthood, and attempts by drought-afflicted nomadic pastoralists to graze their flocks along the fringes of riverine Egypt.

NOT LONG AFTER the deaths of both Joseph and the pharaoh who had been his patron,[23] the good fortune of the Children of Israel suffered a harsh reversal. The severe drought that had impelled their migration from Canaan had long ended, yet they tarried in Egypt. Inertia took hold, and it superseded nostalgia for the ancestral homeland. It must have seemed much easier to remain in the comfort and security attainable in the lush meadows of the eastern fringe of the Nile delta than to return to the arid rangeland of southern Canaan. So, as they had lingered, the Israelites overstayed their welcome. Gradually, tensions developed between them (as they retained their separate identity and alien characteristics) and the indigenous people of Egypt, who eventually reverted to their traditional xenophobia:[24]

A new king arose over Egypt who did not know Joseph. And he said to his people, "Look, the Israelite people are much too numerous for us. Let us deal shrewdly with them, so that they may not increase; otherwise in the event of war they may join our enemies in fighting against us and rise from the ground." So they set taskmasters over them to oppress them with forced labor; and they built storage cities for Pharaoh; Pithom and Rameses. But the more they were oppressed, the more they increased and spread out, so that the [Egyptians] came to dread the Israelites. . . . Ruthlessly they made life bitter for them with harsh labor at mortar and bricks and with all sorts of tasks in the field. (Exodus 1:8–12, 14)

Several facts stand out from this brief description. The practice of enslaving a vanquished or captive people so they could labor for the dominant society was widespread in the ancient Near East. In this case, the public works consisted of building store cities (what nowadays would be called grain-storage silos or depots), which must have become a high-priority undertaking, especially in the wake of the severe and prolonged drought ascribed in the Bible to the time of Joseph.

Much of the construction in Egypt was done with bricks, which were (and still are) made from the alluvial clay deposited by the Nile along its floodplain. The traditional method of making bricks consisted of wetting, kneading, and mixing the clay with straw (for reinforcement) and then pressing the material into rectangular molds to shape the bricks, which were dried in the sun to form mud bricks. Such mud bricks were quite durable in the rainless climate of Egypt. Stronger and more durable bricks could be made by firing the molded mud in a kiln to fuse the grains and harden the material permanently,

Making bricks and constructing buildings in ancient Egypt, tasks such as those performed by the Israelite slaves (Exodus 1:13–14, 5:6–14). This wall painting, in a tomb near Thebes, dates to the fifteenth century B.C.E.

but that process required much fuel and was both laborious and expensive. Monumental buildings were commonly built with bricks and then faced with polished limestone, marble, or granite.

The Israelites multiplied in Egypt (Exodus 1:12). The rate of their population growth apparently exceeded that of the Egyptians, who became concerned lest the Israelites become so numerous as to gain dominance, at least in the eastern delta. The birth rates of nomadic pastoralists was customarily high in order to maintain their population, because the rigors and unsettled conditions of their lives naturally resulted in high infant and maternal mortality rates. (Witness the death of Rachel in childbirth during the clan's migration from Shechem southward.) The stable lifestyle of the sedentary farmers in the riverine domain, on the contrary, probably allowed a larger proportion of the infants to survive. Hence fewer births per family were needed to maintain the population. The birth rates may have been kept deliberately low (perhaps by prolonging the nursing period for each infant) in order to avoid excessive disputes over the inheritance of such limited resources as land and water rights.

To stem the danger of the Israelites' burgeoning population, "Pharaoh charged all his people, saying: 'Every boy that is born [to the Israelites] ye shall throw into the Nile, but let every girl live'" (Exodus 1:22).[25]

NOW BEGINS the story of Moses, and it is in some ways the typical story of an archetypal hero. The hero is born under unusual conditions and then subjected to grave dangers; he is saved miraculously from death, taken away from home, and made to undergo personal trials abroad; finally, he returns to his place of origin to perform great deeds. In other ways, however, the story of Moses is unique.

> There went a man of the house [tribe] of Levi, and took to wife a daughter of Levi. And the woman conceived and bore a son; and when she saw that he was a goodly child, she hid him for three months. And when she could no longer hide him, she took for him an ark of bulrushes, and daubed it with slime and with pitch; and she put the child therein, and laid it in the reeds by the river's bank. And his sister stood afar off, to know what would be done to him. (Exodus 2:1–4)

The way Moses was saved is paralleled in the Egyptian myth of Horus. In the myth, Horus was protected from Seth, the ruler of Egypt, by his mother, Isis, who hid him in the papyrus marshes. Another striking cross-cultural parallel is between the biblical story of Moses and the Mesopotamian legend of Sargon the Great.[26] The idea of the ark of Moses also reprises in miniature the ark of Noah (which also has its Mesopotamian parallel), likewise treated with pitch, designed to save humanity along with all the animals from drowning in the Flood.

The infant Moses was saved, incongruously, by the compassionate daughter of the cruel pharaoh who had decreed the death of every male child born to the Hebrews:

> The daughter of Pharaoh came down to bathe in the Nile. . . . She spied the basket among the reeds and sent her slave girl to fetch it. When she opened it, she saw that it was a child, a boy crying. She took pity on it and said, "This must be a Hebrew child." Then his sister said to Pharaoh's daughter, "Shall I go and get you a Hebrew nurse to suckle the child for you?" And Pharaoh's daughter said, "Yes." So the girl went and called the child's mother. And Pharaoh's daughter said to her, "Take this child and nurse it for me, and I will pay your wages." So the woman took the child and nursed it. When the child grew up, she brought him to Pharaoh's daughter, who made him her son. She named him Moses, explaining, "I drew him out of the water." (Exodus 2:5–10)

The name Moses (or Mose) is evidently Egyptian. It is related to other names (such as Thut-mose and Ah-mose, which imply "child" or "born of") that were, significantly, those of pharaohs. It was given a Hebrew twist by means of a pun on the verb *msh*, which means "drawn out [of water]" (which in the Bible appears only in 2 Samuel 22:17 and Psalm 18:17).

So Moses was raised in the royal household as an Egyptian prince, oblivious to the plight of his own people. When he came of age, however, he ventured out on his own and then witnessed the injustice and cruelty with which the Egyptians treated the Hebrews, whom he apparently knew to be his brethren (Exodus 2:11). Seeing a slave driver flog a slave, Moses killed the oppressor and then had to escape from the wrath of Pharaoh.[27]

Escape where? Leaving the narrow riverine domain of Egypt, Moses could only go into the desert. A silent ancestral call impelled him to go eastward, toward the semiarid pastoral domain of northern Sinai, beyond which beckoned the Negev and Canaan. Somewhere en route, he found a well. And, in the pastoral domain, wherever there was a well of "living water," there were certain to be shepherds.

The arrival of a traveler at a well is a scene described twice before in the Bible: once in connection with Eliezer's mission on behalf of Abraham (Genesis 24), and again in connection with Jacob's escape to Haran (Genesis 29). In both cases, a fortuitous meeting took place at the well, a meeting with a

A slave master scourging a slave, recalling the beating witnessed by Moses
(Exodus 2:11, 5:14). This wall painting in Upper Egypt is dated to
around the fifteenth century B.C.E.

beautiful maiden, a shepherdess: Eliezer met Rebekah, and Jacob met Rachel.
Something very similar happened to Moses:

> Moses fled from Pharaoh. He arrived in the land of Midian; and sat down
> beside a well. Now the priest of Midian had seven daughters. They came to
> draw water, and filled the troughs to water their father's flock. But shepherds
> came and drove them off. Moses rose to their defense and he watered their
> flock. When they returned to their father Reuel,[28] he said, "How is it that you
> have come back so soon today?" They answered, "An Egyptian rescued us from
> the shepherds; he even drew water for us and watered the flock." He said to his
> daughters, "Where is he then? Why did you leave the man? Ask him in to break
> bread." Moses consented to stay with the man, and he gave Moses his daughter
> Zipporah as wife. She bore a son whom he named Gershom, for he said, "I have
> been a stranger in a foreign land." (Exodus 2:15–23)

The name Gershom has multiple connotations: *ger* means "stranger," *garesh*
means "drive away," and *sham* means "over there." So, in a single name, Moses
manifested his initial condition as one driven away from his native country
and made to live as a stranger in a faraway land.

But then a transmutation occurred. Moses took up the life of a shepherd
in the pastoral domain, and the experience reconnected him with the life and
spirit and fate of his ancestors. Thus he no longer was simply a fugitive "Egyptian," but became, in effect, a Hebrew and—being himself free—ready to take
part in the liberation of his oppressed people: "Moses was tending the flock
of his father-in-law, Jethro" (Exodus 3:1),[29] thus repeating the life of and iden-

tifying with Jacob-Israel, the father of the clan that had settled in Egypt. And when Moses was alone in the wilderness, there came to him a startling and momentous revelation:

> He drove the flock into the wilderness, and came to Horeb,[30] the mountain of the Elohim. An angel of Yahweh appeared to him in a blazing fire out of a bush. He gazed, and there was a bush all aflame, yet the bush was not consumed. Moses said, "I must turn aside to look at this marvelous sight; why doesn't the bush burn up?" When Yahweh saw that he had turned aside to look, Elohim called to him out of the bush: "Moses! Moses!" He said, "Here I am."[31] And He said: "Do not come closer; remove your sandals from your feet, for the place on which you stand is holy ground." Then He said: "I am the Elohim of your father, the Elohim of Abraham, the Elohim of Isaac, and the Elohim of Jacob." And Moses hid his face; for he was afraid to look at the Elohim. (Exodus 3:1–6)

A stunning vision, powerfully described. But the passage also contains baffling juxtapositions: first it refers to the mountain of Elohim, then to an angel of Yahweh, then to Yahweh, and finally to Elohim again. We sense that the passage is a synthesis of what were possibly several versions of a mythical story, a synthesis in which the various names or manifestations of God, perhaps originally different perceptions of God, were unified. What strikes us most is the identification of the deity, by whatever name, with fire—that glowing, shimmering, cleansing, fluttering, seemingly alive marvel that had always mesmerized humans. Fire had long served as the agent whereby humans could make offerings to the gods, by which substance was sublimated. Generally, fire is dangerous, a scorcher and devourer of plants and animals, but in Moses's vision this fire did not consume. It was not corporeal but entirely ethereal.

Out of the mysterious fire came the voice, and the voice charged Moses with a mission of destiny:

> Then Yahweh said: "I have marked well the plight of My people in Egypt and have heeded their outcry. . . . I am mindful of their suffering. I have come down to rescue them from the Egyptians and to bring them out of that land to a good and spacious land, a land flowing with milk and honey,[32] the region of the Canaanites. . . . Come, therefore, I will send you to Pharaoh, and you shall free My people, the Israelites, from Egypt." (Exodus 3:7–8, 10)

When Moses asked what name he should use to identify who sent him on his mission, the answer was enigmatic: "Moses said to the Elohim, 'When I come to the Israelites and say to them, "The Elohim of your fathers hath sent me to you"; and they ask me, "What is His name?" What shall I say to them?' And Elohim said to Moses: Eheyeh-Asher-Eheyeh; and He continued: 'Thus shall you say to the Israelites, Eheye sent me to you'" (Exodus 3:13–14).[33] The actual word in Hebrew is *eheyeh*, which is the first-person-singular, future tense of the verb "to be." It may well be the root of the name Yahweh (or Yehovah), which seems to be a compound of the third-person-singular form of the verb in all three tenses (past, present, and future): *hayah, hoveh, yihiyeh* (was, is, and shall be).It implies an invisible, eternal presence.

Moses, apparently baffled by the enigmatic name and awed by the divine presence, hesitated before accepting the charge. He wanted to have tangible signs by which to prove to the skeptical people of Israel, and to the Egyptians, that he was the authentic messenger of God. Even when given such signs, he still hesitated, claiming to be "slow of speech and slow of tongue." But Yahweh overrode his hesitations: "So Moses took his wife and his sons, mounted them on an ass, and went back to the land of Egypt" (Exodus 4:20).

En route to Egypt, an eerie confrontation occurred: "At a night encampment on the way, Yahweh encountered him and sought to kill him. So Zipporah took a flint and cut off her son's foreskin, and touched [Moses's] legs with it, saying, 'You are truly a bridegroom of blood to me!' And when He let him alone, she added, 'A bridegroom of blood because of the circumcision'" (Exodus 4:24–26). The origin of the practice of circumcision is not known with certainty.[34] In the Bible, the ritual is first dated to the time of Abraham and considered a symbolic affirmation of the covenant between Elohim and Abraham (Genesis 17:10–14). As such, it may have been a symbolic—and certainly more humane—ritual substitute for the ancient practice of sacrificing the firstborn son to elicit divine favor. Zipporah's action implies that Yahweh required circumcision and may have been willing to kill anyone (even Moses!) who failed to perform it on his son. The practice was prevalent in Egypt at the

A priest performing a circumcision on a boy, evidently a prevalent practice in ancient Egypt carried out as a rite of passage for adolescent boys. This relief, in a mastaba at Saqqara, is dated to the Sixth Dynasty (ca. 2300 B.C.E.).

time,[35] however, and its adoption by the Israelites may have been one of the lasting influences of Egypt on them.

As the Bible describes him, Moses was uniquely qualified to play the role of intermediary between the Israelites and the Egyptian authorities. To the Egyptian court, he spoke the language of a familiar insider of noble rank who, because of his courageous identification with the Israelites, also carried special moral authority. To the Hebrews, he conveyed the confidence, sophistication, and class of an educated man who, having returned from the pastoral domain, also reminded them of their proudly independent past and offered them a vision of their future as a free nation.

ONCE BACK in Egypt, Moses, aided by his brother Aaron, faced the Children of Israel and announced his mission of liberation. The two then presented themselves to Pharaoh, saying:

> Thus says Yahweh, the Elohim of Israel: "Let my people go, that they may celebrate a festival for Me in the wilderness." But Pharaoh said, "Who is Yahweh that I should heed Him and let Israel go? I do not know Yahweh, nor will I let Israel go." . . . That same day Pharaoh charged the taskmasters and foremen of the people, saying, "You should no longer provide the people with straw for making bricks as heretofore; let them go and gather straw for themselves. But impose upon them the same quota of bricks as they have been making heretofore . . . for they are shirkers. . . . Let heavier work be laid upon the men; let them keep at it and not pay attention to deceitful promises." (Exodus 5:1–9)

Yahweh was the private God of the Israelites, not known to or respected by other peoples and their rulers. So Pharaoh, haughty slave master that he was, acted spitefully to punish the Israelites for the very notion that they might cease working, even temporarily. Then the people of Israel complained to Moses for having worsened their lot, and Moses, in turn, complained to Yahweh. His reply was apparently meant to clarify the confusion over the various names of God: "Elohim spoke to Moses and said unto him: 'I am Yahweh. I appeared to Abraham, Isaac, and Jacob as El Shaddai, but I did not make Myself known to them by My name Yahweh. I also established My covenant with them to give them the land of Canaan'" (Exodus 6:2–4).

To convince the Israelites and the Egyptians that his mission was divinely sanctioned, Moses was instructed to perform supernatural feats. During his first encounter with Yahweh in the desert, he had been told to cast his shepherd's rod to the ground, and it had turned miraculously into a serpent

(Exodus 4:3–4). In Egypt, however, when Moses cast his rod before Pharaoh, it turned into a crocodile (Exodus 7:10). And when the same feat was replicated by the magicians of Egypt, Moses's transformed rod swallowed up theirs. The difference in nuance between a serpent (the most fearsome animal of the desert domain) and a crocodile (the most fearsome animal of the Nilotic riverine domain) is indicative of the context as perceived by the teller of the story, even if the story itself were contrived. Similarly indicative is the next episode in the series of contests of will and power that took place between Moses (representing Yahweh) and Pharaoh, this time on the bank of the Nile:

> Yahweh said to Moses: ". . . Go to Pharaoh in the morning, as he is coming out to the water, and station yourself before him at the edge of the Nile, taking with you the rod that turned into a snake. . . . And say to him: By this you shall know that I am Yahweh. See, I shall strike the water in the Nile with the rod that is in my hand, and it will be turned into blood; and the fish in the Nile will die. The Nile will stink so that the Egyptians will find it impossible to drink the water of the Nile." (Exodus 7:14–18)

With a population variously estimated to range from 1.5 to 2.5 million living on the narrow floodplain of the Nile and using the river as the source of drinking water and fish, the only avenue of navigation, and an outlet for the disposal of waste, the river must have been in constant danger of pollution. From time to time, especially during periods of low flow, spillages of wastes could cause the water to become eutrophic (enriched with nutrients that promote the growth of algae, which make the water turbid, deprived of oxygen, and smelly). Severe pollution would cause massive fish kills—a phenomenon that is prevalent in many rivers around the world that are subjected to pollutants. Moreover, in the early stages of the annual flood in Egypt, the waters rushing downstream from the Blue Nile carried a great load of reddish mud. Both pollution and mud could, on occasion, have caused the waters of the Nile to be so murky as to resemble blood: "All the water in the Nile was turned into blood, and the fish in the Nile died. The Nile stank so that the Egyptians could not drink water from the Nile. . . . And all the Egyptians had to dig round about the Nile for drinking water,[36] because they could not drink the water of the Nile" (Exodus 7:20–22, 24).

Shortly after the water-pollution episode, Egypt reportedly was afflicted with another water-related scourge: a sudden proliferation of frogs. These amphibians thrive in stagnant water, especially if the water is dense with aquatic vegetation that harbors snails and insects. A particularly copious overflow of the Nile would leave behind extensive ponded areas that would subsequently turn into marshes with puddles of eutrophic water that could abound with swarming frogs. When such puddles dried up, myriad frogs would desiccate and die:

> Yahweh said to Moses: "Say to Aaron: Hold out your arm with the rod over the rivers, the canals, and the ponds, and bring up the frogs on the land of Egypt." Aaron held out his arm over the waters of Egypt, and the frogs came up and covered the land of Egypt. But the magicians did the same with their spells, and brought frogs upon the land of Egypt. . . . Then Moses cried out to Yahweh in the matter of the frogs which He had inflicted upon Pharaoh. And Yahweh did as Moses asked; the frogs died out in the houses, the courtyards, and the fields. And they piled them up in heaps, till the land stank. (Exodus 8:1–2, 8–10)

So it seems that at least the first two of the ten plagues that afflicted Egypt may well have been related phenomena.[37] Such occurrences could have been magnified and interpreted afterward as a divine signal or as retribution for the iniquities of the Egyptians toward the Israelites. The next seven scourges reported to have befallen the Egyptians were infestations of lice and of flies (or perhaps mosquitoes, which proliferate in stagnant waters), a disease of animals (referred to as *dever*), an outbreak of boils on humans (perhaps also related to polluted waters), a freak hailstorm, a swarm of locusts, and a period of darkness (perhaps from an eclipse of the sun or a prolonged dust storm). Finally, the Book of Exodus describes the most grievous affliction of all: the death of the firstborn of every family and of every herd of livestock. Although the occurrence of all these plagues in close succession seems exceedingly improbable, each of them was a natural phenomenon likely to have happened in Egypt at one time or another or repeatedly and to have left a lasting impression on the Egyptians and Israelites alike.

An invasion of locusts, for instance, has always been a cause of anxiety throughout the Middle East. Grasshoppers that are indigenous to semiarid parts of eastern Africa and southwestern Asia are usually nonmigratory. At certain times, however, they suddenly band together (especially when their population attains a critical density) to form enormous swarms, which then rise up en masse and migrate to distant agricultural areas in huge wind-driven, sky-blackening "clouds." When they land, the locusts may devour the foliage of a lush field or an orchard in a single day. As quickly as they arrive, the locusts may depart, carried away by a gust of wind. Otherwise, they may remain to lay eggs and produce several generations of voracious "nymphs" that continue to destroy crops:[38]

> Yahweh drove an east wind over the land all that day and all night; and when morning came, the east wind had brought the locusts. Locusts invaded all the

land of Egypt and settled . . . in a thick mass. . . . They hid all the land from view, and the land was darkened. They ate up all the grasses of the field and all the fruit of the trees so nothing green was left of tree or grass of the field, in all of Egypt. . . . Then Pharaoh summoned Moses and Aaron and said, "I stand guilty before Yahweh your Elohim, and before you. Forgive my offense just this once, and plead with Yahweh your Elohim that He but remove this death from me." . . . Yahweh caused a shift to a strong west wind [sea breeze], which lifted the locusts and hurled them into the Sea of Reeds. (Exodus 10:13–19)

Such are the vagaries of the region's weather in general and the winds in particular that a shift of atmospheric pressure distribution may cause an abrupt change from a hot easterly wind to a cool westerly wind or a sea breeze. Such a switch, from one condition to the other, can either bring on or drive off an invasion of locusts.

Even though Egypt receives practically no rain, and the sky is nearly cloudless most of the time, the various winds that visit Egypt can strongly affect the weather. A hot and dry wind from the desert (variously called *sharkiye* [easterly] or *khamsin* [fifty, because it is presumed to occur mostly during the proverbial fifty days of transition from winter to summer and vice versa]) may desiccate the vegetation and blanket the land with so much desert dust as to obscure the sun. A relatively cool and moist breeze from the Mediterranean Sea, though, may bring Egypt occasional relief from the usual dry heat.

As the Israelites prepared to depart from Egypt, they were instructed by Moses to perform a ritual of animal sacrifice that reconnected them with their pastoral ancestry and separated them from the people of Egypt:

Yahweh spoke to Moses and Aaron in the land of Egypt, saying, "Speak to the whole community of Israel and say that on the tenth of this month each shall take a lamb to a family . . . and all the congregation of the Israelites shall slaughter it at twilight. They shall take some of the blood and put it on the two doorposts and the lintel of the houses. . . . They shall eat the flesh that same night, roasted over the fire, with unleavened bread and with bitter herbs. . . . For that night I will go through the land of Egypt and strike down every firstborn. . . . And the blood on the houses where you are staying shall be a sign for you: when I see the blood I will pass over you." (Exodus 12:2–13)

The image of Yahweh in this passage is still quite anthropomorphic: he needs the marking of blood on the doorposts, as he passes from house to house, to distinguish between the homes of the Egyptians and those of the Israelites. The elaborate and symbolic ritual thus prescribed was to be in preparation for the liberation and departure of the people of Israel from their bondage in Egypt. In the process, an official calendar was decreed. Nissan, being the month of spring (corresponding approximately to April), was to begin the calendar. The dusk hour on the fourteenth day of that month (the time of witnessing the first full moon after the vernal equinox) was to begin the festival of Passover. Since the Bible was compiled and canonized after the Babylonian exile, there may well be a Mesopotamian influence in the decision to begin the calendar year in the spring. (In Mesopotamia, the flooding of the Tigris and Euphrates generally takes place following the springtime snowmelt in the upper watersheds of those rivers, and that inundation effectively initiates the new growth of vegetation after a long dry period of dormancy.)

The most storied and picturesque episode of the Exodus is the so-called Parting of the Red Sea. Nowhere in the Scriptures, though, is there any mention of the Red Sea. Rather, the body of water referred to is Yam Soof, which means "reed sea" (Exodus 13:18, 15:4, 22). The word *yamm* in Biblical Hebrew described any body of water, even such a very small one as the water in a ritual vessel (1 Kings 6:23–26; 2 Chronicles 4:6). The body of water referred to in these passages could have been any one of the shallow lakes or marshes that existed along and beyond the eastern edge of the Nile delta. These wetlands were inundated periodically by an eastward-flowing arm of the Nile. Vestigial marshes of this sort, including the Bitter Lakes and Lake Timsah (Lake of the Crocodiles), remained in the area until modern times and are now incorporated into the Suez Canal system. In ancient times, however, they were swamps with a variable depth of brackish water, the shallow sections of which were overgrown with reeds.[39] Bands of fleet-footed fugitives could easily run into the marshes, hide among the tall reeds, and make their way to safety, while a heavy-laden army on horseback and in narrow-wheeled chariots would tend to get bogged down in the slick mud and slippery mire: "The Israelites went into the *yamm* upon dry ground . . . and the Egyptians came in pursuit after them into the *yamm*, all of Pharaoh's horses, chariots, and horsemen. At the morning watch, Yahweh looked down upon the Egyptian army from a pillar of fire and cloud, and threw the Egyptian army into panic. He locked the wheels of their chariots so that they moved forward with difficulty" (Exodus 14:22–25).

To commemorate their close encounter with and deliverance from Pharaoh's army, which their leaders ascribed to a divine miracle, the Israelites (or, rather, their scribes) bequeathed to future generations a greatly exaggerated, supernatural version of their escape from Egypt.[40] To an ecologist familiar with the area, however, it might have been an entirely natural event, a clever and timely use of the terrain by people who, being wide-ranging pastoralists, probably knew the frontier territory, between the Nile delta and the Sinai Peninsula, far better than the river-bound Egyptians.

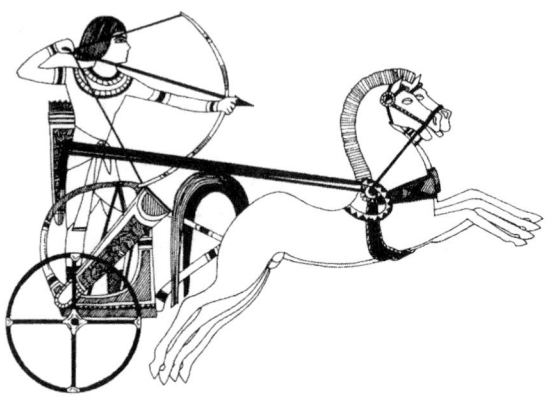

An Egyptian charioteer driving a team of tandem horses, as shown in a wall painting dated to the end of the second millennium B.C.E. The narrow wheels could readily have become bogged down in loose sand as well as in saturated mud.

The image of a pillar of cloud that went before the Israelites by day (Exodus 13:21), guiding them in the desert, may also have been derived from nature. A common sight in Middle Eastern deserts is a phenomenon known as a dust devil. A gust of wind striking an irregular hummock of dry earth may whip up a spiraling pillar of dust in the form of a whirlwind. A dust devil may rise to a considerable height and appear to dance about like a whirling dervish. With a stretch of the imagination, it may fit the biblical description: "The pillar of cloud shifted from in front of them and took up a place behind them, and it came between the army of the Egyptians and the army of Israel. Thus there was the cloud with the darkness, and it cast a spell upon the night" (Exodus 14:19–20). A sustained windstorm in the desert may kick up so much dust as to obscure vision and cast a pall of confusion and disorientation on the land.

The triumphal ode of Moses, following the deliverance in the Sea of Reeds, exults in the prowess of Yahweh:

Then Moses and the Israelites sang this song to Yahweh:

I will sing to Yahweh, for He has triumphed gloriously;
horse and driver He has hurled into the sea.
Yahweh is my strength and might;
He is become my deliverance. . . .
Yahweh the Warrior,
Yahweh is His name!
Pharaoh's chariots and his army He has cast into the sea;

and the pick of his officers are drowned in the Sea of Reeds. . . .
Your right hand, O Yahweh, glorious in power,
Your right hand, O Yahweh, shatters the foe! . . . (Exodus 15:1–6)

One of the stanzas stands out in particular: *Yahweh ish milḥamah, / Yahweh shemo* (Yahweh the Warrior [Man of War], / Yahweh is His name). The notion of Yahweh as a fierce warrior, ready and able to smite the enemies of his chosen people, was to recur on subsequent occasions in the Bible. In each case, the victory in war was not attributed to a human king or leader or to chance circumstances, but to Yahweh. As was the typical style of triumphal poetry in the ancient Near East, this ode to Yahweh magnifies and glorifies the victory for all posterity. It is filled with hyperbole, unrestrained exuberance, and wishful projections into the future[41]—the battles yet to come and the holy Temple yet to be built.

6

THE DESERT DOMAIN

Wanderings in Sinai and the Negev

*Why did you make us leave Egypt to bring us to this wretched
place, a place with no grain or figs or vines or pomegranates?
There is not even water to drink!*

NUMBERS 20:5

THROUGHOUT MOST of history, the desert was regarded as a realm apart,
an extraterritorial domain separate from the principal habitable domains.
The relatively "civilized" residents of the other regions (living in more or less
organized societies) viewed the "wild" people of the lawless desert, few though
they were in number, with fear and hostility, perceiving them to be a threat to
civilization—as, indeed, they were at various times.[1] The desert itself was held
in awe as a place of danger and terror, its mysterious vastness to be entered
only at great risk.

The word "desert" derives from the Latin term *deserere*, which means "to
desert." There are several Hebrew synonyms for the desert, each of which
probably referred originally to a variant of the area we now combine into one
domain. Among those words are *yeshimon, shmamah, ḥoravah, ʿaravah*, and
tsiyah—all of which suggest wasteland, desolation, wilderness, and desicca-
tion. Alas, the nuances originally conveyed by these synonyms apparently
have been obscured by the passage of time. The commonly used word, *midbar*, originally implied pasturage (a place to drive or lay down flocks), so it
probably referred to the semiarid zone (or semidesert) rather than to the fully
arid desert.[2]

The desert has always been a place of great allure and fascination. It is as
much a state of mind as a physical domain. Like a vacuum that draws, so the
very emptiness and enormity of the desert's expanse, the awesome grandeur
of its sculpted landscape, the enigmatic silence that pervades it—all these
qualities and more have intrigued and challenged and inspired people since

Camels resting in a grove of tamarisk trees in a dry creek bed (wadi) in the desert of Sinai.

the beginning of time. In the desert, an upthrust mountain commanding a clear vista all around is a place of spiritual as well as physical elevation, a stage intermediate between Earth and heaven.

A mystical spirit seems to pervade the desert: the spirit sought by hermits and social outcasts as they chose to escape from the noisy world of vain pursuits and oppressive demands. It is the spirit sensed by the solitary prophets who wandered into the desert's stillness to evade the mindless and heartless cruelty of ordinary society and to ponder the ultimate meaning of God's works and of their own higher calling. Here a person can be truly alone and one with nature's elements, free of conventions and unencumbered by imposed duties. Here the unobstructed view of nature's primeval grandeur reveals the greatness of creation and the frailty of humanity, and it inspires a sense of true humility.

ALTHOUGH DESERTS tend to gird the continental sections of the Tropics of Cancer and Capricorn, they occur in more or less distinct masses. The world's greatest by far is the Sahara (an Arabic word that means, quite literally, "desert"), which encompasses the greater part of North Africa. Contiguous with it is the desert of Egypt, divided into the Western Desert and the Eastern Desert by the world's largest oasis, the emerald strip of fertile land along the Nile. The

Eastern Desert of Egypt, in turn, is connected through the deserts of Sinai and the Negev to the deserts of Arabia, Jordan, southern Iraq and Iran, Baluchistan, and northwestern India.

The landscape of these deserts may seem monotonous to a superficial observer, like a time-frozen moonscape stretching on and on to endless, flat or jagged horizons. This seeming uniformity is more illusory than real. The desert is in fact wondrously varied, with dramatic and dazzling landforms. Parts of the desert are covered with dunes—a wavy sea of fluid, shifting sand. In other parts are ranges of craggy mountains, some as high as 13,000 feet (4000 m) above sea level. Cliffs and canyons, upthrusts and faults, sculpted works of scouring rivulets and of blasting sand particles driven by gusty winds, jagged pillars of rock and truncated cones of buttes, maze-like badlands and rutted arroyos, mudflats and salt-encrusted playas—all are on open display, largely bare of soil or vegetation. Dotted irregularly within this sere, varied expanse are incongruous oases, where exotic streams (bringing water from rainfed domains outside the desert) or local concentrations of shallow groundwater emerge at or near the surface, giving rise to anomalous clusters of luxuriant vegetation.

All life in the desert is at once tenuous and tough. To survive the heat and drought, plants and animals (including humans) must rely on special modes of accommodation. Plants adapted to dry conditions, called xerophytes, often have thick or waxy coatings that help minimize evaporation, as well as succulent tissue capable of storing water from periods of plenty to long periods of want. Animals in the desert tend to hide from the heat of the day and become active mainly during the cool nights. Humans, similarly, must learn to find water for themselves and their animals, and to conserve water with maximum efficiency. They must also learn to protect the meager resources available to them from competitors or adversaries. The paucity of resources, the inconstancy of circumstances, and the need to range over large areas to seek subsistence—all create the conditions that foster frequent life-and-death rivalry. Hence desert dwellers must be ready at all times to fight for their own survival.

The movement of people and cargo through the desert always has been physically difficult and hazardous. A typical journey generally lasted for many days, with the travelers walking alongside and directing beasts of burden laden with equipment and commodities. At each way station, the cargo would be unloaded and then reloaded when the journey resumed. Food and water generally were in short supply. Travelers had to endure enervating heat and thirst, sudden flash floods, sandstorms, and rock slides. Moreover, they were ever in danger of being ambushed and robbed by lurking marauders, for whom robbing convoys was a major livelihood.

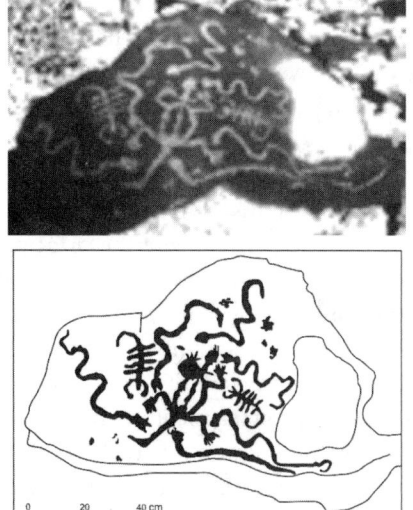

Rock engravings made by sojourners in the desert, by scratching or chipping the dark patina-stained surface of the rock to expose the light-hued limestone beneath. These engravings, showing fearsome creatures—snakes, scorpions, and a *saraf* (venomous lizard)—were found at Mount Karkom, near the border between the Negev and Sinai. Snakes were deified by desert dwellers. A sculpted snake, reportedly made by Moses, was kept in the Temple of Jerusalem as a curative fetish for centuries, until it was removed by King Hezekiah (2 Kings 18:4).

The rugged landscape and oppressive climate and the constant fear of loss of possessions and even death must have cast the desert travelers into a state of constant wariness. To overcome it, they often engaged in religious rites intended to propitiate the protective gods and ward off the evil spirits. The religious impulse was given concrete expression in the construction of monuments and sanctuaries of various sorts, usually along the major roads and at crossroads. These generally were built with carefully selected and fitted stones, arranged in a regular pattern. In some places, stones were inscribed with decorative or symbolic figures, some of which carried religious significance. Beside some of the monuments are the charred remains of animals, such as gazelles or sheep, that were offered as sacrifices to the local deities.

The erection of a monument may have been sanctified by the performance of a ritual, perhaps anointment with oil:

Early in the morning, Jacob took the stone that he had put under his head and set it up as a pillar and poured oil on the top of it. He named that site Bethel [house of God]. . . . Jacob then made a vow, saying: "If Elohim will be with me and protect me on this journey that I am making . . . Yahweh shall be my Elohim. And this stone, which I have set up as a pillar, shall be the abode of Elohim; and all that You give me, I will set aside a tithe for You." (Genesis 28:18–22)

After the domestication of the camel,[3] some desert dwellers became caravaneers. Their function was to convey precious commodities from their sources (such as southern Arabia, India, and the Horn of Africa), across the desert, to the centers of population in the river valleys and along the Mediterranean coast. Among the most valued goods were medicinal herbs, spices and aromatics (incenses and perfumes), ivory, precious stones and other rare minerals, and plant and animal products.

Especially important were medicaments. Long before the advent of scientific medicine, people lived in constant fear of illnesses, not knowing what caused them. From antiquity until modern times, herbs were the principal medicines. The cumulative experience of generations resulted in the selection of particular plants that were believed to be efficacious in the cure of various maladies. Products extracted from those plants were administered to the ill, either orally or topically in the form of oil-based ointments. The places where medicinal herbs grew naturally, or could best be grown agriculturally, became famous: "Is there no balm in Gilead? / Is there no physician there?" asked the prophet, who also advised: "Go up to Gilead, and get balm" and then "Get balm for her wounds, / perhaps she can be healed" (Jeremiah 8:22, 46:11, 51:8).

Spices were also highly prized, far more than at present. In the warm climate of the ancient Near East, without refrigeration or pasteurization, foods spoiled very quickly. Some foods could be preserved by means of fermentation (for example, wine and curdled milk). Unless consumed when entirely fresh, foods that could not be fermented had to be rather heavily spiced to be edible, both to preserve the foods and to mask the disagreeable flavor of rancidity.

Even more valuable were the aromatic extracts of fragrant plants. Many desert plants contain natural products that are strongly aromatic, and they tend to concentrate their saps and resins to a high degree. While many of them are pungent to the smell, some are pleasingly fragrant. When extracted and concentrated still further, the fragrant essences can be stored in sealed vials of alabaster, glazed pottery, or glass and can be transported readily over long distances without losing their distinctive bouquet. Indeed, they were lucrative items of international trade. They served not only as personal cosmetics (especially for the wealthy and titled classes, for whom the fragrances symbolized privilege and refinement), but also—and perhaps primarily—as agents of religious worship.[4] The use of incense in rituals was believed to draw the attention of the gods and delight them, as well as the believers. Aromatic unctions were also applied in Egypt in the embalming process.

Witness the passage pertaining to Joseph's brothers in the vale of Dothan: "Looking up, they saw a caravan of Ishmaelites coming from Gilead, their camels bearing gum, balm, and ladanum to be taken to Egypt" (Genesis 37:25). Years later, when Jacob instructed his sons to go to Egypt with gifts to placate the important man who held the power of life and death over the entire population of that land, the gifts he sent were "a little balm, and a little honey, gum and ladanum, pistachio nuts, and almonds" (Genesis 43:11). Centuries later, the prophet Ezekiel listed the many items of international trade (transported overland through the desert as well as across the sea), including ivory tusks and ebony, spices, carbuncles, purple dye, richly woven work, fine linen, coral, gold, and rubies. In return, sedentary Canaan sold wheat, balsam, honey, oil, and balm (Ezekiel 27:15–17, 22).

Among the numerous clay tablets inscribed in cuneiform that have been found in the Middle East are some that provide price lists for aromatic substances. They attest to the widespread trade in those precious commodities.

THE DESERTS of the Sinai Peninsula and the southern Negev,[5] where the Israelites reportedly wandered for forty years and where their nationhood was forged in isolation (an experience forevermore enshrined in their national memory and lore), together constitute a geographic nexus where the continents of Europe, Asia, and Africa connect. They also form the isthmus where the Indian Ocean (through its arm, the Red Sea) and the Atlantic Ocean (through its offshoot, the Mediterranean Sea) come close together. As such, these deserts have served in history as a narrow land bridge connecting the major centers of sedentary agriculture and urbanized civilization (especially Mesopotamia and Egypt) that had formed on either side of them. Within this intermediate frontier zone wandered various inchoate associations of landless tribes. They subsisted in irregular ways, by hunting such desert animals as the gazelle and ibex, grazing goats on the sparse desert range, transporting goods across the desert from distant sources to the major population centers, and occasionally marauding the settlements on the fringes of those centers. Some of the desert tribes even succeeded from time to time in permanently penetrating and settling in the more favorably endowed domains, either replacing the resident population or assimilating with it. Such was evidently the destiny of the wandering Israelites.

The Sinai Peninsula consists of three greatly contrasting sections: the southern, central, and northern zones.[6] The southern zone is a massif of igneous mountains, mostly granite, incised and crisscrossed by intrusive dikes of congealed magma. The view of these red-hued mountains, rising sheer to great heights, is so spectacular that religiously inspired (mostly European) pilgrims and travelers have tended to assume that here must be the location of Mount Sinai, where Yahweh revealed himself and his Law to Moses. The Monastery of St. Catherine was built here some fifteen centuries ago, at the foot of the highest peak (called Jebel Musa [Mount Moses] by the local Arabs), which the monks believed to be *the* Mount Sinai. Present-day investigators

Evidence of ancient human habitation in the desert: a hunting scene from the Chalcolithic period (ca. fourth millennium B.C.E.), depicted in a rock drawing in the southern Negev. The hunter uses a bow and arrow and is assisted by dogs as he chases a pair of ibexes (wild goats).

consider that attribution to be extremely unlikely, because of the ruggedness of the landscape and the distance from the Nile delta (from which the Israelites had departed) and from the land of Canaan (to which they were headed). Moreover, the turquoise and copper mines in the southwestern part of the Sinai Peninsula were well guarded by Egyptian garrisons, so the Israelites could venture past there only at great risk. Finally, that part of Sinai offers little livelihood to pastoral tribes of any size (except for the possibility of utilizing localized runoff and springwater to grow crops in the wadis that run between the mountains, an undertaking that would have required permanent habitation in the area).

The central zone of the Sinai Peninsula is mostly a series of wide plateaus, strewn with a "desert pavement" of dark flinty gravel (a surface feature called *ḥammada* in Arabic). Although several blocks of sedimentary hills rise above the plateaus, mainly along their northern and western fringes, the topography is mostly flat or undulating. Over their greater extent, the plains support only sparse xerophytic vegetation, but they are dissected by shallow wadis in which grows relatively dense vegetation. The plants consist of several species of trees (for example, acacia, tamarix, and ziziphus), shrubs, and grasses. The wadis receive, absorb, and convey from the desert surface the runoff generated during infrequent rainstorms. Although meager by the standards of the rainfed domain, the vegetation in the wadis can sustain a limited number of grazing animals (particularly camels and goats) and thus provide subsistence to a small population of pastoral nomads. The Arabic name of this section of Sinai, interestingly, is Badiyat et-Tih, which means "the desert of the wanderers."

The northern section of Sinai is mostly an extensive plain of sand dunes. The dunes assume several characteristic shapes, depending on the winds that formed them. Some are crescent-shaped (called *barkhanes*), others are star-shaped, and still others are parallel longitudinal strips (called *seifs*) that may extend for many miles. Here and there rise several isolated blocks of mountains. Along the coast of the Mediterranean are shallow lagoons of brackish water, separated from the sea by more or less continuous sandbars.

The Negev desert, smaller than the Sinai, also consists of three principal zones. The southern part is igneous, with mountains of varicolored granite and gneiss. North of it lie the Central Negev Highlands (reaching a maximal height of 3000 feet [900 m] above sea level), composed of mainly sedimentary limestone, chalk, and sandstone. The topography is wave-like, consisting of a series of parallel alternating ridges (anticlines) and vales (synclines), with a southwest–northeast axis. In several places, the anticlines were breached and gouged out by erosional processes that created spectacular craters, or cirques. The northern parts of the Central Negev Highlands tend to receive more rain than the southern Negev and provide a fair amount of pasturage, especially along the wadis. Runoff farming was practiced in this subdistrict, especially during the Nabatean,[7] Roman, and Byzantine eras (fourth century B.C.E.–seventh century C.E.), based on the collection of runoff from sloping areas and its utilization in small bottomland plots. Finally, the relatively humid northern plains consist of sand- and loess-covered expanses with vegetation that allows extensive grazing. Dryland farming is also practiced in this area, albeit on a marginal basis.[8]

NUMEROUS SCHOLARS and amateurs have tried to decipher the passages in the Books of Exodus, Leviticus, Numbers, and Deuteronomy that describe the wanderings of the Israelites in Sinai in order to trace the actual path followed by the tribes from Egypt to Canaan.[9] Yet the entire chronicle remains an enigma. Those who believe in the literal truth of the biblical account have encountered great difficulty in identifying locations that are mentioned but hardly described in the Scriptures. Only a few of the listed place-names fit sites that are recognizable today. Although circumstantial evidence has been claimed here and there, no definite traces of the Israelites' sojourn in or passage through Sinai have been found to date.

The biblical account of the long trek through Sinai was apparently written many centuries after the purported event, so the story seems to have been modified and perhaps even radically reconstructed from ancient oral traditions by writers whose primary aim was not to achieve historical veracity but to promote religious doctrine. The random roaming of a motley band of nomads may have been interpreted retrospectively as a purposeful journey toward an ultimate destination. No wonder, then, that various investigators (some of whom were motivated by personal, ideological, religious, or national aims of their own) have suggested no fewer than nine locations where the "Red Sea" or the Sea of Reeds was crossed and no fewer than fourteen sites for Mount Sinai.[10] (The proposed sites of Mount Sinai have varied from western Sinai to northwestern Arabia and, in between, to the southern Negev.) So much depends on the particular interpretation of the ambiguous Writ and of the tantalizing landscape.

What matters in the end is not the exact route followed by those fugitives from Egypt through the trackless wilderness more than three thousand years ago, but the essence of their experience in the desert, inasmuch as it affected the consciousness and lore of their cultural descendants. So strong and frequent are the references in subsequent parts of the Bible to the sojourn in the desert and so indelibly etched is the memory of it that we are driven to believe that the lore must have arisen out of a real event. The same experience led the Israelites to sanctify certain landmarks in the desert, particularly prominent mountains, where they would—at first in actuality and later in imagination—worship and from which they would draw inspiration.

The belief that the Israelites (or at least an influential nucleus among them) lived for a time in the desert, survived its austerity, and sublimated the experience spiritually as an encounter with the Almighty seems more plausible than the counter-belief that the entire story is pure invention. Other groups that did not share the experience in Sinai (except vicariously) may have joined at later times with those who had trekked through that desert to constitute the larger Israelite nation. The early Israelites may well have been a blend of various groups with different backgrounds. The attribution of a common ancestry and shared experiences in Egypt and Sinai may have been a later claim aimed at promoting national and religious unity.

The shortest possible route for the Israelites to have traversed from the Nile delta to Canaan was along the coast of the Mediterranean Sea or parallel to it through the plain of sand dunes in the northern Sinai Peninsula. Indeed, that was the path followed by the Egyptian armies in their campaigns of conquest throughout the Levant, and the same path in the opposite direction was followed by the armies of their adversaries who periodically sought to invade Egypt. Hence the stretch of northern Sinai along the coast was heavily fortified and garrisoned much of the time by Egyptian forces. That may well have been the precise reason why the Israelites avoided that route: "When Pharaoh let the people go, Elohim did not lead them by way of the land of the Philistines, although it was nearer; for Elohim said: 'The people may have a change of heart when they see war, and return to Egypt.' So Elohim led the people roundabout, by way of the wilderness at the Sea of Reeds" (Exodus 13:17–18).

Relating the military threat to the Philistines may have been the idea of a later writer or writers, on the basis of the situation that existed at their own time. There is some evidence, however, that the Egyptians deliberately settled some of the Sea Peoples (who had become their mercenaries and who included the Philistines) along the southeastern coast of the Mediterranean in order to protect the approaches to Egypt. In any case, that passage accords with the evidence that the coastal route from Egypt to Canaan was well protected at the time. As it happened, however, the Israelites could not avoid war even while taking the detour through the central Sinai, although the irregular forces they had to face there were probably less powerful than the organized army that was arrayed along the northern route.

If, indeed, the Exodus and the trek across Sinai did occur (albeit on a scale much more modest than that related in the Scriptures), the most probable initial leg of the route would have been from the eastern fringes of the Nile delta, through the depression known today as Wadi Tumeilat, toward the Bitter Lakes.[11]

A site where the Israelites may have crossed the Sea of Reeds was the juncture between the Great Bitter Lake and the Small Bitter Lake. An earthen ridge that is a couple of miles wide but only a few feet high could have offered relatively easy passage. At various times, the ridge may have been covered by a shallow depth of water, whereas at other times it may have been exposed. Even in relatively dry periods, however, the high water table (fed by seepage from the twin lakes) would have made the soil soggy and promoted the profuse growth of phreatophytes such as phragmites and papyrus reeds. So it seems fairly reasonable that the light-footed Israelites could have crossed there on "dry land," while the heavy-laden Egyptian army, mounted on horses and riding in narrow-wheeled chariots, could have become mired in the soft mud.

Now the Children of Israel, newly freed of slavery in Egypt, reportedly found themselves in the great and terrible desert, knowing not what to do and how to survive. Slavery in Egypt was humiliating and arduous, but it was secure. Freedom in the desert seemed glorious at first, but terribly insecure. The most elemental imperative was water:

Moses caused Israel to set out from the Sea of Reeds. They went on into the wilderness of Shur; they traveled three days in the wilderness and found no water. They came to Marah, but they could not drink the water of Marah because it was bitter; that is why it was named Marah. And the people grumbled against Moses, saying, "What shall we drink?" So he cried out to Yahweh, and Yahweh showed him a piece of wood; he threw it into the water and the water became sweet.... And they came to Elim, where there were twelve springs of water and seventy palm trees; and they encamped there beside the water. (Exodus 15:22–25, 27)

Several explanations can be offered for Moses's feat. One is that the water of Marah was in a stagnant pool that was covered with a foul-smelling and foul-tasting scum of algae and animal waste. Casting a tree into the water may have stirred it up sufficiently to bring to the surface better-quality water. Another possibility is that the tree that Moses cast into the water, being a desert plant, exuded salt from its leaves, as does tamarisk. If so, the tree may have been salty enough to cause the suspended matter in the water to coagulate and settle, thereby rendering the water clear of the offensive turbidity. Moses may have known how to clarify fouled waters from his earlier experience as a shepherd in the semiarid domain, at the desert's edge.[12]

On another occasion of thirst, Moses was instructed to extract water from a rock by striking it with his rod:

From the wilderness of Sin the whole Israelite community continued by stages as Yahweh would command. They encamped at Rephidim, and there was no water for the people to drink. The people quarreled with Moses, "Give us water to drink," they said.... Then Yahweh said to Moses ... "Strike the rock and water will issue from it, and the people will drink." And Moses did so in the sight of the elders of Israel. (Exodus 17:1–6)

Obtaining water by striking a rock is not necessarily as fantastic as it first seems, provided that it is the right kind of rock. In some places in the desert (although they may be far apart), shallow (perched) groundwater oozes toward the surface; as it evaporates, it forms an encrustation of precipitated calcite (calcium carbonate) and gypsum (calcium sulfate). This porous and brittle rock-like formation, called evaporite or travertine, is recognizable to anyone familiar with the desert (as was Moses). By breaking the crust in some places, Moses could have released the water percolating below it.[13]

As they were led into the desert and made to embark on the rigorous trek through the waterless wasteland, the people of Israel soon began to long for the security and abundance they had left behind in Egypt, and they forgot the humiliation and hard labor to which they had been subjected there: "In the wilderness, the whole Israelite community grumbled against Moses and Aaron. The Israelites said to them, 'If only we had died by the hand of Yahweh in the land of Egypt, when we sat by the fleshpots, when we ate our fill of bread! For you have brought us out into this wilderness to starve this whole congregation to death'" (Exodus 16:2–3).

According to the Book of Exodus, Yahweh then provided the Israelites with two kinds of food: manna and quail:

> Yahweh said to Moses: "I will rain down bread[14] for you from the sky, and the people shall go out and gather each day that day's portion." . . . In the evening quail appeared and covered the camp; in the morning there was a fall of dew about the camp. When the fall of dew lifted, there, over the surface of the wilderness, lay a fine and flaky substance, as fine as frost on the ground. When the Israelites saw it, they said to one another, *man hoo* [What is it?]—for they did not know what it was. And Moses said to them, "That is the bread which Yahweh has given you to eat." (Exodus 16:4, 13–18)

Various explanations have been offered for the manna—whose name derives from the Israelites' question—none of them convincing. One conjectured candidate for the title is the sweet whitish sap exuded by the foliage of a species of tamarisk tree after being stung by an insect. But the amount and nutritional value of this sap would be completely insufficient to serve as food for any number of people. The story of manna, therefore, must be regarded as completely mythical.

The story of the quails, however, has at least some basis in reality. Each year, these birds migrate in large flocks across the Mediterranean Sea, to Africa in the autumn and back to Europe in the spring. They tend to fly very low, hovering above the ground. At the end of a long flight across the Mediterranean, they are so exhausted as to drop to the ground, where they can be picked up by hand. The Bedouin of northern Sinai have traditionally strung up vertical nets along the seashore to capture the birds by the droves as they descend on the coast.[15] So the account of the quail being borne by the wind from the sea (Exodus 16:13; Numbers 11:31–32) has some basis in reality, although it is hard to imagine the supply being steady enough to feed a large tribe wandering across the Sinai Peninsula for all of forty years. As many as 600,000 men, as well as women and children, are reported to have left Egypt with Moses (Exodus 12:37), but the desert of Sinai certainly could not have supported so large a group. Far more plausible is that a limited number of people (no more than a few thousand) subsisted in Sinai by grazing goats and by hunting. Indeed, there are several indications in the relevant books of the Pentateuch that the Israelites in Sinai had flocks of goats and sheep.

The desert through which the Israelites passed on their journey toward the promised land was not entirely empty. There were tribes of desert dwellers, engaged in extensive grazing of the dry range, hunting, and occasional marauding. Such groups naturally resented intruders into their domain who might compete with them for the area's exceedingly meager resources. One such tribe was called Amalek, and it became a particularly bitter enemy of the Israelites:

> Amalek came and fought with Israel at Rephidim. Moses said to Joshua, "Pick some men for us, and go out and do battle with Amalek." . . . And Joshua overwhelmed the people of Amalek with the sword. Then Yahweh said to Moses, "Inscribe this in a document as a reminder, and read it aloud to Joshua: I will utterly blot out[16] the memory of Amalek from under heaven!" And Moses built an altar and named it Yahweh-nissi [Yahweh my banner, or miracle]. . . . Yahweh will be at war with Amalek throughout the ages. (Exodus 17:8–16)

Feuds between rival tribes in the desert, fateful to the very survival of each, always tended to be very extreme and to last for many generations—until one or the other side prevailed.

In contrast to the Amalekites, the Midianites (among whom Moses found refuge when he first escaped from Pharaoh) became, for a time, allies of the Israelites. And when Moses's father-in-law, Jethro, the priest of Midian, heard of all that had happened to Moses and to the Israelites, he came to greet Moses in the wilderness: "Jethro rejoiced over all the kindness that Yahweh had shown Israel when He delivered them from the Egyptians. Jethro said . . . 'Now I know that Yahweh is greater than all the *elohim*.'[17] . . . And Jethro, Moses' father-in-law, brought a burnt offering and sacrifices for Elohim; and Aaron came with all the elders of Israel to partake of the meal before Elohim with Moses' father-in-law" (Exodus 18:9–12).

Each nation had its own god or set of gods. So while Jethro acknowledged the greatness of Moses's god, he did not thereby accept him as the only god. Note that the word *elohim* (being in the plural form) may mean "gods" in a generic sense, or it may mean a particular, singular God. As such, the name Elohim probably represents a stage in the transition from a belief in multiple gods to a gradual acceptance of the one and only God: Yahweh.

THE BOOK of Exodus describes the ultimate defining event for Moses and the Israelites: the singular and momentous revelation on Mount Sinai, including the confirmation of the Holy Covenant, and conferral of the Ten Commandments:

> On the third new moon after the Israelites had gone forth from the land of Egypt . . . they entered the wilderness of Sinai and encamped . . . in front of the

THE HOSTILITY BETWEEN
THE ISRAELITES AND THE AMALEKITES

The vehement hostility between the Israelites and the Amalekites illustrates the grim fight-to-the-death rivalry between nomadic tribes over territorial rights in the desert domain, the rights to sparse pastures and meager water supplies. No permanent peace arrangement could be made between nomads, whose locations and circumstances were ever shifting, depending on the vagaries of their drought-prone environment. Settlers along the edge of the desert domain were always in danger of the hit-and-run attacks of the desperate nomads. Consequently, the self-preservation instinct of the Israelites, themselves in the early stages of the transition from nomadism to sedentarism in a marginal environment, was to strike their perennial enemies, the Amalekites, mercilessly, wherever they could. As ancestral memories tend to persist beyond their original contexts, the hatred felt by the Israelites toward the Amalekites (as representatives of desert raiders in general) remained etched in Israelite lore and was even given a religious sanction: "Then Yahweh said to Moses: 'Inscribe this in a document as a reminder, and read it aloud to Joshua: I will utterly blot out the memory of Amalek from under heaven!' And Moses built an altar and named it Yahweh-nissi [Yahweh my banner, or miracle]. He said, 'It means, hand upon the throne of Yah! Yahweh will be at war with Amalek throughout the ages'" (Exodus 17:14–16).

This lethal enmity toward the raiding nomads persisted for many generations, long after the Israelites had ceased being wanderers in the desert domain, having settled in the land of Canaan and having coalesced to form a united kingdom. Even then, Samuel saw fit to instruct the newly anointed King Saul: "I am the one that Yahweh sent to anoint you over His people in Israel. Therefore, listen to Yahweh's command! Thus said Yahweh Tsevaot [Lord of hosts, God of war]: I am exacting the penalty for what Amalek did to Israel, for the assault made upon them on the road, on their way up from Egypt. Now go, attack Amalek, and proscribe all that belongs to him. Spare no one, but kill alike men and women, infants and sucklings, oxen and sheep, camels and asses!" (1 Samuel 15:1–3). Later, when Saul pitied Agag, the chief of the Amaleks, and spared his life, as well as the animals in some of the Amalekites' flocks, the fierce prophet Samuel condemned Saul publicly and then proceeded personally to slay the Amalekite chief (1 Samuel 15:33).

The Israelites' ruthless mores of the desert were moderated slightly when it came to the sedentary Canaanites, whose land the Israelites came to possess:

When you approach a town to attack it, you shall offer it terms of peace. If it responds peaceably and lets you in, all the people present there shall serve you at forced labor. If it does not surrender to you but would join battle with you, you shall lay siege to it; and when Yahweh your Elohim delivers it into your hands, you shall put all its males to the sword. You may, however, take as your booty the women, the children, the livestock, and everything in the town—all its spoil—and enjoy the use of the spoil of your enemy, which Yahweh your Elohim gives you. (Deuteronomy 20:10–14).

In the conduct of war, the early Israelites were hardly, or only marginally, more humane than all the other warring peoples of the ancient Near East. War has never been a gentle activity. Nowhere is that cruel truism more apparent than on the wall of a temple at Medinet Habu in Egypt, where Rameses III commemorated his victory over invaders from Libya. Displayed among the trophies brought from that battle were large piles of the severed penises of the slaughtered enemies.

mountain, and Moses went up to the Elohim. Yahweh called to him from the mountain, saying, "... If you will obey Me faithfully and keep My covenant, you shall be My treasured possession among all the peoples. Indeed, all the earth is Mine, but you shall be to Me a kingdom of priests and a holy nation." (Exodus 19:1–6)

This was to be a unique moment, unlike any previous or subsequent encounter with God. For once, Yahweh's message would come not as a vision in a dream, not as a whisper in a dark cave, not as a flickering light in a small burning bush seen by one man, but as a thundering voice emanating from a thick cloud and resounding from the mountaintop for an entire assemblage to hear and tremble:

> As morning dawned, there was thunder, and lightning, and a dense cloud upon the mountain, and a very loud blast of the horn, and all the people who were in the camp trembled. . . . Now Mount Sinai was all in smoke, for Yahweh had come down upon it in fire; the smoke rose like the smoke of a kiln, and the whole mountain trembled violently. The blare of horn grew louder and louder. As Moses spoke, Elohim answered him in thunder. Yahweh came down upon Mount Sinai, on the top of the mountain, and Yahweh called Moses to the top of the mountain and Moses went up. (Exodus 19:16–20)

No natural event fits that description. Whoever believes the veracity of this, the most awesome of all occasions, needs no naturalist explanation. Those who seek a natural cause can find none. There are no active volcanoes in Sinai.

A mountain in southern Sinai, thought by some to be the mountain on which Moses received the Ten Commandments.

An earthquake is a possible event, but a rare one, and not specific to any particular mountain.

What is significant is the biblical writers' association of God's revelation with the desert and, more particularly, with a mountain in the desert. Even today, with all the means of easy access to (and departure from) once-remote places, there remains an awe-inspiring quality to the desert landscape. The raw exposure of the elements, the upthrust cliffs amid the soft sands, the shattered boulders fallen from rugged slopes, the hidden caverns and shadowy ravines, the unlidded sunlight and distant vistas by day and the crystal-clear skies and piercing stars of night—all bespeak the primeval power of nature and of its creator. Above all reigns a pervasive stillness and a sense of expectancy, as though having once heard the voice of the Almighty, the desert dare not break the silence as it awaits that voice to issue from within and resound once again for all the world to hear.

Alas, as the narrative continues, the assembled Israelites failed to rise to the momentous occasion. While Moses was on the mountain for forty days and nights, they grew restive and demanded that Aaron create for them a tangible statue that could serve as a symbol or an object of worship. Familiar as he was with the multifarious idols of the Egyptian pantheon, Aaron chose a figure that most represented his people's pastoral tradition:

> When the people saw that Moses was so long in coming down from the mountain, the people gathered against Aaron and said to him, "Come, make us a god who shall go before us, for that man Moses, who brought us from the land of Egypt—we do not know what has happened to him." Aaron said to them, "Take off the gold rings that are on the ears of your wives, your sons, and your daughters, and bring them to me." . . . This he took from them and cast in a mold, and made it into a molten calf. And they exclaimed, "This is your god, O Israel, who brought you out of Egypt!" When Aaron saw this, he built an altar before it. . . . The people offered up burnt offerings. . . . They sat down to eat and drink, and then rose to dance. (Exodus 32:1–6)

The calf god could have been either a pastoral or a rainfed-farming icon (as the Canaanite god El was traditionally represented as a bull). So it seems that the generations of sojourning in Egypt had not eliminated the pastoral lore and the Canaanite orientation of the Israelites. Nor was the radical notion of an abstract, invisible Yahweh yet fixed in their consciousness. Thus when left alone, they reverted to the ancient cult of animal-god worship and to the pagan rite of orgiastic dancing, and they would tend to do so over the ensuing generations, practically until the time of the exile in the sixth century B.C.E.

A Canaanite (or early Israelite) version of the bull god, crafted in bronze in the twelfth century B.C.E. The bull may resemble the golden calf reportedly worshiped by the recalcitrant Israelites in Sinai during the prolonged absence of Moses (Exodus 32).

In response to their early betrayal of the Covenant, Yahweh—so it is written—considered destroying the people of Israel. But he was restrained by Moses, who pleaded with Yahweh on behalf of his people. Here the role of God and man were momentarily reversed: Yahweh was portrayed as impulsive and mercilessly vindictive, while Moses represented a higher conscience and compassion:

> Yahweh said to Moses, "I see that this is a stiffnecked people. Now, let Me be, that My anger may blaze forth against them and that I may destroy them, and make of you a great nation."[18] But Moses implored Yahweh his Elohim, saying, "Let not Your anger, O Yahweh, blaze forth against Your people. . . . Let not the Egyptian say, 'It was with evil intent that He delivered them, only to kill them off in the mountains and annihilate them from the face of the earth.'" . . . And Yahweh renounced the punishment He had planned to bring upon His people. (Exodus 32:9–14)

So important was the principle of justice tempered by mercy that it could be demanded of even the Almighty!

But then Moses himself was seized with wrath as he meted out his own punishment against the very people on whose behalf he had pleaded for mercy from Yahweh: "Moses stood up in the gate of the camp and said, 'Whoever is for Yahweh, come here!' And all the Levites rallied to him. He said to them, '. . . Each of you put sword on thigh, go back and forth from gate to gate throughout the camp, and slay brother, neighbor, and kin.' The Levites did as Moses had bidden, and some three thousand of the people fell that day" (Exodus 32:26–28). The inconsistency in Moses's attitude and behavior is a reflection of the inner struggle within the evolving faith of the Israelites, between the contradictory notions of a harsh God demanding absolute obedience and a compassionate God promoting patience and forgiveness along with justice.

After punishing the people for their transgression, Moses returned to the mountain to receive the Ten Commandments and to bring the Law to his people. When he descended from his intimate encounter with Yahweh, his face radiated with light:

> As Moses came down from the mountain bearing the two tablets . . . [he] was not aware that the skin of his face was radiant, since he had spoken with Him. Aaron and all the Israelites saw that the skin of Moses' face was radiant; and they shrank from coming near him. But Moses called to them, and Aaron and all the chieftains in the assembly returned to him, and Moses spoke to them. Afterward all the Israelites came near, and he instructed them concerning all that Yahweh had imparted to him on Mount Sinai. And when Moses had finished speaking with them, he put a veil over his face. (Exodus 34:29–33)

Having spent forty days alone on the mountaintop, Moses may have had such a sunburned and shiny face that it appeared to be eerily aglow. He then had to cover his face to protect it from further exposure. That is a mundane explanation, but perhaps a plausible one.[19]

The drama of the revelation on Mount Sinai, so vividly described, was designed to impress on all future generations of Israelites the supreme importance of the Mosaic Law and the principles it embodies. These are conveyed most succinctly in the Ten Commandments:

> I am Yahweh your Elohim, who brought you out of the land of Egypt, the house of bondage.
>
> You shall have no other gods besides Me.
>
> You shall not make for yourself a sculptured image, or any likeness of what is in the heavens above, or on the earth below, or in the waters under the earth. You shall not bow down to them or serve them. . . .
>
> You shall not swear falsely by the name of Yahweh your Elohim; for Yahweh will not clear one who swears falsely by His name.
>
> Remember the Sabbath day and keep it holy. Six days you shall labor and do all your work, but the seventh day is a Sabbath of Yahweh your Elohim; you shall not do any work—you, your son or daughter, your male or female slave, your cattle, or the stranger who is within your settlements. . . .
>
> Honor your father and your mother that you may long endure on the land that Yahweh your Elohim is assigning to you.
>
> You shall not murder.
>
> You shall not commit adultery.
>
> You shall not steal.
>
> You shall not bear false witness against your neighbor.
>
> You shall not covet your neighbor's house; you shall not covet your neighbor's wife, or his male or female servant, or his ox or his ass, or anything that is your neighbor's. (Exodus 20:2–14)

The Ten Commandments remain the fundamental tenets of ethical monotheism, representing a ringing declaration of high principles expressed in concise practical terms.[20] That statement has never been surpassed, or even equaled, in power, clarity, and brevity. Even more than it specifies explicitly, it conveys implicitly. Why obey its edicts? Because it is mandated by the supreme universal moral authority. Why not worship other gods? Because there is only one God. Why grant rest to all working humans and animals on the Sabbath? Because God himself rested on that day. Why honor parents, and why refrain from killing and adultery and stealing and bearing false witness and coveting? Because humans are required by God, who put them on Earth and who granted them strength and wisdom, to act justly and humanely.

The requirement to rest on the Sabbath, in particular, was given a cosmic rationale. It was presented as being not only a humane act, granting rest to the weary, or a spiritual act, providing time for contemplation and prayer, but an emulation of God's own act, "for in six days Yahweh made heaven and earth, and on the seventh day He ceased from work and rested."[21] That made the requirement absolute (Exodus 31:12–17). The Sabbath thus was a unique holy day. All other holidays were somehow linked to the cycles of nature or to historical events. The Sabbath stands alone as a living symbol of God's transcendence.[22]

The principles underlying the Ten Commandments were represented as having been handed down in one fell swoop on one memorable day from the summit of one holy mountain. It is much more likely that they evolved gradually over many generations of trial and tribulation and much error and anguish and agony and soul-searching. It is noteworthy that the Ten Commandments specify no fixed ritual of worship, especially no form of human or animal sacrifice, which had long served as the principal means of worship and supplication in the pagan religions as well as in the early stages of the Israelite religion. Nor do they require a tithe or an annual pilgrimage to a holy temple or obedience to any human authority in whom is vested the exclusive power to represent God or to serve as his intermediary. Such requirements, and many more, were added in numerous corollary rules. Yet the paramount code of behavior remained a set of simple, universal, and absolute principles meant to guide individual and communal behavior, not just on particular occasions and at specific places, but at all times and in all locations.

As far as we know, later generations of Israelites never sought to build a temple on Mount Sinai, nor did they tend to revisit it on pilgrimage. In the end, the actual place faded into myth and became immaterial, compared with the principles claimed to have originated there but meant to be applied everywhere.

In the desert, not only the abode of the people was transient, but also that of their Yahweh. The maker of heaven and Earth ruled everywhere, yet made his tangible presence among the people of Israel wherever they moved. The last five chapters of Exodus (36–40) provide a highly detailed description of *ohel hamoed* (Tent of Assembly, or Tabernacle). It was ornate, with fine artistic and symbolic decorations. It housed the Holy Ark (probably containing the Tablets of the Law) and was, in effect, a portable Temple.[23] As such, it served for a tribe of nomadic pastoralists the same function as did a permanent cult center for a sedentary society. Being a universal God, Yahweh was present everywhere, of course, yet was believed to be centered in the Tent of Assembly—but that was merely one of many ambiguities or inconsistencies of the evolving religious faith. Another discrepancy was the inclusion within the Tabernacle of sculpted cherubim with spread wings (Exodus 37:7–9), despite the absolute prohibition of graven images, in the Ten Commandments and elsewhere.

Incidentally, worship in the Tent of Assembly required the attendees to wash their hands and feet, as was the traditional custom of the pastoralists. This and other details suggest that the Israelites never became (or intended to become) permanent dwellers in the desert domain. Rather, they were—consciously and in fact—mere wayfarers on their journey to their promised land, which they fantasized to be a place of lush pastures and bountiful fields and fruitful orchards. Yet the Israelites remained in—or in close proximity to—the desert long enough to absorb elements of its lore and to assimilate them into their holistic worldview.

THE NOSTALGIC MEMORY of Israel's formative experience in the desert was repeatedly invoked in later periods of the nation's history, long after the settlement in Canaan.[24] The prophet Elijah, when persecuted and dejected, sought solace and nearness to God in the solitude of the desert. After wandering in the vast and desolate landscape for the proverbial forty days and forty nights, he finally experienced the presence of God on Mount Horeb:

> Frightened, he fled at once for his life[25] . . . into the wilderness. He came to a broom bush and sat down under it, and prayed that he might die. . . . Suddenly, an angel touched him and said to him, "Arise and eat." He looked about, and there, beside his head, was a cake baked on hot stones and a jar of water![26] . . . He walked forty days and forty nights as far as the mountain of the Elohim at Horeb.[27] There he went into a cave. . . . Then the word of Yahweh came to him.
> (1 Kings 19:3–9)

The next passage, describing Yahweh's revelation to Elijah, is surely one of the most sublimely beautiful and inspiring in the entire Bible:

> "Come out," He called, "and stand on the mountain before Yahweh." And lo, Yahweh passed by. There was a great and mighty wind, splitting mountains and shattering rocks by the power of Yahweh; but Yahweh was not in the wind. After the wind—an earthquake; but Yahweh was not in the earthquake. After the earthquake—fire, but Yahweh was not in the fire. And after the fire—a sound of thin silence.[28] When Elijah heard it, he wrapped his mantle about his face and went out and stood at the entrance of the cave. (1 Kings 19:11–13)

Many generations before the time of Elijah, Yahweh was said to have demonstrated his presence to the assembled people by means of a powerful sound-and-light display of fireworks (Exodus 19:9, 16–18). Evidently, something had changed in the interim in the way the Deity was to be perceived. Although no less powerful in his ability to rend mountains and smite enemies, Yahweh became increasingly subtle and spiritual, a mover of the human spirit no less than a mover of mountains.

The prophet Jeremiah summoned up memories of the nation's past and evoked longings for the purity of faith and clarity of vision originally manifested in the desert, as he expressed Yahweh's love for his people:

> Thus said Yahweh:
> I remember . . . the devotion of your youth,
> your love as a bride—
> How you followed me in the wilderness,
> in a land not sown.
> Israel was holy to Yahweh,
> the first fruit of His harvest. (Jeremiah 2:2)

But then, feeling rejected and dejected, as had Elijah before him, Jeremiah too sought personal solace in the desert as he pleaded wistfully:

> Oh, to be in the desert,
> at an encampment for wayfarers!
> Oh, to leave my people,
> to go away from them—
> For they are all adulterers,
> a band of traitors. (Jeremiah 9:1)

The prophet Isaiah envisioned a time of universal salvation, in which God would redeem the desert as well as all humanity:

The arid land shall be glad, the wilderness shall rejoice
and shall blossom like a rose.
It shall blossom abundantly,
it shall also exult and shout.
It shall receive the glory of Lebanon,
the splendor of Carmel and Sharon.
They shall behold the glory of Yahweh,
the splendor of our Elohim. . . .
For waters shall burst forth in the desert,
streams in the wilderness.
Torrid earth shall become a pool;
parched land, fountains of water. . . .
And the ransomed of Yahweh shall return
and come with shouting to Zion,
crowned with joy everlasting.
They shall attain joy and gladness,
while sorrow and sighing flee. (Isaiah 35:1–10)

7

THE RAINFED DOMAIN

Settlement in the Hill District of Canaan

Yahweh your Elohim is bringing you into a good land, a land with streams and springs issuing from plain and hill; a land of wheat and barley, of vines, figs, and pomegranates, a land of olive trees and honey; a land where you may eat food without stint, where you will lack nothing.

DEUTERONOMY 8:7–9

THE RAINFED DOMAIN extends in a wide arc from the Zagros Mountains in the east; through the so-called Armenian and Taurus Mountains in the north, and the parallel Lebanon Mountains and Anti-Lebanon Range in the west; to the hill districts of Galilee, Carmel, Samaria, and Judea in the southwest. This is a region of highlands, hills, and fertile intermontane valleys.

The land of Canaan lies on the southwestern edge of the arc. Its northern part consists of five parallel, ecologically distinct, districts, only three of which are humid enough to be considered part of the rainfed domain. The first of these three, on the west, is the coastal plain, wide in the south and tapering toward the north (where Mount Carmel juts into the Mediterranean Sea, delimiting the coastal plain to the south and forming a wide bay to the north). The coastal plain is underlain by calcareous sandstone, and its soils are mostly sandy but in places are alluvial clay loams. To the east, the land rises to a piedmont district, called the Shefela. The undulating foothills are underlain by chalk and soft limestone, and its soils are loams and clay loams. East of the Shefela, in turn, rises a range of low mountains that constitutes the spinal column of the country. These highlands are interrupted by intermontane valleys that divide them into subdistricts with the biblical names of Galilee in the north, Samaria in the center, and Judah (or Judea) in the south.

The largest of the intermontane valleys lies across the northern part of the country, forming the boundary between the subdistricts of Galilee and Samaria. This is the Valley of Jezreel, through which flow the stream of Kishon westward, along the edge of Mount Carmel toward the Mediterranean Sea,

The orographic effect, whereby the windward side of a mountain range forces moisture-laden air to rise, condense its vapor to form clouds, and precipitate the vapor as rain. The leeward slope causes the moisture-deprived air to descend, compress, warm up, and desiccate the land. Thus the west-facing slopes of Canaan are in the rainfed domain, while the east-facing slopes are in the desert domain.

and the stream of Harod eastward, through the Beisän Valley to the Jordan Rift Valley. The Valley of Jezreel was cultivated for grain and had meadows of grass,[1] along with a series of marshes that were periodically inundated by the overflow of the streams. Along the rim of the valley were the settlements of Megiddo, Ta'anakh, Harosheth, Endor, Jezreel, and—on its eastern edge, where it links with the Jordan Valley—Beth-Shean. These settlements lay astride the region's major trade routes that connected Canaan with all the neighboring lands.

The topography of the central highlands is asymmetrical: the relatively humid west-facing slopes descend gently toward the Mediterranean Sea, but the east-facing slopes drop steeply toward the Jordan Valley. Being in the rain shadow of the cloud-bearing westerly winds of winter, the east-facing slopes constitute a desert domain that is incised by steep canyons.

The geologic trench known as the Jordan Rift Valley is a riverine domain all its own. It consists of deep deposits of marl, which is a mixture of clay and fine particles of lime. These deposits are residual sediments from the floor of an inland sea that has long since receded and dried up, leaving only vestigial lakes. In the northern section of the valley were two freshwater lakes, through which flowed the swift and exuberant stream of the upper Jordan. Coming out of the larger of the two lakes (Sea of Galilee), the lower Jordan wended its sinuous course southward for about 60 miles (100 km), only to die a tired death in the thick steamy brine of the Dead Sea, in the desolate valley of Sodom and Gomorrah.[2]

The southern part of Canaan is called the Negev (literally, "dried up"). Its northern section is in the pastoral domain, whereas its southern section lies

View of the upper Jordan River. The name Jordan (Hebrew, Yarden) denotes "descender" in recognition of the river's steep descent from its source at the foot of Mount Hermon to the Sea of Galilee.

in the desert domain. Thus, its small size notwithstanding, the land of Canaan exhibits a great diversity of climate and terrain, and hence of ecological conditions.

THE TYPICAL Mediterranean climate of the country is characterized by an annual cycle of contrasting seasons: a cool rainy winter and a warm dry summer. The rains typically begin in October or November and cease in April. The annual precipitation ranges from some 16 inches (400 mm) at the southern fringe of the rainfed domain, to more than 40 inches (1000 mm) in the relatively humid uplands in the north, and generally exceeds the potential evaporation rate during the winter. In the summer, however, the potential evaporation rate is well above 40 inches (1000 mm), so, in the absence of irrigation, crop plants must depend entirely on the residual soil moisture from the preceding winter. When soil moisture is deficient, plants wilt and crops may fail entirely.[3]

Although rainfall in the uplands is not always abundant and not ideally distributed in time and space, it is generally sufficient to support a mixture of native trees,[4] shrubs, and grasses. Originally, the more humid areas were covered with dense forests, whereas the less humid areas gave rise to savannah-like open woodlands, with widely spaced trees (or clumps of trees) and interspersed patches of shrubs and grasses. In the spring, the countryside is carpeted with and perfumed by a profusion of flowers, including lilies, anemones, cyclamens, poppies, and hollyhocks.

THE PHYSIOGRAPHY OF CANAAN

To understand the ecology of the land of Canaan, we must consider the features of its geography and climate. First let us envision an elongated sea intruding itself like a wedge, 2500 miles (4000 km) from west to east, into the midst of a huge landmass, and call that sea the Mediterranean. North of the sea lies the humid continent of Europe; south of it lies the arid region of North Africa. The sea contributes moisture to the air currents swirling from the west and delivering rain to the lands lying to the north and east of its shores. Now notice a small country flanking the southeastern coast of the sea, with part of it extending above the sea's southern edge (hence being in the path of rain-bearing winds from the sea) and the other part extending below that edge (hence abutting the Sahara and the Arabian Deserts). Even knowing nothing about the topography, we may surmise that rainfall in that country is maximal in the north and west (along the sea) and diminishes toward the east and south (away from the sea and into the desert).

Assume, moreover, that the land rises from the coast, from sea level to an elevation of, say, 3000 feet (900 m) at a distance of 30 miles (50 km) or so from the sea, after which it levels out to form a plateau. Such a topography produces an orographic effect, forcing the clouds coming from the sea to rise over the land and to drop greater rainfall on the higher ground. In such a country, all the streams conveying runoff tend to flow from the plateau and the west-facing slopes toward the sea.

Now let us imagine the improbable—that the plateau was torn asunder in relatively recent geologic history, forming a deep, narrow trench on a north–south line parallel to the coast and some 50 to 60 miles (80 to 100 km) from it. Suppose, even more improbably, that the trench forms the deepest valley on the continental surface of the Earth, a chasm whose lowest point lies more than 1300 feet (400 m) below sea level. The presence of such a rift in the Earth's crust would divide the country in two and likely change its entire hydrological regimen. As the moisture-laden air from the sea is forced upward by the west-facing slopes, its vapor condenses and precipitates there as rain. Then, as the air descends into the rift valley, it becomes drier and warmer. As a result, the east-facing slopes of the highlands together with the rift valley itself, being in the rain shadow of the central highlands, constitutes a desert. And the streams from both sides of the divided plateau and from the mountains to its north rush into the deep valley, creating a river that runs through it and a saline lake at its landlocked lowest point (where water evaporates continuously but salts accumulate).[1]

To complete our exercise in imaginary geography, we now call that river the Jordan (Yarden [river "that descends"]) and its terminal lake, the Dead Sea (Yam Hamelaḥ [Sea of Salt]). In so doing, we have divided our hypothetical country into two entities: the western part, consisting of modern Israel and the Palestinian West Bank, together termed Canaan in the Bible; and the eastern part, consisting of the modern kingdom of Jordan (formerly called Transjordan, which reflects its biblical name, 'Ever ha Yarden).

1. D. Nir, *Geomorphology of Eretz Israel* (Hebrew) (Jerusalem: Akademon, 1989).

When the native tree cover was cleared to form arable land—an activity that took place over many centuries—the combination of rain and fertile soil could support the cultivation of perennial fruit-bearing trees and of annual crops, grown in either the winter or the summer. (Growing a summer crop under rainfed [unirrigated] conditions required fallowing the land during the preceding winter in order to collect and conserve soil moisture for crop

THE GEZER AGRICULTURAL CALENDAR

Gezer was one of the largest cities during the Canaanite and Israelite periods.[1] Its importance was due to its proximity to the fertile coastal plain and to two important trade routes (that along the plain from north to south and that through the Valley of Ayalon from east to west). A small tablet of soft limestone was discovered at Gezer by R. A. S. Macalister in 1908. On it was an inscription, in Biblical Hebrew (or the nearly identical Canaanite), listing the sequence of agricultural activities performed during the twelve months of the year. Dated to the third quarter of the tenth century (ca. 925 B.C.E.), the calendar is considered to be the earliest discovered record of ancient Hebrew writing.[2] It has been variously analyzed and interpreted.[3] A plausible possibility is that it was written by a young person as a school exercise (cast in verse, a kind of mnemonic ditty):

Its two months are (olive) harvest
Its two months are planting (grain)
Its two months are late planting
Its month is hoeing up of flax
Its month is harvest of barley
Its month is harvest and feasting
Its two months are vine-tending
Its month is summer fruit[4]

The Gezer calendar. Its inscription in ancient Hebrew, or Canaanite, specifies the annual cycle of agricultural activities.

The calendar appears to begin in the autumn. Olives are usually harvested in late September and October. The sowing of winter grains (wheat and barley) is usually carried out in November and December. The "late planting" of spring-growing crops is done in January and February. Row crops are hoed to control weeds mainly in the late winter and early spring, in the month of March, and flax is harvested in April. The harvesting of barley may begin in April and continue into May, when wheat is also harvested and threshed. Grapevines are tended in June and July, and summer fruits—grapes, figs, and pomegranates—are gathered in August and early September.

1. M. Avi-Yonah and E. G. Kraeling, *Our Living Bible* (Chicago: Biblical Publications, 1972).

2. The Gezer calendar consists entirely of consonants, except for the final *vav* (*waw*) attached to the repeated word *yrhw*, which appears to represent the vowel *o*, signifying "its month."

3. *Biblical Encyclopedia* (Jerusalem: Bialik, 1954), 2:471–474; *Encyclopaedia Hebraica* (Jerusalem: Encyclopaedia, 1955), 10:579–581; Y. Feliks, *Agriculture in Eretz-Israel in the Periods of the Bible, the Misnah, and the Talmud: Halakkah [Doctrine] and Practice in Basic Agricultural Work* (Hebrew) (Jerusalem: Mas, 1990).

4. "The Gezer Calendar," in J. B. Pritchard, ed., *The Ancient Near East: An Anthology of Texts and Pictures* (Princeton, N.J.: Princeton University Press, 1958), 1:209.

THE EVOLUTION OF THE HEBREW CALENDAR

In Mesopotamia, the calendar year began in the spring, when the Tigris and Euphrates Rivers were in spate after having received the fresh snowmelt from the eastern Taurus and western Zagros Mountains. After the nearly rainless winter, the arriving flood initiated the growing season. The Mesopotamian names for the months were retained in the Hebrew calendar.[1] The name of the first month in the Hebrew calendar, Tishrei, is apparently a loanword from the Akkadian *shurru* (to begin). Marheshvan, the second month in the Hebrew calendar but the eighth in the Mesopotamian calendar, is evidently derived from the Akkadian *warhu samnu* (eighth month). Kislev is similarly of Akkadian origin. So are Tebeth, which implies "month of sinking in mud," and Shevat, from *shabatu*, which implies "[month of] beating rain." Adar seems to come from the Akkadian *iddar*, which means "[month of the] threshing floor." Nissan is, likewise, from the Akkadian *nisannu*, as is Iyar from *aaru*. Sivan evolved from the Akkadian *simanu* (fixed date, time). Tammuz is named after the Babylonian and Assyrian god Dumu-zi (the son who rises, the faithful son). Av (Ab) is related to the Akkadian *abu* (father, master). Finally, Elul corresponds to the Akkadian *elulu*, which signifies "harvest time."

Generations later, when the Israelites were established in the rainfed domain of Canaan, they moved the beginning of the year without changing the names of the months. In the rainfed domain, the biological and agricultural awakening begins in the autumn. After the long dry summer, the onset of the rainy season (generally in October) brings about the germination and renewed growth of both native and crop plants. Hence *rosh hashana* (head [beginning] of the year) was pegged to the new moon nearest to the autumnal equinox. Explicit recognition of Rosh Hashana, however, is absent in the Bible, which refers to the holiday as "the first day of the seventh month" (Numbers 23:24). Clearly, the festival was set as the beginning of the Jewish calendar year some time afterward. The religiously sanctioned impulse to self-reflection gave Rosh Hashana a special meaning as the start of a ten-day period of repentance.

The practical need to establish a consistent calendar of activities was combined with the ancient rite of spring and the religious injunction to celebrate the liberation from slavery in the "month of spring."[2] That, as well as the agricultural imperative, necessitated reconciling the original lunar-based calendar with the solar cycle.

The original calendar of the pastoralists was based on the waxing and waning of the moon. The lunar cycle is 29.5 days long. A year consisting of 12 lunar months therefore totals 354 days—about 11 days short of the solar year, which has about 365 days. The Arabs, who continued as pastoralists longer than the Israelites, have maintained the unadjusted lunar calendar. Hence their holidays may fluctuate from summer to winter in different years. A purely lunar calendar may be adequate for a nomadic society, but is not sufficient for a settled society dependent on farming. All agricultural activities must necessarily relate to the seasons, which are determined by the solar cycle. A people that was once nomadic and became agricultural may adhere to the ancient tradition of lunar months, but must adjust it to serve the need of farmers to schedule their annual tasks to accord with the expectable onset of the rainy and dry seasons. The Israelites adjusted their lunar calendar by adding an extra lunar month just before Nissan (the Passover month) in seven years out of nineteen. So about one year of every three in the Jewish calendar consists of thirteen months.[3]

---

1. E. Klein, *A Comprehensive Etymological Dictionary of the Hebrew Language* (Jerusalem: Carta, 1987).

2. "You shall observe the Feast of Unleavened Bread . . . at the set time in the month of [spring], for in it you went forth from Egypt" (Exodus 23:15, 34:18); "Observe the month of spring, and keep the passover to Yahweh your Elohim, for in the month of spring Yahweh your Elohim brought you forth out of Egypt by night" (Deuteronomy 16:1).

3. Such a year is called *shana meuberet* (pregnant year).

use during the rainless summer.) The principal crops listed in the Bible were wheat, barley, grapes, figs, pomegranates, and olives. The marginal threshold rainfall for annual cropping is about 12 inches (300 mm); less rain than that would generally result in crop failure, and more rain would give proportionately higher yields.

The effectiveness of rain in sustaining crops depends on the presence of a receptive and retentive soil. To be productive, the soil must be able to absorb the rainwater rather than shed it and to store the moisture in the rooting zone of the crops to be grown. The typical soil of the Mediterranean uplands is called *terra rossa*. It is a loam or a clay-loam, formed on hard limestone, and is noted for its red color and favorable structure. Another prevalent soil, called *rendzina*, is a darkish or grayish loam formed on soft limestone and on chalky or marly bedrock. Both soils are relatively receptive to rain. The amount of water that any soil can retain, however, depends on its depth. And herein is the problem. The residual upland soils tend to be rather shallow, covering the hillsides to a depth that seldom exceeds 2 feet (0.6 m), and on steep slopes is much less than that. Only on flat plateaus and particularly in intermontane valleys, where sediment is deposited by gravity and inflowing water, does the soil attain much greater depth.

The typically intense rainstorms that periodically lash the hills of the Levant during the winter are highly erosive. Especially vulnerable are soils on steep slopes that have been denuded of their vegetative cover and have been pulverized as a result of tilling or of trampling by grazing animals. Water trickling off such slopes scours and leaches the topsoil, thereby depriving the remaining soil of nutrients and reducing the thickness of the rooting zone and its capacity to absorb and store moisture. Ironically, the beneficent rain that is so eagerly awaited by farmers can become a voracious monster, gnawing at the soil and wearing away the land. Hence the greatest problem of rainfed agriculture on sloping ground is how to control the erosive power of rain and promote the penetration of water into the soil rather than its escape as surface runoff. Unless this problem is solved, there can be no sustainable agriculture on the rainfed uplands (just as there can be no sustainable agriculture in the irrigated river valleys unless the twin problems of salination and waterlogging are solved). So rainfed farming in the Levant, which began in the flat intermontane valleys where erosion was not a serious problem, came to depend on specialized methods for conserving soil and water as it gradually expanded—under pressure of increasing population—onto the hillsides.[5]

THE BOOK of Joshua describes the entry of the Israelites into Canaan from the east[6] and the conquest of the land by a series of battles with the city-states there.[7] In this campaign, the Israelites were reportedly led by Moses's successor, Joshua, the son of Nun. His symbolic first act as leader was to direct the Israelites into Canaan by crossing the Jordan River "on dry land" (Joshua 3:14–17). In so doing, he had, in effect, repeated Moses's act of crossing the Sea of Reeds.[8] However, Joshua was not primarily a spiritual leader, neither a priest nor even a member of the tribe of Levi (who were granted special religious status). Rather, he was a member of the tribe of Ephraim, descended from Joseph, and was appointed by Moses to serve as a practical military commander. In that role, he did not promulgate new laws but affirmed and applied those handed down by his mentor, Moses. His main task was to wrest the land from its pagan inhabitants and then allocate territories to the various tribes of Israelites. In the process, he oversaw the transformation of a motley collection of wandering tribes into a permanently settled (albeit still heterogeneous) national federation. The Israelites thus underwent the fundamental transition from nomadic pastoralism in the arid domain of northern Sinai, through quasi-permanent pastoralism in the semiarid domain of southern Canaan, and rainfed farming in central and northern Canaan, to—eventually, and in part—settled living in the urban domain of the country's major cities.

Many scholars believe that the conquest was not a concerted campaign but a series of separate, small-scale events that were later collected and combined into a unitary narrative.[9] The Israelites' takeover of Canaan, as best can be discerned from available evidence, was not a single victorious march from east to west and from south to north until "all the land" was conquered. Instead, it seems to have taken place gradually over several centuries. Israelite tribes apparently infiltrated into Canaan and settled wherever they could find room between the existing city-states. Eventually, they absorbed those city-states, either by peaceful assimilation or by violent action, or by both means. Some of the battles are described too vividly and realistically to be discounted totally. However, the scale of the battles and the successes attributed to Joshua may well have been greatly exaggerated in the telling.

Quite apart from the military conquest, the process of settlement on the land in the rainfed domain of Canaan could not have occurred all at once, but took place in stages:

> I will not drive them out before you in a single year, lest the land become desolate and the wild beasts multiply to your hurt. I will drive them out before you little by little, until you have increased and possess the land. I will set your borders from the Red Sea to the Sea of the Philistines, and from the wilderness to the river; for I will deliver the inhabitants of the land into your hands, and you will drive them out before you. You shall make no covenant with them and their gods. (Exodus 23:29–31)

CLASHES ACROSS DOMAIN BOUNDARIES

The marginal, semiarid areas between the rainfed domain and the desert domain became an arena in which two opposing movements took place simultaneously or in alternating sequence. One was the inward movement of nomads from the sparsely populated desert domain, driven by drought, dwindling resources, and destitution. The other was the outward movement of people from the fertile, densely settled rainfed domain, driven by overpopulation. Both groups of migrants were in the category now referred to as environmental refugees. Each group carried with it its own culture, traditions, and preferred mode of subsistence, which it tried to apply in or adapt to the frontier region. The inevitable clash between the migrants could result in the ascendancy of one group over the other and the expulsion or even the annihilation of the vanquished group. A less cruel alternative was the gradual blending of the two societies by assimilation.

Such may have been the experience of the Israelites during their prolonged encounter with the "indigenous" peoples who had preceded them in the pastoral domain of southern Canaan and the rainfed domain of central and northern Canaan. An initially violent struggle to the death with the Amalekites continued with other Canaanite tribes, only to resolve itself gradually in ethnic amalgamation and cultural synthesis. One example of an amicable relationship was that between the Israelites and the pastoral Kenites. The ultimate blending was facilitated by the ancestral relationship of the Canaanites and the Israelites, who had spoken practically the same language from the outset. Over the long course of time, the monotheistic religion and ancestral mythology of the Israelites prevailed, but only after a long internal religious struggle and the absorption of elements of Canaanite tradition (along with elements of other traditions) into what later—much later—came to be known as Judaism.

Examples of ethnic assimilation between the Israelites and the local populace are the marriages of Judah's sons with the Canaanite woman Tamar (later consummated by Judah himself [Genesis 38:6–24]), of Moses with Midianite (Exodus 2:21) and Cushite (Numbers 12:1) women, and of Boaz with the Moabite woman Ruth (Ruth 4:13).

Since communities in each domain had their own gods, and newcomers were expected to accept the local gods, the Israelites had to be admonished, lest they, too, be tempted to adopt local practices: "You shall not bow down to their gods in worship or follow their rituals, but shall tear them down and smash their pillars to bits. You shall serve Yahweh your Elohim, and He will bless your bread and your water" (Exodus 23:24–25). Worship of the local gods would be not merely a religious transgression, but indeed a threat to the nation's future. Religious compromise could have led quite naturally to cultural and then physical assimilation and loss of the Israelites' distinct identity. Therefore, religious exclusivity became a national interest as well.

To the destitute and desperate nomads emerging from the deserts of the Negev and Sinai at the beginning of the Iron Age (ca. 1200 B.C.E.), the promised land must have seemed like a veritable paradise. They had to be made aware that rainfed farming differed fundamentally not only from nomadic pastoralism but also from the type of riverine farming that they had witnessed in Egypt: "The land you are about to enter and possess is not like the land of Egypt from which you have come. There the grain you sowed had to be watered by your labors, like a vegetable garden; but the land you are about to cross into and possess, a land of hills and valleys, soaks up water from the rains of heaven. It is a land that Yahweh your Elohim looks after . . . from year's beginning to its end" (Deuteronomy 11:10–12).[10] In reality, however, Canaan was very possibly the least convenient location for a small nation to try to establish itself. Not only did it lie between the two competing regional centers of power, Mesopotamia and Egypt—the hammer and the anvil, as it were—but it lay at the climatically unstable transition zone between the arid and the humid zones.

There were great ecological differences even within the rainier parts of Canaan. In the hill districts, variations in the direction and steepness of slopes create variations in the density and composition of vegetation. Even where rainfall is practically uniform, differences in the influx of solar radiation can be decisive. South-facing slopes may be subject to an evaporative demand twice as intense as that of north-facing slopes; thus the former tend to be much drier than the latter. The important city of Shechem, for example, is nested between two mountains: Mount Gerizim and Mount Ebal. Viewed from the valley, the north-facing slope of Gerizim appears luxuriant compared with the sere south-facing slope of Ebal.[11] Hence "when Yahweh your Elohim brings you into the

The city of Shechem, nestled between Mount Ebal in the north (*left*) and Mount Gerizim in the south (*right*). The south-facing slope of Ebal appears barren, whereas the north-facing slope of Gerizim abounds with vegetation. The contrast, due to differential exposure to the sun, may have been the reason why the Bible refers to Ebal as the Mount of the Curse and to Gerizim as the Mount of the Blessing (Deuteronomy 11:29, 27:13).

land that you are about to enter and possess, you shall pronounce the blessing at Mount Gerizim and the curse at Mount Ebal" (Deuteronomy 11:29).

The most desirable areas of Canaan were already well populated at the time of the supposed entry of the Israelites. The fertile valleys were especially densely occupied, so the Israelites had no recourse but to settle in the central mountain range.[12] Here they faced new and unfamiliar challenges: a variable and sometimes capricious pattern of rainfall; limited dependable sources of water (springs or perennial streams) for domestic needs and practically none for irrigation; shallow, stony, and erodible soils; a rugged terrain with practically no flat land; and a thicket of oaks, pines, and dense shrubs that the settlers had to clear away. The need to clear the forests was expressed in the injunction to the brother-tribes of Ephraim and Manasseh (together called the Sons of Joseph) by Joshua, who, while himself an Ephraimite, was charged with leading the entire coalition of twelve tribes:[13]

> The Josephites complained to Joshua, saying, "Why have you assigned as our portion a single allotment and a single district, seeing that we are a numerous people whom Yahweh has blessed so greatly?" "If you are a numerous people,"

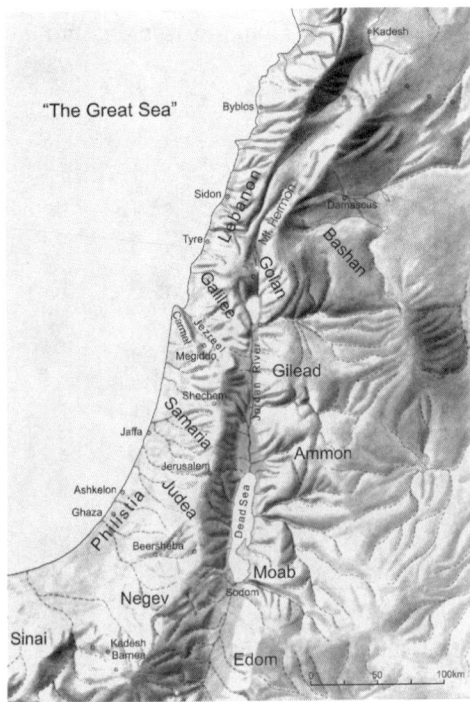

Relief map of the land of Canaan and its districts.

Joshua answered them, "go up to the forest country and clear an area for your-selves there. . . ." "The hill country is not enough for us," the Josephites replied, "and all the Canaanites who live in the valley have iron chariots, those in Beth-shean and its dependencies and those in the Valley of Jezreel." But Joshua declared to the House of Joseph, to Ephraim and Manasseh, ". . . The hill country shall be yours as well; true, it is forest land, but you will clear it and possess it to its farthest limits.[14] And you shall also dispossess the Canaanites, even though they have iron chariots and even though they are strong." (Joshua 17:14–18)

The chariots of iron that the Canaanites possessed were the ancient equiv-alent of the tanks of modern warfare. However, they were effective only in the plains (and only if the earth was dry and firm), but were useless on the steep and rocky hillsides. And so it was that the Israelites, not out of choice but out of sheer necessity, took up residence in the rugged hill country of Canaan and became highlanders rather than plainsmen. (Only much later, as they gained in numbers and technology, were the Israelites able to descend to the plains and not only take over the Canaanite city-states there but even vanquish the powerful and technologically advanced Philistines.)

AFTER THE DEATH of Joshua, the loosely federated Israelite tribes tried to continue the process of settlement even without a central leader: "In those days there was no king in Israel; everyone did as he pleased" (Judges 21:25). Just how incomplete the Israelites' occupancy of Canaan was at that time is indicated in the long list of places that remained for some generations in the hands of the original inhabitants (Judges 1:21, 27–35). Indeed, the report of land annexation in the Book of Judges seems more tentative and incremental, and therefore more realistic, than the grandiose account of military conquest in the Book of Joshua.

What the Israelites did, it would seem, was to adjust to the ways of the land in any way they could. That accommodation included a tendency to emulate the native peoples by worshiping their idols (chief among whom were Ba'al, the god of sky and rain, and Ashera, the goddess of earth and fertility).[15] That, however, did not prevent their neighbors from trying repeatedly to overcome and subjugate the Israelites. The Judges who arose at various times to mobilize opposition to those oppressors were local, charismatic leaders and warriors. Although called Judges, they served more as military chieftains than as magis-trates. Each of the Judges was concerned about just one tribe or, at most, a co-alition of a few tribes, periodically threatened by the still-dominant Canaanite city-states or by bands of plundering nomads, such as the Midianites from the pastoral or desert domain:

> The hand of the Midianites prevailed over Israel; and because of Midian, the Israelites provided themselves with refuges in the caves and strongholds of the mountains. After the Israelites had done their sowing, Midian, Amalek, and the Kedemites [Sons of the East] would come up and raid them; they would attack them, destroy the produce of the land all the way to Gaza, and leave no means of sustenance in Israel, not a sheep or an ox or an ass. For they would come up with their livestock and their tents, swarming as thick as locusts; they and their camels were innumerable. Thus they would invade the land and ravage it. Israel was reduced to utter misery by the Midianites. (Judges 6:2–6)

On this occasion, the Israelites were led by Gideon, a young farmer from the tribe of Manasseh (Judges 6–7). With just three hundred volunteers, Gideon surprised the encamped Midianites in the night and caused them to flee in panic.

Only seldom, however, did several of the Israelite tribes band together under jointly agreed leadership to wage a common war. One of the greatest of the Judges was Deborah, who rose to defend her people against the dominant Canaanites forces headed by the king of the powerful city-state of Hazor:

> The Israelites once again did what was offensive to Yahweh.... And Yahweh surrendered them to King Jabin of Canaan, who reigned in Hazor. His army commander was Sisera, whose base was Harosheth-goiim. The Israelites cried out to Yahweh; for he [Sisera] had nine hundred iron chariots, and he had oppressed Israel ruthlessly for twenty years.
>
> Deborah, wife of Lappidoth, was a prophetess; she led Israel at that time. She used to sit under the Palm of Deborah, between Ramah and Bethel in the hill country of Ephraim, and the Israelites would come to her for decisions. She summoned Barak son of Abinoam, of Kedesh in Naphtali, and said to him, "Yahweh, the Elohim of Israel, has commanded: Go, march up to Mount Tabor, and take with you ten thousand men of Naphtali and Zebulun. And I will draw Sisera, Jabin's army commander, with his chariots and his troops, toward you up to the Wadi Kishon; and I will deliver him into your hands." ...
>
> Sisera was informed that Barak ... had gone up to Mount Tabor. So Sisera ordered all his chariots ... and all his troops to move ... to Wadi Kishon.... Barak charged down Mount Tabor, followed by ten thousand men, and Yahweh threw Sisera and all his chariots and army into a panic before the onslaught of Barak. Sisera leaped from his chariot and fled on foot.... All of Sisera's soldiers fell by the sword. (Judges 4:1–16)

So great was the victory that Deborah composed a triumphal ode in celebration. In it, she related the fortuitous environmental circumstances that had enabled Barak's infantrymen to vanquish Sisera's charioteers:

> The kings of Canaan fought
> at Taanach, by Megiddo's waters. . . .
> The stars fought from heaven,
> from their courses they fought against Sisera.
> The torrent Kishon swept them away,
> the raging torrent, the torrent of Kishon. . . .
> Then the horses' hoofs pounded
> as headlong galloped the steeds. (Judges 5:19–22)

What evidently happened was that a flash flood (such as often follows an intense rainstorm) occurred in the brook of Kishon, which then overflowed its banks. The entire section of the valley along the brook, normally firm and easily traversable, turned into soft mire, in which the chariots were bogged down.[16] The Israelite foot soldiers could then catch the hapless charioteers, as they were stuck in the mud behind frenzied horses, and smite them one by one. Sisera himself fled on foot and sought refuge in the tent of Yael, of the pastoral Kenite tribe, who fed him, lulled him to sleep, and then killed him.[17]

In her celebratory ode, Deborah also castigated the tribes of Israel that had evaded the battle, because of either complacency or fear, as she decried the lack of solidarity among the tribes of Israel:

> Among the clans of Reuben
> were great qualms of heart.
> Why then did you stay among the sheepfolds
> and listen as they pipe for the flocks? . . .
> Gilead tarried beyond the Jordan;
> and Dan—why did he linger by the ships?
> Asher remained at the seacoast
> and tarried at his landings. (Judges 5:15–17)

None of the Judges held sway over Israel as a whole or was a true forerunner of the united monarchy. Moreover, the long intervals between the appearance of successive Judges shows that there was no continuity of authority. Thus the Book of Judges offers not a seamless account but a collection of separate episodes, the chronological sequence of which is uncertain. One curious aspect is the repeated reference to periods of precisely *forty* years, a magical number that recurs frequently in the Bible,[18] such as in connection with the duration of the Israelites' sojourn in Sinai (forty years) and Moses's stay on Mount Sinai (forty days and nights).

AS AN essential condition of their adjustment to the hilly terrain of Canaan's rainfed domain, the Israelites had to learn ways of soil and water conservation: how to hew out and plaster cisterns in order to collect and store rainwater for the dry season, and how to carve out arable fields on steep slopes. They overcame this challenge by gathering stones from the ground and using them to

ON FALLOWING AND FESTIVALS

Although purportedly handed down in the desert of the Sinai Peninsula, many of the Mosaic laws pertain directly to the ecology and mode of life in the riverine, pastoral, and—especially—rainfed domains. Outstanding examples are the injunctions to fallow the land every seven years and to celebrate three major nature-based, agricultural festivals:

> Six years you shall sow your land and gather in its yield; but in the seventh you shall let it rest and lie fallow. Let the needy among your people eat of it, and what they leave let the wild beasts eat. You shall do the same with your vineyards and your olive groves. (Exodus 23:10–11)

> Three times a year you shall hold a festival for Me: You shall observe the Feast of Unleavened Bread—eating unleavened bread for seven days as I have commanded you—at the set time in the month of Abib [spring], for in it you went forth from Egypt; . . . and the Feast of the Harvest, of the first fruits of your work, of what you sow in the field; and the Feast of Ingathering at the end of the year, when you gather in the results of your work from the field. Three times a year all your males shall appear before the Sovereign [*adon*], Yahweh. (Exodus 23:14–17)

The practice of fallowing apparently originated in Mesopotamia, where farmers learned from experience that "resting" the land periodically can help control, or at least slow, the process of groundwater rise and salination. During the fallow period, the farmers refrained from irrigating their fields and allowed the natural process of internal drainage to take place. The weeds that grew freely during the fallow period extracted water from the water table and thus helped lower it. The residue of crops and weeds, left in place and grazed by livestock during the fallow period, also contributed nutrients to the soil, thus helping to restore its fertility.

Fallowing was not similarly practiced in Egypt, simply because there it was unnecessary. The annual pulsation of the Nile provided automatic drainage and leaching of salts, and the increment of rich silt brought by the flood served to augment soil fertility. However, fallowing became relevant again in the rainfed land of Canaan, where soil cultivation and crop production, especially on sloping land, caused the erosion of soil, depletion of organic matter, and gradual loss of fertility. The remedies were the retention of stubble, application of animal manure, strict conservation of soil (by such means as terracing), and—once again—periodic fallowing of the land.

erect retaining walls on the contours, thus dividing the slopes into a series of flat terraces. Behind the terrace walls, which kept the soil from eroding, they planted olives, grapes, figs, pomegranates, and almond and other fruit trees. In the valleys, they sowed field crops (vegetables, wheat, and barley), whereas the unterraced portions of the hillsides were for pasture. Patches of the native vegetative cover—forest trees and shrubs—remained on the steepest and rockiest slopes, which could not be utilized otherwise. To maintain or restore soil fertility in the cultivated areas, the Israelite farmers were instructed to leave the land fallow for one year out of seven, a sort of sabbatical for the land known as *shmitah.*

The prophets were familiar with the lives of farmers and shepherds, and they drew many of their parables and allegories from the rural lore. The back-breaking work that was required to establish a vineyard on a stony hillside and the risk that it would not be fruitful are described most poignantly in Isaiah's outcry.[19] While his plaint was directed against the failure of the people of Israel to obey Yahweh, he used the failure of a vineyard to yield grapes as a metaphor that must have been very familiar to his intended audience:

> Let me sing for my beloved
> a song of my lover about his vineyard.
> My beloved had a vineyard
> on a fruitful hill.
> He broke the ground, cleared it of stones
> and planted it with choice vines.
> He built a watchtower inside it;
> he even hewed a wine press in it;
> for he hoped it would yield grapes.
> Instead, it yielded sour grapes. . . .
> Now I am going to tell you
> what I will do to my vineyard:
> I will remove its hedge,
> that it may be ravaged;
> I will break down its wall,
> that it may be trampled.
> And I will make it a desolation.
> It shall not be pruned or hoed,
> and it shall be overgrown with briers and thistles. (Isaiah 5:1–6)

The diligence and the competence of an exemplary farmer, who performs his varied tasks in sequence, using the methods and implements needed for each job, are lauded by Isaiah:

A terraced slope in the hills of Judea. The terraces may well date to biblical times.

Does he who plows to sow plow all the time,
breaking up and furrowing his land?
When he has smoothed its surface,
does he not rather broadcast black cumin and cumin,
or set wheat in a row,
barley in a strip,
and emmer in a patch?
For He teaches him the right manner,
his Elohim instructs him.
So, too, black cumin is not threshed with a threshing board,
nor is the wheel of a threshing sledge rolled over cumin,
but black cumin is beaten out with a stick
and cumin with a rod.
It is cereal that is crushed.
For even if he threshes it well
and the wheel of his sledge and his horses overwhelm it,
he does not crush it. (Isaiah 28:24–28)

Finally the object of the metaphor is revealed:

That, too, is ordered by Yahweh Tsevaot;
His counsel is unfathomable,
His wisdom marvelous. (Isaiah 28:29)

The threat that the rains would be withheld and that drought would scourge the inhabitants of the land was a specter looming over the practice of farming, as well as of pasturing, especially in the southern, marginal sections of the rainfed domain. No less important than the total amount of rain was its distribution within each season. Timely early rains were essential to promote germination and early seedling growth. Midseason rains were needed for subsequent crop development, particularly where the amount of moisture in the root zone was limited by the shallowness or the coarse texture of the soil. Late rains were important for the filling of the ripening grain and fruit to ensure a good harvest. The most favorable rainfall regimen provided gentle rains, well distributed throughout the growing season. Excessively intense and untimely rains, conversely, could damage the crops during their most vulnerable stage of growth, as well as accelerate soil erosion. The worst condition, however, was a prolonged dry spell that could cause total crop failure.[20]

That is why the divine promise was couched in terms of the provision and timeliness of the rains: "If you obey the commandments that I enjoin upon you this day, loving Yahweh your Elohim and serving Him with all your heart and soul, I will grant the rain for your land in season, the early rain and the late. You shall gather in your new grain and wine and oil—I will also provide grass in the fields for your cattle—and thus you shall eat your fill" (Deuteronomy 11:13–15). That promise was immediately followed by a clear threat: "Take care not to be lured away to serve other gods and bow to them. For Yahweh's anger will flare up against you, and He will shut up the skies so that there will be no rain and the ground will not yield its produce; and you will soon perish from the good land that Yahweh is assigning to you" (Deuteronomy 11:16–17).

The primal fears of the people of Israel (drought, pestilence, invasion, and exile) were expressed by King Solomon, on the occasion of dedicating to Yahweh the Temple that he had built in Jerusalem:

Should Your people Israel be routed by an enemy because they have sinned against You, and then turn back to You and acknowledge Your name, and they offer prayer and supplication to You in this House, O, hear in heaven and pardon the sin of Your people Israel, and restore them to the land that You gave to their fathers. Should the heavens be shut up and there be no rain because they have sinned against You . . . and then repent their sins . . . send down rain upon the land which You gave to Your people as their heritage. So, too, if there is famine in the land, if there is pestilence, blight, mildew, locusts or caterpillars, or if an enemy oppresses them in any of the settlements of the land . . . O, hear in Your heavenly abode, and pardon and take action! (1 Kings 8:33–39)

The utter dependence on timely rains was shared by the newly settled Israelites with the indigenous Canaanites, whose principal deities were the rain god, Baʻal, and the earth goddess of fertility, Ashera. Hence the Israelites were naturally inclined to adopt the lifestyle and lore of the local populace. Against this tendency the prophets of Yahweh exhorted and railed repeatedly, often vehemently.

The prophet Samuel sought to prove the rainmaking prowess of Yahweh: "'Now stand by and see the marvelous thing that Yahweh will do before your eyes. It is the season of the wheat harvest.[21] I will pray to Yahweh and He will send thunder and rain; then you will take thought and realize. . . .' Samuel prayed to Yahweh, and Yahweh sent thunder and rain that day, and the people stood in awe of Yahweh and of Samuel" (1 Samuel 12:16–18).

The prophet Elijah set out to prove the supremacy of Yahweh over Baʻal to King Ahab, who had adopted pagan ways, and to the people of Samaria.[22] To do so, he had to show that Yahweh was superior to Baʻal in bringing the coveted rain at a time of severe drought. The background to the dramatic contest was described thus: "Ahab son of Omri reigned over Israel in Samaria for twenty two years. Ahab . . . did what was displeasing to Yahweh . . . he took as wife Jezebel daughter of King Eth-Baʻal of the Phoenicians, and he went and served Baʻal and worshipped him" (1 Kings 16:29–31). Therefore,

> Elijah the Tishbite, an inhabitant of Gilead, said to Ahab: "As Yahweh lives, the Elohim of Israel whom I serve, there will be no dew or rain except at my bidding." The word of Yahweh came to him: "Leave this place, turn eastward and go into hiding by the Wadi Cherith, which is east of the Jordan. You will drink from the wadi, and I have commanded the ravens to feed you there." He proceeded to do as Yahweh had bidden. . . . After some time the wadi dried up, because there was no rain in the land. (1 Kings 17:1–7)

So severe was the drought that Ahab ordered his minister Obadiah: "Go through the land, to all the springs of water and to all the wadis. Perhaps we shall find some grass to keep horses and mules alive, so that we are not left without beasts" (1 Kings 18:5). So Yahweh sent Elijah to confront Ahab: "The word of Yahweh came to Elijah: 'Go, appear before Ahab; then I will send rain upon the earth'" (1 Kings 18:1). Before bringing rain, Elijah challenged the prophets of Baʻal to a contest of fire (1 Kings 21–39), which he won.[23] Elijah then smote the prophets of Baʻal. But still, his triumph was not complete. To establish Yahweh's total power, Elijah undertook the supreme test—to bring rain:

> Then Elijah said unto Ahab: "Go up, eat and drink, for there is a rumbling of [approaching] rain." And he said to his servant: "Go up and look toward the sea." He went up and looked and reported, "There is nothing." Seven times [Elijah] said, "Go back," and the seventh time [the servant] reported, "A cloud as small as a man's hand is rising in the west." ... Meanwhile the sky grew black with clouds; there was wind, and a heavy downpour fell. (1 Kings 18:41–45)

The unpredictable appearance of rain is a result of the shifting patterns of air masses along the eastern Mediterranean during the winter. To the ancients, who had no weather charts and were unaware of changing high-pressure and low-pressure air masses, the sudden onset of a rainstorm seemed to be a purposeful divine act.

After Elijah's reported ascension to heaven (2 Kings 2:1, 11–12), his protégé, Elisha, also had occasion to conjure up water by invoking the name of Yahweh. It happened in a desert creek bed during the joint campaign of the kings of Judah and Israel against Mesha, the king of Moab. (The sister kingdoms of Judah and Israel were more often than not at odds with each other, but cooperated occasionally.) The story is told as follows:

> Now King Mesha of Moab was a sheep breeder and he used to pay as tribute to the king of Israel a hundred thousand lambs and the wool of a hundred thousand rams. But when Ahab died, the king of Moab rebelled against the king of Israel. So King Jehoram promptly set out from Samaria and mustered all Israel. At the same time, he sent this message to King Jehoshaphat of Judah: "The king of Moab has rebelled against me; will you come with me to make war on Moab?" He replied, "I will go." ... And he asked, "Which route shall we take?" [Jehoram] replied, "The road through the wilderness of Edom." (2 Kings 3:4–8)

The shorter and easier route from Samaria and Jerusalem to Moab led across the Jordan River, just north of the Dead Sea, but that route was likely to have been well fortified. Wishing to avoid those defenses and to surprise the Moabites, the Israelites chose the more circuitous route, going by way of Edom and, in order to surprise Moab, attacking it from the south. But the land of Edom is very mountainous, and the approaches to it require a journey through the hyper-arid desert of Aravah (the southern section of the Jordan Rift Valley, between the Dead Sea and the Red Sea) and then up steep and rugged canyons:

> So they marched for seven days ... and there was no water left for the army or for the animals that were with them. "Alas!" cried the king of Israel, "Yahweh has brought ... [us] together only to deliver ... [us] into the hands of Moab."

ELISHA THE MIRACLE MAKER

Several of the miracles attributed to Elisha may have natural explanations. One pertains to the curative qualities of the springs that flow into the Jordan River:

> Na'aman, commander of the army of the king of Aram . . . though a great warrior, was a leper. . . . So Na'aman came with his horses and chariots and halted at the door of Elisha's house. Elisha sent a messenger to him, "Go and bathe seven times in the Jordan, and your flesh shall be clean." But Na'aman was angered and walked away. "I thought," he said, "he would surely come out to me, and would stand and invoke Yahweh his Elohim by name, and would wave his hand toward the spot, and cure the affected part. Are not Amanah and Pharpar, the rivers of Damascus, better than all the waters of Israel? I could bathe in them and be clean!" And he stalked off in a rage. . . . [But after his servants mollified him] he went down and immersed himself in the Jordan seven times, as the man of the Elohim had bidden; and his flesh became like a little boy's, and he was clean. (2 Kings 5:1, 9–14)

Na'aman's malady may not have been leprosy, but psoriasis or another form of eczema—an allergenic skin disease characterized by redness, blistering, scaling, and itching that is exacerbated by excessive sweating and secondary infections. The waters of several saline hot springs that flow into the Jordan River seem to alleviate some of the symptoms, and Elisha was evidently aware of that. Today, no less than in centuries past, these springs attract many sufferers and are believed to be efficacious in the treatment of skin and other maladies.

Several other miracles were attributed to Elisha. One of his purported deeds was to purify foul water by casting salt into it (2 Kings 2:19–22). The stagnant water may have been turbid with mud and suspended organic matter. The salt would have coagulated the suspension and clarified the water. (This deed is reminiscent of the act by Moses at Marah [Exodus 15:25].) On another occasion, Elisha revived an unconscious child by administering mouth-to-mouth respiration (2 Kings 4:32–35), much as his mentor, the prophet Elijah, had done before (1 Kings 17:17–23).

But Jehoshaphat said: "Isn't there a prophet of Yahweh here, through whom we may inquire of Yahweh?" One of the courtiers . . . said, "Elisha son of Shafat, who poured water on the hands of Elijah,[24] is here." (2 Kings 3:9–11)

Elisha abhorred King Jehoram (the son of Ahab and Jezebel), whom he considered an apostate. But he did approve of King Jehoshaphat. So, reluctantly, Elisha came to the rescue:

The hand of Yahweh came upon him, and he said: "Thus said Yahweh: . . . You shall see no wind, you shall see no rain, yet the wadi shall be filled with water; and you and your cattle and pack animals shall drink. And this is but a slight thing in the sight of Yahweh, for He will also deliver Moab into your hands." In the morning, when it was time to present the meal offering, water suddenly came from the direction of Edom and the land was covered by water. (2 Kings 3:15–20)

What could have happened was that a sudden rainstorm over the mountains of Edom, east of the desert of Aravah, sent a flash flood hurtling down a rainless stretch of the wadi through which the Israelite expedition was attempting to wend its way toward the Moabite heartland. Such an occurrence is well known among desert dwellers, who are always careful not to encamp in a streambed (but at some distance from its banks), lest a sudden torrent sweep away both them and their possessions. Evidence that the arriving water was runoff from Edom is provided in the description: "Next morning, when they arose, the sun was shining over the water, and from the distance the water appeared to the Moabites as red as blood" (2 Kings 3:21). Characteristic of Edom is the red hue of the outcropping sedimentary formation known as Nubian Sandstone,[25] overlying the granite bedrock. Rainwater scouring the soft sandstone and trickling off it is often tainted with suspended silt, made red by the presence of iron oxides, including the blood-red mineral hematite. Hence the description of the water in that passage is fairly realistic.

As the Bible describes the battle between the Israelites and the Moabites, the latter were decisively defeated and their land was cruelly devastated: "The Israelites attacked the Moabites, who fled before them ... and they destroyed the towns. Each man threw a stone into each fertile field, so that it was covered over; and they stopped up every spring and felled every fruit tree" (2 Kings 3:25).[26] In desperation, the king of Moab made the supreme sacrifice to his god, and that deed reportedly caused the Israelites to end their siege and return to their own land: "So he [King Mesha] took his firstborn son, who was to succeed him as king, and offered him up on the wall as a burnt offering. A great wrath came upon Israel, so they withdrew from him" (2 Kings 3:27).[27]

THE ISRAELITES were able to maintain autonomy in Canaan for less than 1000 years of their 3500-year history. Starting around 1200 B.C.E. and continuing for the next two hundred years, the twelve tribes functioned separately, maintaining only a loose connection with one another. During part of that time, one or several of the tribes came under the domination of hostile nations, so they had to fight repeatedly to maintain or regain their independence.

Around the beginning of the first millennium B.C.E., the imperative of self-defense drove the tribes to unite and form a kingdom, first under Saul and then under David and his son Solomon. Soon, however (ca. 900 B.C.E.), the united kingdom split as a consequence of the rivalry between the two dominant tribes: Judah in the south and Ephraim in the north. Judah, the largest tribe, assimilated the tribes of Simeon and Levi and retained Jerusalem, the capital of the Davidic kingdom. The other ten tribes were governed from Samaria, the capital of the northern kingdom.

The rift had disastrous consequences: it resulted in internecine fighting and weakened the nation as a whole. Neither of the two semi-kingdoms had the power to withstand the external forces arrayed against it. The harsh political and climatic realities of life in Canaan were never commensurate with the dream cherished by, and the bright promise believed to have been made to, the wandering Hebrew pastoralists who initially settled there. And yet that tenuous sliver of land became for the Israelites and their surviving descendants, the Jews, their one and only homeland, the focus of their nationhood, their Land of Israel, to which they clung physically as long as they could—and spiritually always.

8

THE MARITIME DOMAIN

Interactions with Philistines and Phoenicians

The ocean sounds, O Yahweh,
the ocean sounds its thunder,
the ocean sounds its pounding.
Above the thunder of the mighty waters,
more majestic than the breakers of the sea,
is Yahweh, majestic on high.

PSALM 93:3–4

THE EASTERN SHORE of the Mediterranean was divided between Phoenicia in the north and Canaan in the south. The two contiguous coasts differ physiographically. The northern coast is rugged, as the steep slopes of the Lebanon Mountains approach the sea and occasionally jut into it, leaving a very narrow coastal plain or none at all. Hence there is little room for agricultural development along the shore. But the coast is endowed with numerous peninsulas, islets, coves, and estuaries that provide excellent sites for harbors. Consequently, the societies that inhabited this part of the Levant turned seaward to engage in fishing and in seafaring trade. Shipbuilding was facilitated by the availability of suitable wood in the lush forests of cedars, pines, cypresses, and junipers growing on the sea-facing slopes of the Lebanon Mountains. Those wood resources became highly desired commodities in the regionwide trade that developed during the Bronze and Iron Ages.[1] The Bronze Age cities of Ugarit[2] and Byblos were important trading centers in their heyday and bequeathed us a rich legacy of records in a West Semitic language that is very similar to early Hebrew.[3]

Farther south, in Canaan, the mountain ranges recede from the Mediterranean Sea (except at Mount Carmel), thus opening up a coastal plain that becomes wider toward the southern section of the country. In contrast with the coast of Lebanon, that of Canaan is, for most of its length, straight and sandy, with few natural harbors. During the Bronze Age, the main havens for coast-hugging ships that had to escape the occasionally violent storms were in the shallow estuaries of the perennial streams, but they, like harbors, are

View from the north of the estuary of the Qishon River, at which lay the ancient anchorage now known as Tel Abu Hawam, with Mount Carmel in the background.

few and far between (since most of the country's streams are dry during the greater part of each year). In the Early Iron Age, the technology of constructed harbors was developed, and thereafter several sites on the coast of Canaan—including Acre, Ashkelon, Athlit, Dor, and Jaffa—became prominent ports of call for seafaring trade and its connection with the agricultural hinterland and beyond.

SAILORS WHO NEGOTIATED the restless waters of the Mediterranean were faced with the capricious, occasionally violent, moods of the sea and the weather—the waves, currents, and winds. They also had to cope with the ruggedness of the coast, its submerged offshore rocks and sandy shoals, which presented hazards to navigation. While the open sea offered luring potentials for exploration, communication, and lucrative trade, it also posed physical and psychological challenges, hardships, and dangers. Venturing into the open sea was always an act of courage, impelled and sustained by faith in a benevolent power able to ensure the safety and success of the maritime enterprise. And so it was that the seafarers of the Mediterranean conceived religious beliefs and practices that helped them cope spiritually with the rigors and hazards of their occupation.

Having descended from land-based cultures, the mariners of the Levant retained some of the traditions of their agricultural ancestors, especially those that were meaningful in their new domain. For example, they continued to worship the Canaanite storm god, Ba'al, who provided the rains needed to grow crops and pasture on land.[4] They believed that he could send the beneficent winds that filled the sails and facilitated the success of a voyage at sea, or could instigate the violent tempests, lashing hail, and thunderous lightning that could cause its disastrous end. To placate the pent-up fury of that god and to appeal to him as a guardian spirit, they carried cultic objects, decorated their vessels with symbolic figures, and offered sacrifices on portable altars carried on board or at selected locations on land. They built temples on promontories visible from the sea and dedicated them to Ba'al. Their rites included recitations, prayers, divination of omens, and taboos—all designed to elicit the assistance of the benevolent gods or to ward off or appease the malevolent ones.

Especially important in their pantheon were Ba'al Haddu and his variants: Ba'al Samem (Ba'al of the Sky), Ba'al Malage or Malahu (Lord of the Sailors), Ba'al Rosh (Lord of the Summit), and Ba'al Sapon or Tsafon (Ba'al of the North), who was presumed to reside atop a high mountain, somewhere in Lebanon, that was visible from the sea over a great distance and probably used as a landmark and navigational aid.[5] (The Bible mentions Ba'al Tsafon as a location on the eastern boundary of Egypt, near the Sea of Reeds [Exodus 14:2, 9; Numbers 33:7], thus raising the possibility that the name was used for several places dedicated to the god along the coast of the eastern Mediterranean.) Indeed, the coastal-maritime pantheon encompassed a bewildering array of gods who were constantly changing—merging, separating, and proliferating. Included among the mariners' deities were manifestations of the goddesses Ashera and Tanit (who may have been variants of the same goddess, as both were represented by a lunar disk and/or crescent).

The seafarers of the Levant recognized mythical sea creatures, in addition to the various manifestations of Ba'al. Within the deep and murky waters of the raging sea there seemed to lurk many voracious monsters, including Yamm (creature of the sea), Leviathan or Lotan (an enormous sea creature), and Tanin (a giant dragon-like or serpent-like creature), each of which could roil the waters and threaten to devour sailors cast overboard or even swallow entire ships. Hence Canaanite sailors prayed to their guardian god, Ba'al, to suppress or defeat the malevolent Yamm and the other brutes of the deep. One or another of those imaginary creatures apparently metamorphosed into the Greek god of the sea, Poseidon (the Roman god Neptune), depicted as riding on a winged sea horse.

Several of the sea monsters are mentioned in the Bible, no doubt as a consequence of the land-based Israelites' interactions with the peoples of the maritime domain. The Israelites generally regarded the sea as an untamed, thrusting, and threatening environment that must be constrained by a powerful and benevolent God

who closed the sea behind doors
when it gushed forth out of the womb,
when I clothed it in clouds,

when I made breakers My limit for it,
and set up its bar and doors,
and said, "You must come so far and no farther,
here your surging waves will stop." (Job 38:8–11)[6]

A number of the biblical prophets praised the power of Yahweh to subdue the creatures of the sea:

Are you wroth, O Yahweh, with Neharim [rivers], . . .
Your rage against Yamm—
that You are driving Your steeds,
Your victorious chariot? . . .
A torrent of rain comes down;
loud roars the deep,
the sky returns the echo,
sun and moon stand still on high,
as Your arrows fly in brightness,
Your flashing spear in brilliance. (Habakkuk 3:8–11)

Isaiah proclaimed Yahweh's superiority over all the marine gods: "In that day Yahweh will punish with His great, cruel, mighty sword Leviathan the Elusive Serpent—Leviathan the Twisting Serpent; He will slay the Dragon of the Sea" (Isaiah 27:1). Amos exalted Adonai's ability to ferret out his enemies wherever they may hide and send forth His own monsters to subdue the untamed ones: "If they conceal themselves from My sight at the bottom of the sea, / there I will command the serpent to bite them" (Amos 9:3).

In the Psalms, the monsters of the deep are both the enemies of Yahweh and his subservient creations:

It was You who drove back the sea with Your might,
who smashed the heads of the monsters in the waters;
it was You who crushed the heads of Leviathan,
who left him as food for the denizens of the desert;
it was You who released springs and torrents,
who made mighty rivers run dry. (Psalm 74:13–15)

How many are the things You have made, O Yahweh! . . .
The earth is full of Your creatures.
There is the sea, vast and wide,
with its creatures beyond number. . . .
There go the ships,

and Leviathan that You formed to sport with. . . .
Open Your hand, they are well satisfied;
hide Your face, they are terrified. . . . (Psalm 104:24–29)

Psalm 148:7 commands the sea monster Tanin and the ocean depths to praise
Yahweh. The descriptions in the Bible of Yahweh subduing the sea creatures
parallel those in the Ugaritic Ba'al cycle.[7]

IN THE Early Iron Age (ca. 1150 B.C.E.), the southern coast of Canaan was colonized by the Philistines.[8] They were one of a varied group of seafaring tribes
known collectively as the Sea Peoples. Their places of origin are still uncertain,
but may well have been the Ionian Islands, as well as the coasts of Asia Minor,
Greece, Crete, or even Sicily and Sardinia.[9] The Sea Peoples have been called
the nomads of the sea for their tendency to venture far from their homelands
in search of trade or plunder.[10] (In some ways, their lifestyle resembled that of
the Vikings, who ranged along the coasts of northwestern Europe some two
millennia later.)

Occasionally, the Sea Peoples raided the coastal towns around the Mediterranean Sea and, at other times, invaded the lands of the inhabitants and
tried to settle permanently. One of their raids was made famous by the triumphal wall carving in the mortuary temple of Rameses III (r. 1198–1167 B.C.E.)
at Medinet Habu in Egypt.[11] In this carving, the Sea Peoples are shown in
their ships, fighting a losing naval battle with the Egyptian forces. Some of
the Sea Peoples entered into the service of the various states of the region, as
either mercenary soldiers or hired sailors. Among them were the Peleste, who

An Egyptian representation of the battle between the forces of Rameses III
and the invading Sea Peoples in the twelfth century B.C.E. This detail is from
a carving on a wall of the mortuary temple of Rameses III at Medinet Habu.

settled on the coast of Canaan and became the Philistines of the Bible. They may have done so under their own initiative or at the behest of the Egyptians, who hoped thereby to protect the northeastern frontier of their empire. In the course of time, the Philistines seem to have turned their backs to the sea to become primarily land-based farmers and urbanites.

The Philistines (whose name may be related to the Hebrew word *polshim* [invaders]) happened to enter Canaan from the west and settled in the southern part of its coastal plain at about the same time that the Israelites entered the country (as the Bible tells us) from the east and settled in the central highlands. (Having earlier tried, and failed, to enter Canaan directly from the south, the Israelites had been forced to detour around the provinces of Edom and Moab and to finally cross the Jordan River at Jericho.)

Soon the two peoples met and, quite inevitably, clashed. At first, the Philistines had the upper hand. They appear to have been more cohesive, organized, and battle-ready than the Israelites. In addition, they had a technological advantage. They were skilled at metallurgy and toolmaking. For a time, therefore, the Israelites were militarily inferior to the Philistines and dependent on them, and the Philistines pressed their advantage forcibly as well as profitably:

> No smith was found in all the land of Israel, for the Philistines were afraid that the Hebrews would make swords or spears. So all the Israelites had to go down to the Philistines to have their plowshares,[12] mattocks, axes, and coulters sharpened. The charge for sharpening was a *pim* for plowshares, mattocks, three-pronged forks, and axes, and for setting the goads. Thus on the day of battle, no sword or spear was to be found in the possession of the troops with Saul and Jonathan. (1 Samuel 13:19–22)

With their technological superiority—based on the use of iron as well as bronze tools,[13] weapons, and, perhaps, chariots—the Philistines terrorized the tribes that dwelt in the lowlands and foothills. Lacking the ability to forge metal implements, the Israelites were at first unable to make effective weapons with which they could ward off the invading Philistines. The Israelites therefore retreated to their mountain strongholds, where skill in archery and cunning use of the rugged terrain and its thicket of trees and shrubs gave the highlanders a defensive advantage over the plainsmen.[14]

Three of the tribes of Israel reportedly tried to settle along the coast. Of Dan and Asher, the Judge Deborah said:

> Dan, why did he linger by the ships?[15]
> Asher remained on the seacoast
> and dwelt at its bays. (Judges 5:17)

And Zebulun, Jacob prophesied,

> shall dwell by the seashore;
> he shall be a haven for ships,
> and his flank shall rest on Sidon. (Genesis 49:13)

Of the three coastal tribes, Dan was nearest to the Philistines and suffered most from their aggression. Not surprisingly, the first recorded resistance against the Philistines was carried out by a man of the tribe of Dan, named Samson, whose legendary exploits have ever since been cited as deeds of an admirable (albeit, morally flawed) hero.[16] But Samson's daring struggle was in the nature of a one-man guerrilla operation rather than an organized national campaign. The people of the tribe of Dan were actually forced to relinquish their original abode along the coast just north of the Philistine heartland and to seek an alternative settlement site in the Upper Jordan Valley, in a place ever since called Dan.

For a time, the Philistines were even able to push inland from their coastal-plain redoubts (the five principalities of Gath, Gaza, Ashdod, Ashkelon, and Ekron), take control of Beth-Shean in the eastern Valley of Jezreel, and con-duct periodic forays into the central highlands. The clashes between the two nations continued for several generations. At one point, the priests leading the people of Israel even tried—in desperation—to summon the direct interven-tion of Yahweh, the Lord of hosts, by bringing the Ark of the Covenant to the battlefield:

> Israel marched out to engage the Philistines in battle . . . and Israel was routed by the Philistines, who slew about four thousand men. . . . Then the elders of Israel [said] . . . "Let us fetch the Ark of the Covenant of Yahweh from Shiloh; then He will be among us and will deliver us from the hands of our enemies." S the troops sent men to Shiloh; there the sons of 'Eli [the high priest], Hophni and Phineas, were in charge of the Ark of the Covenant of the Elohim, and they brought down from there the Ark of the Covenant of Yahweh Tsevaot [Lord of hosts] Enthroned on the Cherubim. (1 Samuel 4:1–4)

But the Ark availed them not: "The Philistines fought, Israel was routed, and they all fled to their homes. The defeat was very great, thirty thousand foot soldiers of Israel fell there. The Ark of Elohim was captured, and 'Eli's two sons, Hophni and Phineas, were slain" (1 Samuel 4:10–11).

There followed a fancifully reported confrontation between the god of the Philistines, Dagon, and the Ark of Yahweh (1 Samuel 5:1–4), which seems to imply that Dagon was himself a living god who, by prostrating himself, accepted the superiority of Yahweh. We do not know the form of the idol Dagon, but the name (probably derived from *dag* [fish]) suggests that he was a maritime god, inherited by the Philistines from their seafaring ancestors.[17] The Philistines, fearing further acts of revenge by the invisible Yahweh, chose to play it safe by returning the Ark to the Israelites.

On the Israelite side, it became clear that something far more organized had to be done to protect the nascent nation from total domination and eventual annihilation by the Philistines. The loose association of squabbling tribes had to combine its forces in order to ward off the Philistines. The external threat induced the tribes to form a centralized kingdom, capable of waging a more effective war against the aggressive enemy. The first king, Saul, dared to challenge the Philistines and—although he won some skirmishes—was eventually defeated and killed in battle on the slopes of Mount Gilboa, overlooking Beth-Shean.[18] The predicament of the Israelites ended only after the Philistines' technological monopoly was broken in the days of Saul's successor, King David.

The Israelites eventually prevailed against the Philistines.[19] And when they did, they attempted to take to the sea by forging an alliance with the region's quintessential seafarers: the Phoenicians. Although the Israelites never became a maritime power in the Mediterranean Sea, the Jerusalem-based kings—Solomon (1 Kings 9:26–28) and Jehoshaphat (1 Kings 22:49)—reportedly built ships at the Red Sea port of Elath[20] and attempted to trade with southern Arabia and eastern Africa. And the Israelites lived near enough to the sea to be aware of its moods, its phenomena, its dangers, and its mysteries. Hence the Bible abounds with references to the sea. Indeed, the word *yamm* (which generally meant "sea," but could also refer to any body of water) appears almost four hundred times.

AFTER VANQUISHING the Philistines and consolidating their own state, the Israelites came into contact with the Phoenicians, the preeminent seafarers of the eastern Mediterranean. The Phoenicians had come into prominence in the Early Iron Age (ca. 1150 B.C.E.). Based on the northern coast of the eastern Mediterranean, they were skilled at utilizing the resources from their forests to construct houses and temples, as well as ships. In addition, they excelled at making glass, using coastal sand, and dyes, using a substance produced by a type of sea snail called murex. Indeed, it was the dye they produced and traded that gave the Phoenicians the name by which they came to be known outside their own country. Its reddish-purple color reminded the Greeks of the mythical firebird, phoenix, so they called its suppliers Phoenicians. They,

View of Tyre, one of the two
(along with Sidon) principal
seaports of the Phoenicians.

however, referred to themselves as Tyrians or Sidonites, after their two principal coastal cities and seaports: Tyre and Sidon.

From their home ports, the Phoenicians ranged around the perimeter of
the Mediterranean Sea (and beyond, reaching Britain, western Africa, and
possibly even South America). Moreover, they established colonies and maritime trading centers on the island of Malta; in Carthage, on the shore of North
Africa; in Cartagena, on the east coast of Iberia; and in various sites in Sicily and the Appennine (Italian) Peninsula.[21] Along with the goods they conveyed, the Phoenicians disseminated the alphabet that had been developed by
Semitic traders who, in the course of their contacts with Pharaonic Egypt,
simplified a few of the numerous hieroglyphic symbols in order to represent
the consonants of their Hebrew-like language.[22]

The Bible provides a fascinating account of the arrangements made between the Israelite king Solomon of Jerusalem and the Phoenician king Hiram
of Tyre concerning the shipment by sea of cedar and cypress wood for the
construction of Solomon's Temple and palace:

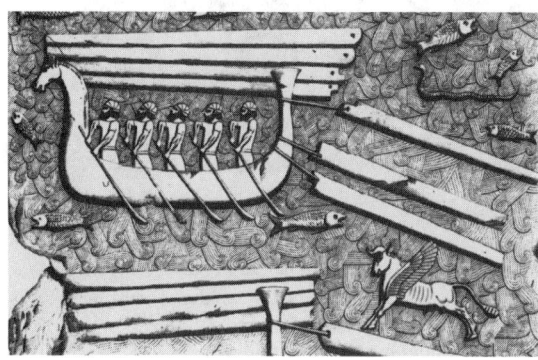

A simplified depiction
of Phoenician ships hauling
logs of wood. This relief is
from the palace of Sargon II
(r. 721–705 B.C.E.) at
Khorsabad.

Solomon sent this message to Hiram: "You know that my father David could not build a house for the name of Yahweh his Elohim because of the enemies that encompassed him. . . . But now Yahweh my Elohim has given me respite all around. . . . And so I propose to build a house for the name of Yahweh my Elohim. . . . Please, then, give orders for cedars to be cut for me in Lebanon. My servants will work with yours, and I will pay you any wages you may ask for your servants; for as we know, there is none among us who knows how to cut timber like the Sidonians." . . .

So Hiram sent word to Solomon: ". . . I will supply all the cedar and cypress logs you require. My servants will bring them down to the sea from [Mount] Lebanon; and at sea I will make them into floats and [deliver them] to any place that you designate to me. There I shall break them up for you to carry away. You, in turn, will supply the food I require for my household." So Hiram kept Solomon provided with all the cedar and cypress wood he required, and Solomon delivered to Hiram 20,000 *kors* of wheat as provisions for his household and 20 *kors* of beaten oil. Such was Solomon's annual payment. (1 Kings 5:3–6, 8–11)

The passage implies clearly that the Phoenicians of Tyre were well endowed with timber resources and were skilled at logging and transporting them by

A remnant grove of cedars in Lebanon. In ancient times, the entire mountain range of Lebanon was covered with a dense growth of cedars, cypresses, and junipers. The Phoenicians began to cut the forest trees systematically, for constructing boats and houses, and—being astute traders—they exported the timber. Over the ensuing centuries, the forests were gradually cleared and the denuded mountainsides exposed to erosion.

sea, but lacked a productive farming hinterland for self-sufficiency in food supplies. The Israelites, on the contrary, had become enterprising farmers capable
of producing surplus grain and olive oil (at least in years of normal rainfall).
So great was the Tyrians' need for farmland that King Solomon offered King
Hiram a section of land in return for continuing supplies of timber and gold.

The offer, however, was not entirely satisfactory:

> King Hiram of Tyre had supplied Solomon with all the cedar and cypress tim-
> ber and gold that he required—King Solomon in turn gave Hiram twenty towns
> in the region of Galilee. But when Hiram came from Tyre to inspect the towns
> that Solomon had given him, he was not pleased with them. "My brother," he
> said, "what sort of towns are these you have given me?" So they were named
> the land of Kabul, as is still the case. However, Hiram sent the king one hun-
> dred and twenty talents of gold. (1 Kings 9:11–14)

The word *kabul* apparently implied "unproductive land." It may have referred
to a coastal district in western Galilee that was particularly ill-drained and
swampy, perhaps in the floodplain of the Naaman River. In modern Hebrew,
the word *kabul* refers to peat or a mucky soil typical of marshes. The district
of Kabul, incidentally, is also mentioned in Joshua 19:27, in connection with
the same area (allocated to the tribe of Asher).

King Solomon's cooperation with the Phoenicians extended to seafaring as
well as to exchanging timber for agricultural products and marginal real estate. The Israelites sought and applied Phoenician expertise in building a fleet
of ships to sail the Red Sea. By so doing, Solomon in effect extended the reach
of his kingdom southward to the arm of the Red Sea now known as the Gulf
of Aqaba, thus taking control of the entire Negev. The fleet was launched successfully from Etzion-geber, "which is near Eloth [elsewhere in the Bible called
Elath]" (1 Kings 9:26–28). The expedition is reported to have reached the land
of Ophir—perhaps at the southern tip of the Arabian Peninsula [23] or on the
Horn of Africa, from where it brought back much gold.[24]

A similar naval venture was attempted a few generations later by one of
Solomon's successors, King Jehoshaphat of Judah. This time, however, the
effort ended in disaster, as the ships were wrecked at Etzion-geber (1 Kings
22:49). The failure may have been due to Jehoshaphat's determination to go it
alone. He did not enlist the expert assistance of the Phoenicians and pointedly
refused to cooperate with the northern kingdom of Israel (1 Kings 22:50).

The interaction between the Israelites and the Phoenicians was not only
economic, but also social, cultural, and religious—much to the dismay of the
Yahwist writers of the Bible. The kingdom of Israel, based in Samaria, was
especially influenced by its northern neighbors, the Phoenicians of Tyre and

Sidon. King Ahab of Israel, for instance, married the Phoenician princess Jezebel,[25] daughter of Eth-Ba'al, king of the Sidonites, and "he went and served Ba'al in the temple of Ba'al which he built in Samaria. Ahab also made a sacred post. Ahab did more to vex Yahweh, the Elohim of Israel, than all the kings of Israel who preceded him" (1 Kings 16:31–33). Ahab's successor, Ahaziah, "followed in the footsteps of his father and his mother. . . . He worshipped Ba'al and bowed down to him; he vexed Yahweh, the Elohim of Israel, just as his father had done" (1 Kings 22:53–54).

The cultural and religious influence of the Phoenicians persisted and even penetrated the southern kingdom of Judah, based in Jerusalem. King Jehoram, son of King Jehoshaphat, "followed the practices of the kings of Israel—whatever the House of Ahab did, for he had married a daughter of Ahab [and Jezebel] and he did what was displeasing to Yahweh" (2 Kings 8:18). His half-Sidonite wife, Athaliah, evidently influenced their son Ahaziah, who succeeded Jehoram to the throne. Like his father and mother, "he walked in the ways of the House of Ahab and did what was displeasing to Yahweh . . . for he was related by marriage to the House of Ahab" (2 Kings 9:27).

After the fall of the kingdom of Israel to the Assyrians in 721 B.C.E., efforts were made in the kingdom of Judah to purge those pagan influences, especially by Kings Hezekiah and Josiah.

ONE BOOK in the Bible that does describe the rigors and hazards of seafaring very explicitly is the Book of Jonah. It begins with Yahweh's charge to the prophet Jonah, whose name means, literally, "dove" (reminiscent of Noah sending the dove to seek dry land following the Great Flood), to go to the great Assyrian city of Nineveh and castigate the people for their iniquity. (Nowhere else in the Bible was an Israelite or a Jewish prophet sent to preach to the gentile heathens.) Jonah tried to evade the onerous assignment by escaping across the sea to faraway Tarshish (perhaps the city of Tarsus, later the home of the Christian apostle Paul, on the coast of Asia Minor): "The word of Yahweh came to Jonah son of Amittai: 'Go at once to Nineveh, that great city, and proclaim judgment upon it; for their wickedness has come before Me.' Jonah, however, . . . went down to Joppa[26] and found a ship going to Tarshish. He paid the fare and went aboard to sail . . . away from the service of Yahweh" (Jonah 1:1–3).

Perhaps Jonah thought that the jurisdiction of Yahweh (being the God of the land-based Israelites) was confined to the terrestrial realm and did not extend to the maritime domain. In this notion, however, the book that bears

Jonah's name shows him to have been mistaken. Nor was he apparently aware of the potential hazards of a sea voyage on the stormy Mediterranean:

> Then Yahweh cast a mighty wind upon the sea, and such a great tempest came upon the sea that the ship was in danger of breaking up. In their fright, the sailors cried out, each to his own god; and they flung the ship's cargo overboard to make it lighter for them. Jonah, meanwhile, had gone down into the hold of the vessel where he lay down and fell asleep. The captain went over to him and cried out, "How can you be sleeping so soundly! Up, call upon your god! Perhaps the god will be kind to us and we will not perish." (Jonah 1:4–6)

The ship's crew evidently was a mixed lot, coming from different locations and cultures, so each man had his own god or gods, and each tried to placate his god by throwing gifts into the waters and by praying. But their gifts and prayers still did not stop the storm. So, in the common belief that a storm must be caused by an angry god, the frightened sailors sought to identify the person who had provoked that anger. And they did so by casting lots, on the assumption that the results of such lots are not random but determined by the gods, whose purpose may thus be revealed: "The men said to one another, 'Let us cast lots and find out on whose account this misfortune has come.' They cast lots and the lot fell on Jonah. They said to him, 'Tell us, you who have brought this misfortune upon us, what is your business? . . . What is your country?' 'I am a Hebrew,' he replied. 'I worship Yahweh, the Elohim of Heaven, who made both sea and land'" (Jonah 1:7–9).[27]

In the mind-set of the pagan believers, the gods of all domains and peoples were valid within their respective contexts. And since, in this case, the sailors had established by drawing lots that the Hebrew in their midst had displeased his god, Yahweh, and thus had caused the storm, they now wished to know how they might placate that angry god and so save their own lives:[28] "The men were greatly terrified. . . . They said to him, 'What must we do to you to make the sea calm around us?' For the sea was growing more and more stormy. He answered: 'Heave me overboard, and the sea will calm down'" (Jonah 1:10–13). So the men cast Jonah overboard, and, lo, "the sea stopped raging" (Jonah 1:15). And then they did what all the ancient Near Eastern cultures did to express their gratitude to whichever god they believed had saved them: "They offered a sacrifice unto Yahweh, and made vows" (Jonah 1:16).

The rest of the story is too well known to require repetition in detail. Jonah was swallowed by a "great fish" (Jonah 2:1), from whose innards he prayed to Yahweh, and was miraculously delivered back to safety. After his misadventures, Jonah accepted the divine mission that he had tried to evade and, as instructed, proclaimed the demise of the city of Nineveh. But when the people of that city repented, Yahweh, in compassion, decided to spare Nineveh (unlike what he had done to the iniquitous cities of Sodom and Gomorrah). So Jonah was unhappy, feeling that his pronouncement had been negated. There followed a symbolic act by Yahweh Elohim, who first grew a *kikayon* over Jonah's head, to shade him from the sun. On the morrow, however, Elohim[29] caused the plant to wilt. As Jonah fainted in the heat and rued the demise of the shady plant,[30] Yahweh explained: "You cared about the plant, which you did not work for and which you did not grow, which appeared overnight and perished overnight. And should not I care about Nineveh, that great city, in which there are more than a hundred and twenty thousand persons who do not yet know their right hand from their left, and many beasts as well!" (Jonah 4:10–11).

Thus it was that a tale of one man's attempt to shirk God's mission turned out to be an object lesson in morality and humane compassion. The Book of Jonah thus marks an important step in the transformation of Hebrew monotheism from a strictly national religion to a universal one directed toward—and compassionate to—all peoples.

ALTHOUGH THE Israelites never became mariners in a major way, the proximity of the Mediterranean Sea eventually influenced their culture profoundly. It affected their language in terms of imagery, metaphor, scope, and vocabulary. It expanded their perception of the world and of God the creator, and hence of their own place in the natural order. It brought them into intimate contact with maritime cultures that were, in turn, connected by ancestry and trade to the entire Mediterranean world. That connection served to foster communication across ethnic and cultural boundaries and to widen the scope of the Israelites' own frame of reference, which initially had been mainly pastoral and largely confined to the pastoral and riverine domains. Access to the sea forged links with advanced and diverse civilizations, which helped transform an insulated, inward-looking, exclusivist culture into a more open, out-reaching, universalist civilization. In effect, it initiated a process that in time made possible the rise and spread of Christianity and the wide acceptance of the Bible as one of the fundaments of Western civilization as a whole.

9

THE URBAN DOMAIN

Convergence of King and Cult in Jerusalem

I rejoiced when they said to me,
"We are going to the House of Yahweh."
Our feet stood
inside your gates, O Jerusalem,
Jerusalem built up,
a city knit together. . . .
There the thrones of judgment stood,
thrones of the house of David. . . .
May there be well-being within your ramparts,
peace in your citadels.

PSALM 122:1–7

E XPOSURE TO each of the varied ecological domains in the ancient Near East and to the multifarious cults of their resident societies preconditioned the Israelites to the realization of nature's (hence God's) overarching unity, which is the essential premise of monotheism. A complex history created a palimpsest of acquired cultural elements ripe to be synthesized. However, the preconditioning did not result automatically in a sudden collective epiphany. What brought the heterogeneous ethnocultural elements into synergetic fusion was not so much a spontaneous coalescence of ideas as a purposefully directed process. The doctrine that eventually emerged (complete with commandments, codes, and rituals) was formulated by a religious leadership and then adopted and promoted as policy by an established monarchy.

In other words, the early experiences of the Israelites provided the constituents, as it were—the primordial mix—by which monotheism could arise. What was needed was a catalyst to bring about the reconstitution of those elements into a new and cohesive national-religious ethos.

Significantly, the fusion did not take place in any one of the various ecological domains: not in the lush gardens of the riverine domain, not in the rangelands of the semiarid pastoral domain, not in the farming communities of the rainfed domain, not along the sandy or rocky beaches of the coastal domain, not even in the austere fastness of the desert domain.[1] The separate strands of culture and ideology and faith and ritual were finally made to interweave and merge only in the synthetic environment of the city, in the urban domain, induced to do so by a dynastic regime whose basic political purpose was to promote national unity and religious cohesion.

That seminal development began to take shape at a particular time and place. The time was the consolidation of the Davidic kingdom, which first ruled over the unified tribes of Israel as a whole (ca. 1000–900 B.C.E.), but later was confined to Judah and its associated tribes. It was a time of transformation, during which an originally loose assemblage of clans and tribes engaged in herding and farming strove to attain national consolidation. The main place was the city of Jerusalem. It was in that urban center that the disparate elements were brought together and Judaic monotheism was promulgated and fostered as the official state religion.

The motley collection of squabbling tribes, such as the Israelites had been during the preceding period (described in the Book of Judges), was altogether too vulnerable to the periodic incursions of desert nomads, to the organized forces of the neighboring Canaanite city-states, and—ultimately—to the massed armies of the great regional empires: Mesopotamia and Egypt. Most immediately pressing, at the dawn of the Iron Age, was the conflict with the technologically superior Philistines. To defend against such formidable enemies, the Israelite tribes were forced to unite under a single military leadership, which ultimately assumed the form of a monarchy. Such a monarchy, in turn, required the establishment of a centralized administration, complete with a court, an aristocracy, a standing army, an enforceable set of laws, and—of course—a system of mandatory taxation.

To elicit the loyalty of the people, the instituted leaders had a vested interest in inculcating the belief in a common national ideology, ancestry, history, and destiny. They used the radical ideal of monotheism as the centralizing principle and the negation of prevailing polytheism as the rallying cry that would draw and forge the Israelites, designated as the chosen people of the one and only God, into a distinct and cohesive entity, in contrast to and apart from the surrounding cultures.

THE FIRST EFFORT at establishing a united monarchy was made by Saul. He came from the tribe of Benjamin, which was small but centrally positioned between the large rival tribes of Judah in the south and Ephraim in the north. Saul achieved partial success in his campaigns against some of the Israelites' enemies (including the long-hated Amalekites), but finally failed and was killed, along with his son Jonathan, in the decisive battle against the Philistines. Saul's major shortcoming, apart from his unstable temper, was his failure to reconcile the fledgling temporal authority of the king with the established ecclesiastical authority of the priests (represented at the time by the powerful personality of Samuel). Nor did Saul consolidate the diffuse modes of religious worship, which were practiced in various locations and whose forms tended to imitate the ways of the native Canaanites. What Saul's reign lacked most was a spiritual vision that could inspire an embattled and inchoate nation.

After the deaths of Samuel and Saul, David assumed the throne—first over his own tribe of Judah (with its capital of Hebron, site of the cave of Machpelah) and then over the reunified nation of Israel. One of his most important acts, undertaken around 995 B.C.E., was to establish a capital in the buffer zone between the territories of the southern and northern tribes (an early Israelite version of the selection, nearly three millennia later, of the District of Columbia as a neutral site between the North and the South for the capital of the United States).[2] In that no-man's-land, fortuitously, stood the old city-state of Jerusalem, still occupied by a native Canaanite group called the Jebusites (who evidently had held out against the invading Israelites since the time of Joshua).[3] David captured that stronghold (2 Samuel 5:6–9) and made it the capital of his united kingdom.

David's even more astute and symbolic act was to bring the Ark of the Covenant to Jerusalem from the rural location where it had languished after its capture by the Philistines and its return to the Israelites, some decades earlier (1 Samuel 4:11, 6:1–12). The transfer was carried out with great pomp (2 Samuel 6:2–15), with dancing and shouting and sounding the horn. By placing the Ark in Jerusalem, David established a combined political and religious focus for the nation. And by thus signifying the presence (and hence the imprimatur) of Yahweh, David sought to sanctify the earthly authority of the monarchy and to make the priesthood subservient to it. The presence of the Ark, followed by the erection of an altar atop the hill overlooking the city and, later,

A view of Jerusalem from the south, showing the Temple Mount in the center right, the vale of Qidron below it, and the vale of Hinnom on the left. In the upper right, across the Qidron, are the Mount of Olives and Mount Scopus.

DAVID'S FATEFUL TRANSGRESSION

Except for Abraham and Moses, David is the most admired man in the Bible—his character lauded, his exploits praised, and his accomplishments exalted most lavishly. David's eventful life—his early years as an obscure shepherd, his purported duel with the Philistine giant Goliath,[1] his entry into the royal court as King Saul's personal attendant, his subsequent banishment and resort to banditry, his refuge among the Philistines, his assumption of kingship over the tribe of Judah and, later, accession to kingship of all Israel, his conquest of Jerusalem and its establishment as the country's capital, his military triumphs, and his establishment of a permanent Judean dynasty—is described with such romantic extravagance that some scholars have come to regard the entire saga as a work of fiction.

There is at least one incident in the life of David that is anything but laudatory. It reveals David to have been less than perfect, vulnerable to temptation and prone to treachery in the effort to cover up his own misdeed. The story itself is so well known that we need to recount only its bare outline (2 Samuel 11–12).

Alone one day, on the roof of his palace, David observed the enticing figure of a woman bathing nearby. He inquired and learned that she was Bathsheba, wife of Uriah—a Hittite soldier in the king's service. Knowing that her husband was away fighting the Ammonites, David sent for the woman and lay with her. Consequently, Bathsheba informed David that she had conceived by him. David then contrived to have Uriah released and brought home in the expectation that he would lie with his wife and assume that he was the father. But Uriah, out of loyalty to his battlefield comrades, refused to indulge in the comforts of home as long as the war continued. So David sent him back into battle, secretly instructing his commander to put Uriah in harm's way and arrange to get him killed. When that was done, David took the widowed Bathsheba to be his own wife.

But then the prophet Nathan intervened. He came to David and told him an allegorical story about a rich man with much livestock of his own who took a poor man's only lamb and served it to his guest. Hearing that, David flew into a righteous rage and cried that the man who so mistreated his poor neighbor deserved to die. At that moment Nathan pointed to the king and declared: "That man is you!" (2 Samuel 12:7). David then bowed to Nathan's reprimand and repented, but he was not absolved. The incident became a turning point in David's theretofore charmed career. It was followed by a series of tragedies and reversals of fortune, leading to David's gradual decline.

That dramatic moment of truth, when the king was confronted and called to task by the authority of a divinely inspired moral law, is a high point in the evolution of ethical monotheism.[2] That the biblical writers made no allowances for exposing his sinful act testifies to their courage and honesty. If the biblical account were merely a sycophantic paean to the greatness of the king—as many of the written records of ancient kings indeed were (for example, in Mesopotamia and Egypt)—it surely would have omitted the embarrassing story of David's egregious sin. As it is, the story—being so believably human—lends credence to the biblical account (which, although undoubtedly embellished, still seems to contain a kernel of truth).

---

1. An alternative version, however, attributes the slaying of Goliath to David's cousin Elhanan (2 Samuel 21:19).

2. A similar confrontation occurred generations later between the prophet Elijah and King Ahab in the vineyard of Naboth: "Would you murder and also take possession?" (1 Kings 21:19).

by the construction of Solomon's Temple there, imparted an aura of sanctity to the dynastic throne of David and his successors. David himself was represented to the people not merely as a secular king, but as God's anointed one (literally, *mashiah* [messiah]). That aura helped ensure the continuity of the Davidic dynasty in Jerusalem—even after the secession of the northern tribes—in contrast to the frequent changes of kingship in Samaria, capital of the northern kingdom of Israel.

What lent additional sanctity to Jerusalem was the subsequent, apparently mythical, association of the hill abutting Jerusalem on the north with Mount Moriah,[4] where, according to Genesis 22, Abraham was prepared to sacrifice his son Isaac to Elohim, only to be stopped by Yahweh's angel. Many centuries after the time of Abraham, King David, having conquered the city, purchased the threshing floor of the Jebusite farmer Araunah (2 Samuel 24:18–25) atop the hill, there to build an altar to Yahweh.

The same site was chosen by David's son and successor, Solomon, for building the Temple (1 Kings 6–7), in which he placed the Holy Ark (1 Kings 8:1–9) containing the Tablets of the Law, which Moses had brought down from Mount Sinai.[5] The Temple then became the focus of the nation's religion, the destination of obligatory thrice-yearly pilgrimages and the site of gift offering, where elaborate ceremonies were performed by priests and Levites and where commoners and noblemen and kings alike worshiped Yahweh. The practice of religion was no longer a spontaneous, individual act but a formalized, centralized, obligatory, and regulated public ritual.[6]

THUS AT a crucial stage in the history of the Israelites, as the desperate tribes struggled with external enemies as well as with one another, the cult of the one God Yahweh came to serve an important national purpose. It became the unifying principle and rallying cry—indeed, the spiritual and cultic cement—that bonded the strife-ridden tribes into a single nation. The stories of Abraham, Isaac, and Jacob were resurrected out of the dim past to emphasize the ancient roots of the national affinity, claimed to derive from one family. Equally powerful was the ethos of Israel's role as Yahweh's chosen nation and its destiny of glorious redemption, conditional on only the people's strict obedience to his commandments.

The formulation and formalization of monotheism and its adoption as the official Jerusalemite religion did not happen all at once. Evidently, it had to evolve from idolatry (the worship of idols) through monolatry (the worship of one major god, in preference to all existing minor gods) or henotheism (the worship of one god without denying the existence of other gods) to pure monotheism (the worship of an absolutely singular God). Even when monotheism was adopted as the state religion, it was not completely free from remnants of earlier beliefs. Nor was it readily accepted by all the people of the kingdom, many of whom continued to hedge their bets, as it were, between the earthy Canaanite nature gods and the ethereal Yahweh. However, since it was mainly the monotheistic doctrine that has been bequeathed to us by the writers and compilers of the Bible, we can learn about the persistence of other currents of thought and forms of worship among the people from only the passing mention of them or the denunciation of them by the puritanical prophets. At times, even the preachings of the Yahwistic prophets betray vestigial polytheistic images.

The ideology of monotheism, however, was not simply contrived by a ruling urban elite to serve its own purposes and then imposed on a reluctant populace. In fact, the tenets of ethical monotheism were fervently believed and championed by their foremost adherents, especially the forceful and eloquent Jerusalemite prophets Nathan, Isaiah, Jeremiah, Amos, and others. Yet it was undoubtedly true that the faith and practice they advocated served the historical purpose of uniting and rallying the nation. Otherwise, the nation could not have survived the trials to which it was subjected under the prevailing environmental and geopolitical circumstances. Without the bond of religious faith and practice, the nation might have disintegrated and eventually disappeared, as indeed have nearly all the nations and cultures that were contemporary with the Israelites in the ancient Near East.

The process of urbanization that began in the ninth and eighth centuries B.C.E. was accompanied by increasingly widespread literacy. Whereas in earlier generations only specialized scribes could read and write, now the written scrolls that gradually came to compose the Torah became accessible to a wider public, not only to an exclusive priesthood. This, too, served to promote the general acceptance of monotheism.

THE EARLIEST MENTION of Jerusalem seems to have been made in Egyptian writings of the Twelfth Dynasty (ca. 1900 B.C.E.) in the form of Aushamen. In later Akkadian and Assyrian writings, the city is called Urushalima (the city of the god Shalem).[7] The same name, Shalem, appears in the Bible in connection with an episode in the life of Abraham (Genesis 14:18–20). Returning from battle, Abraham was offered bread and wine by the king of Shalem, Melchizedek (king of justice), who was said to have been "priest of the supreme El." The city is also mentioned in the Amarna Letters, which are dated to the fourteenth century B.C.E.

The original Canaanite settlement was on a spur projecting southward from the higher ridge that later came to be known as the Temple Mount. The spur was marked on the east by the vale of Qidron and on the west by the vale known in Hellenic times as Tyropoeon (vale of the cheese makers). The two vales converge at the southern tip of the spur, where they meet in the vale of Hinnom. In subsequent centuries, the city expanded both westward and northward as both its importance and its population grew greatly.

The location of the original settlement was advantageous because it abutted the major north–south route that ran along the crest of the country's mountain range. That path connected the southern part of Canaan's hill country, the district of Hebron, with the northern part of the hill country, the district of Shechem. The mountain range is deeply incised by ravines on both its western flank, which descends toward the Mediterranean Sea, and, especially, its eastern flank, which drops steeply toward the Jordan Rift Valley. Only the very spine of the range, its watershed divide, is conveniently traversable.

The second advantage of the site was that it also lay astride an important east–west artery of traffic that connected the Jordan Valley (with its ancient settlement center of Jericho, just north of the Dead Sea) and the districts to the east (across the Jordan River) with the Mediterranean coast and its major seaports: Jaffa and Ashkelon. The path that ascended from Jericho to Jerusalem and then descended to the coast is more readily traversable than some alternative routes across that section of the country's central mountain range.

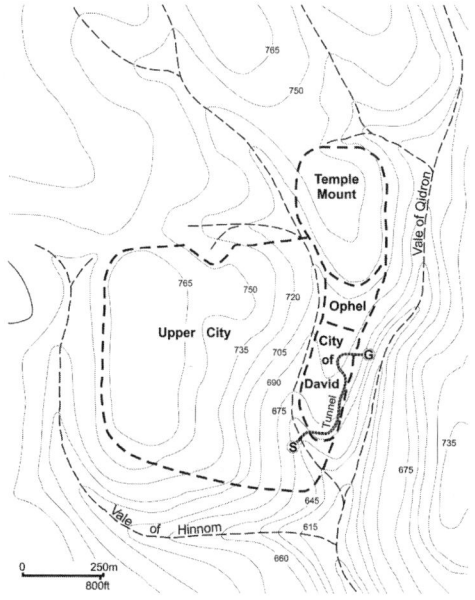

Topographic map of ancient Jerusalem and vicinity. The City of David was built on the site of a Jebusite town under the fortified hill called Ophel, and the tunnel built during the reign of Hezekiah (2 Chronicles 32:30) diverted water from the Gihon (G) to the pool of Siloam (S).

THE ORIGINAL GEHENNA

Some words of biblical derivation are used in common parlance without awareness of their original meaning or context. One such word is "gehenna," which means "a place of suffering, the abode of condemned souls."[1] The Hebrew word is *geihinnom*, a contraction of the phrase *gei Ben-Hinnom*, literally "valley of the son of Hinnom" (presumably, a tract named after its one-time owner). The valley lies to the south of Jerusalem, below the spur of the City of David and curving around the western and southern sides of a high hill (now known, erroneously, as Mount Zion). During the First Temple period, it apparently was an unsavory place, the town dump to which the lepers and other outcasts of society were relegated, and where despicable pagan practices were conducted (Jeremiah 7:31–32, 19:1–6, 32:35; 2 Kings 23:10; 2 Chronicles 28:3, 33:6), including the burning of children as sacrifices to Molech.[2] So pungent and abhorrent was the place that its name came to represent purgatory, or hell, synonymous with Sheol—the underworld where the dead dwelt in darkness (Deuteronomy 32:22 and some sixty-four other passages).

Another word for hell is *'azazel*, apparently after the name of a cliff in the Judean Desert. From that cliff, a goat (*sa'ir*) was released into the wilderness to serve as the bearer of sins and thus to provide the people vicarious expiation (Leviticus 16:8, 10, 26). From that custom, we retain the term "scapegoat."[3]

---

1. *Oxford English Dictionary* (Oxford: Oxford University Press, 2000), s.v. "gehenna"; *American Heritage Dictionary* (Boston: Houghton Mifflin, 1985), s.v. "gehenna."

2. The custom of sacrificing a beloved child to win the favor of the gods was widely prevalent in the ancient world. Consider, for example, Agamemnon's willingness to sacrifice his daughter Iphigenia to induce the gods to provide favorable winds. A parallel case in the Bible was the deed of Jephthah, who sacrificed his daughter in thanksgiving to God for having granted him victory over the Ammonites (Judges 11:34–40).

3. The original meaning of the word *'azazel* is somewhat obscure. Perhaps the simplest interpretation is "removal" or, more specifically, "expulsion of the goat." See E. A. Klein, *A Comprehensive Etymological Dictionary of the Hebrew Language* (Jerusalem: Carta, 1987), 468. The word used for "he-goat" in this context is not *tayish*, but *sa'ir*, which is a euphemism. It means, literally, "hairy" and hence, by implication, "hairy one"—a he-goat. By further extension, it eventually came to imply a satyr or demon. Another interesting linguistic extension of the word refers to a thin rain shower (Deuteronomy 32:2), perhaps because the black rain clouds are reminiscent of black he-goats. (In contrast, the word for copious rain is *revivim*.) The association of a he-goat with wandering in the wilderness derives from the lore of pastoralists in the semiarid zone. For a thorough discussion of sacrificial rites, including the offering of children to Molech and the release of the "scapegoat," see W. Burkert, *Creation of the Sacred: Tracks of Biology in Early Religions* (Cambridge, Mass.: Harvard University Press, 1996).

The third advantage of Jerusalem's location was its access to a guaranteed supply of water, in the form of a karstic spring known as the Gihon (surging) or Shiloah (sending forth).[8] The crest of the mountain range has few such springs, since most springs emerge lower down in the incised wadis. So vital was the spring that Jerusalem probably owed its very existence to it.[9] The Gihon emerged naturally just below the eastern edge of the spur on which the city had been built. Natural fissures or shafts hewn into the rock may have been used by David's men as a way to enter the city stealthily and take it over (2 Samuel 5:6–8).

The opening to the spring of Gihon (Siloam).

Centuries after the time of David and Solomon and eight years after they de-stroyed the kingdom of Israel, the Assyrians, led by King Sennacherib, threat-ened to besiege Jerusalem and deny its access to the Gihon.[10] In preparation for the attack, King Hezekiah hastily ordered the diversion of the spring into the walled city proper: "It was Hezekiah who stopped up the spring of water of Upper Gihon, leading it downward west of the City of David" (2 Chronicles 32:30). In order to complete the 1740-foot (530-m) tunnel before the enemy arrived, the workers excavated it from both ends. Where they met, deep inside the bedrock underlying the city, they left an inscription—written in classical Hebrew, with words containing vocalic letters—expressing joy at the success-ful completion of their difficult and dangerous task.[11] To rally the entire nation in the face of the impending Assyrian onslaught, Hezekiah had to reaffirm the

The inscription, in ancient Hebrew, found in 1880
inside the diversion tunnel of Siloam.

spiritual bond linking the nation to its past. To do so, he also reasserted the centrality of Jerusalem as the focus of Yahweh's presence: "Hezekiah sent word to all Israel and Judah; he also wrote letters to Ephraim and Manasseh to come to the House of Yahweh in Jerusalem to keep the Passover for Yahweh the Elohim of Israel" (2 Chronicles 30:1). Hezekiah narrowly survived the attack by Sennacherib, who lifted the siege and withdrew his forces without breaching the walls of Jerusalem. This deliverance was hailed by the prophet Isaiah as a miracle wrought by Yahweh (2 Kings 18:32–37).[12]

Even greater unification and centralization of faith and ritual were instituted during the reign of Josiah (640–609 B.C.E.), who launched a religious revival and reformation in Jerusalem. It was in the course of repairs ordered by King Josiah in the Temple that the high priest Hilkiah discovered, as though by chance, a scroll of the Torah. He gave the scroll to the royal scribe Shaphan, who then read it to the king: "When the king heard the words of the scroll of the Torah, he rent his clothes" (2 Kings 22:11). Josiah then summoned all the elders of Judah and Jerusalem:

> The king went up to the House of Yahweh, together with all the men of Judah and all the inhabitants of Jerusalem, and the priests and the prophets—all the people, young and old. And he read to them the entire text of the covenant scroll which had been found in the House of Yahweh. The king stood by the pillar and solemnized the covenant before Yahweh: that they would follow Yahweh and observe His commandments, His injunctions, and His laws with all their heart and soul; that they would fulfill all the terms of this covenant as inscribed upon the scroll. And all the people entered into the covenant.
>
> Then the king ordered . . . to bring out of the Temple of Yahweh all the objects made for Ba'al and Asherah and all the hosts of heaven. He burned them outside Jerusalem in the fields of Kidron, and he removed the ashes to Bethel. He suppressed the idolatrous priests whom the kings of Judah had appointed to make offerings at the shrines in the towns of Judah and in the environs of Jerusalem. . . . He tore down the cubicles of the male prostitutes in the House of Yahweh, at the place where the women wove coverings for Asherah. . . . He also defiled Topheth, which is in the valley of Ben-Hinnom, so that no one might consign his son or daughter to the fire of Molech. . . . He burned the chariots of the sun. . . . He tore down the altars made by [prior] kings. . . . The king also defiled the shrines . . . which King Solomon had built for Ashtoreth, the abomination of the Sidonians, for Chemosh, the abomination of Moab, and for Milcom, the abomination of the Ammonites. (2 Kings 23:2–13)

We learn much from these and other passages (2 Chronicles 34–35) about the deeds of Josiah. We also learn much about the pagan rituals that had been adopted by his predecessors, who apparently hedged their commitment to

WORSHIP OF THE GODDESS ASHERA

In the pantheon of Canaanite nature gods, the reigning goddess was Ashera, the feminine counterpart of the principal god, Ba'al or El. She was the goddess of the terrestrial realm—the land and its fertility and the reproduction and well-being of all the Earth's creatures. She was worshiped throughout the ancient Near East, particularly in the rainfed domain. Her importance is manifested by numerous references to her in tablets and inscriptions found at such sites as Ugarit and Ebla, in present-day Syria.

There is much evidence that the cult of Ashera was prevalent among the early Israelites as well as the Canaanites (Judges 3:7; 1 Kings 14:23; 2 Kings 17:10), notwithstanding the vehement denunciation of it by the prophets of Yahweh. That they felt the need to castigate the cult attests to its prevalence and persistence. The Bible contains no fewer than forty direct references to Ashera and various indirect ones, including nine mentions of Astoreth, evidently an alternative appellation for the Canaanite goddess. Evidently, it was a seductive cult, complete with sexual practices (including the services of female and male prostitutes [2 Kings 23:7; Hosea 4:14]). Hence many of the people who accepted the supremacy of the masculine Yahweh were also attracted to the worship of the feminine Ashera.

Worship of the goddess was promoted even by King Solomon, who had built the Temple dedicated to Yahweh in Jerusalem (1 Kings 11:1–5). The subsequent division of the united kingdom was later attributed by the biblical writers to Yahweh's punishment of Solomon's transgression (1 Kings 11:31–33). After that schism, the northern kingdom came increasingly under the cultural influence of the Phoenicians. The worship of Ashera was especially promoted in Samaria by the Phoenician-born Queen Jezebel, wife of King Ahab. Ashera appears to have been worshiped alongside Yahweh for a considerable period of time.[1]

In Jerusalem, Queen Ma'akhah had "an abominable" image made for Ashera (1 Kings 15:13; 2 Chronicles 15:6). King Jehoshaphat made an early effort to remove the idols, including Ashera, from the capital (2 Chronicles 17:6), but his attempt apparently failed. After the destruction of the northern kingdom of Israel by the Assyrians, however, a strongly puritanical monotheistic movement began in the southern kingdom of Judah, aimed at affirming the faith in Yahweh alone. King Hezekiah began to remove the idols, including Ashera, from the Temple (2 Chronicles 29:5, 16, 30:14, 31:1), but his son Manasseh (2 Kings 21:7; Chronicles 33:3) returned them. A more decisive action was taken by King Josiah to purge Judah of the idols (2 Chronicles 34:3–4).

The cult of Ashera did not disappear entirely, it seems, until the destruction of Jerusalem and the exile of the Judean elite to Babylonia. That traumatic defeat was regarded by the nation's religious leaders as Yahweh's retribution for the persistence of paganism in the kingdom of Judah, just as it had existed in the previously destroyed kingdom of Israel.

---

1. Z. Meshel, "Kuntilat 'Ajrud, an Israelite Site on the Sinai Border" (Hebrew), *Qadmoniot* 9 (1976): 119–124; S. Scham, "The Lost Goddess of Israel," *Archaeology* 58 (2005): 36–40; W. G. Dever, *What Did the Biblical Writers Know and When Did They Know It?* (Grand Rapids, Mich.: Eerdmans, 2001).

Yahweh and practiced a kind of equivocal monotheism while maintaining altars and allowing the offering of sacrifices (including child sacrifices) to the various gods of the neighboring peoples.

King Josiah then convened the people and conducted a communal Passover celebration (2 Chronicles 35:1). So complete was his commitment to the doctrine of the Torah and his effort to eliminate non-Yahwistic practices and sanctuaries that the biblical writer paid him the highest accolade: "There was no king like him before who turned back to Yahweh with all his heart and soul and might, in full accord with the Teaching of Moses, nor any like him after him" (2 Kings 23:25).

AFTER ALL his great deeds in establishing monotheism as the absolute religion of the state, Josiah met a sudden and tragic death on the battlefield. Unfortunately, he involved himself, perhaps unnecessarily, in the war that was being waged between the Egyptians and the Assyrians, the two regional titans of the time. As King Neco of Egypt marched through Canaan toward Carchemish on the Euphrates, he warned Josiah to stay out of the fray. Josiah did not heed the advice but chose to engage Necho in the plain of Megiddo.[13] There he was shot by archers and died on his return to Jerusalem (2 Chronicles 35:20–24).

Scholars believe that the scroll of the Torah found in the Temple and adopted by King Josiah was none other than the Book of Deuteronomy, or at least the law code in it (Deuteronomy 12–26).[14] Although Deuteronomy is presented as Moses's farewell speech before his death, and it is purported to be set on the plateau of Moab, before the Israelites crossed the Jordan River to enter the promised land, it appears to have been written centuries later, during the First Temple period in Jerusalem, for its warnings and tenets are pertinent to that period. It is a selective, modified reprise of the accounts and laws stated in the Books of Exodus, Leviticus, and Numbers. It differs from them, however, in its distinctive style and in important features of content.[15]

The law code in Deuteronomy decrees that there is to be only one place for sacrifice, one central altar where "Yahweh sets His name." It therefore required the destruction of the variously located "high places" (generally, prominent hilltops) at which sacrifices had been offered in earlier times. Significantly, the Books of Samuel and Kings do not mandate the centralization of worship. The prophet Samuel sacrificed in several places, as did Kings Saul and David. Clearly, then, one of the major aims of Josiah's reform (based on the claimed rediscovery of the Torah) was the absolute centralization and standardization of worship, to be instituted as state policy.

The abolition of provincial worship appears to have caused a profound change in the entire practice of the religion. Denied the right to perform sacrifices and other rituals anywhere but at the Temple, those unable to attend the Temple sought alternative forms of worship. The new modes of religious expression that came into being included communal prayers and readings of the Torah. Gradually, therefore, Judaic religion underwent a metamorphosis,

ON COOKING A KID IN ITS MOTHER'S "MILK"

This culinary rule is stated three times in the Bible: "You shall not cook a kid in its mother's milk" (Exodus 23:19, 34:26; Deuteronomy 14:21). It was extended by later rabbinical interpreters to prohibit the mixing of any meat products with any dairy products and, as such, has long been accepted as one of the most important dietary laws of Orthodox Judaism. No explanation for the original edict is given in the Bible. Maimonides held that this law prohibits a pagan rite, although no such specific rite is known.[1] Other explanations are based on humane grounds. However, there is an entirely different possibility.

The original Hebrew Scriptures were rendered without vowels. Centuries later, vowels were added to facilitate the correct reading of the Torah. In this dietary injunction, the vowels applied to the three-consonant word *ḤLV* render it *ḥalev* (the milk of), as in the phrase *ḥalev imo*, which means "the milk of its mother." But what if a slightly different vowelization is applied: *ḥelev imo?* Then the phrase means "the fat (tallow) of its mother." A small change in pronunciation, a major change of meaning.

The second possibility makes sense, since it is consistent with various other Torah prohibitions regarding *ḥelev*, including the following: "It is the law for all time throughout the ages, in all your settlements: you must not eat any *ḥelev* [fat] or any blood" (Leviticus 3:17). The same edict is then restated and emphasized: "Yahweh spoke to Moses, saying: Speak to the people of Israel thus: You shall eat no *ḥelev* of ox or sheep or goat. *Ḥelev* from an animal that died or was torn by beasts may be put to any use, but you must not eat it" (Leviticus 7:22–24). A possible explanation is implied in the following: "The *ḥelev* [fat] of My festal offering shall not be left lying until morning" (Exodus 23:18). Clearly, animal fat was liable to spoil (turn rancid, reek, perhaps even become toxic) very quickly in the prevailing heat of the ancient Near East, absent any method of refrigeration or alternative preservation. If eating *ḥelev* was prohibited, then, obviously, so should have been the consumption of a kid seethed in its mother's *ḥelev*.[2]

The prohibition of seething a kid in its mother's fat can also be explained on humanitarian and practical grounds, since doing so would require slaughtering both the kid and its mother (Leviticus 22:28; Deuteronomy 22:6–7).

The purported prohibition of mixing meat and dairy products seems altogether inconsistent with what we know about the lifestyle of seminomadic pastoralists, for whom meat and milk constitute the principal foods. Given their need to move repeatedly, the Israelites would have had no way to separate those staples. Such separation, and the consequent use of different dishes and utensils, was possible for only a sedentary (for example, urban) society, especially if it wished to set itself apart from the prevailing customs of its neighbors. There are passages in the Bible, in fact, that describe the mixing of milk and meat products without reservation (contradicting the principle of separation). Abraham served three guests (later revealed to be angels!) a meal of meat and milk: "Abraham ran to the herd, took a calf, tender and choice, and gave it to a servant-boy, who hastened to prepare it. He took curds and milk and the calf that had been prepared and set these before them and waited on them under the tree as they ate" (Genesis 18:7–8). Another passage extols the beneficence of Yahweh: "He set him [Israel] atop the highlands, to feast on the yield of the earth; He fed him honey from the crag, and oil from the flinty rock, curd of kine and milk of flocks; with the best of lambs, and rams of Bashan, and he-goats" (Deuteronomy 32:13–14). On one occasion, David and his loyal troops are described as feasting on curds and mutton (2 Samuel 17:29).

However plausible this alternative version—*ḥelev* rather than *ḥalev*—may be, the *ḥalev imo* (milk of its mother) rendition was accepted long ago and is entrenched, as though no other interpretation of the injunction were possible.

The proscription on the eating of swine is another indication of the Israelites' nomadic and pastoral background, the formative early experiences that shaped their most durable

continued

traditions. Pigs—in contrast with sheep, goats, and even cows—cannot be herded and taken to distant pastures in a hot, dry environment. Therefore, they were kept by only settled farmers, mostly in pens where they were allowed to wallow in puddles and were fed household leftovers. Moreover, the animals so kept tended to accumulate fat, which, unless consumed immediately after an animal was slaughtered, could putrefy in the warm climate. To nomadic pastoralists, pigs seemed unclean and repulsive. This attitude was so ingrained that it persisted even after nomadic societies themselves settled down and took up farming. The Israelites' abhorrence of the pig was accentuated by its having been a major source of meat for their mortal enemies, the Canaanites. When that revulsion was given religious sanction, the taboo was retained long after it had lost its original rationale.

1. Maimonides (1135–1204), whose full name was Moshe ben Maimon, was the most influential Jewish theologian and philosopher of the Middle Ages. In his work *Mishneh Torah* (*Review of the Law*), he defined and explained the rules of the Halakah (Jewish religious law). For a modern interpretation of this and other Torah edicts, see D. L. Lieber, ed., *Etz Hayim: Torah and Commentary* (New York: Rabbinical Assembly, 2001).

2. The consumption of blood was prohibited because, in the perception of the ancient Hebrews, blood represented the essence of life. Eating blood therefore would be tantamount to consuming a live animal, "for the life of the flesh is in the blood" (Leviticus 17:11). Hence blood must be drained from a carcass, and the remainder drawn out by means of salts, before the flesh of an animal can be prepared for food.

becoming less a religion of traditional sacrifices and more a religion of prayer and study. Ultimately, this change enabled the religion to survive total detachment from the Temple and even from the homeland and to persevere during prolonged exile.

THE DISADVANTAGE of the site of Jerusalem was its strategic vulnerability to attack from the higher ground nearby. Its location just to the west of the Gihon placed it at a topographic disadvantage. Indeed, the city was surrounded by mountains and was especially approachable from the north, which explains the warnings by Jeremiah: *mitsafon tipatah hara'a*, or "From the north shall disaster break loose" (Jeremiah 1:14), and "For evil is appearing from the north, and great disaster" (Jeremiah 6:1). To overcome the vulnerability, great walls were constructed, only to be breached by enemies and reconstructed by defenders repeatedly over the ensuing centuries.

The Bible portrays the reigns of David and his son Solomon as a golden age of military and economic ascendancy, during which Jerusalem became a great capital, with imposing palaces and public buildings. In recent decades, archaeologists have searched for remains of that golden age but have so far found only meager evidence of it. Some of them have come to believe that the biblical account, written long after the time of David and Solomon, is delib-

LIFE IN THE URBAN DOMAIN

A city, particularly a capital city, was a densely populated and hence closely knit, organized community. The urban domain developed its own characteristic social and economic structure, a set of loyalties and a hierarchy no longer based entirely on the traditional clan system. Along with a defensive wall and public facilities came a new set of relationships and precepts—indeed, an entirely different culture that included new modes of worship and religious rites, legal standards, behavioral mores, and pastimes. Most city dwellers no longer engaged in traditional farming and its directly related enterprises, as they had in their ancestral rural encampments and villages. Instead, they engaged in industrial, artisanal, mercantile, clerical, cultic, and service professions. Among the specialized craftsmen were weavers, butchers, leather tanners, and metalworkers—some working as entrepreneurs and some as menial laborers.

The commanding presence of the king and his administrators and the formalized ritual of the temple were manifested in imposing buildings that were generally located on higher ground and were visible from every point in the city and even from afar. Members of particular social classes and professions lived in more or less distinct clusters of houses or neighborhoods, separated by alleyways or streets. An open square usually was located just inside the city gate, and it provided a location where the people of the city could assemble and where public discussions and trials were held (for example, 2 Samuel 15:2). It was also in the city square that returning heroes were celebrated and visiting strangers were either welcomed or turned away.[1] Finally, and very importantly, the city provided its denizens with a measure of security against attack, to which the villagers in the open countryside were generally exposed.

Because of the many advantageous features and amenities of Jerusalem, its inhabitants often identified with the city, took pride in its beauty, and were ever concerned for its fate. In the Song of Songs, the lover compares the beauty of his beloved to that of the city: "You are comely . . . , my love, / comely like Jerusalem" (Song of Songs 6:4).

Despite the attractions of urban life, the inhabitants of the crowded city were much more vulnerable to contagious diseases than were their rural neighbors: a fatal epidemic broke out in several Philistine cities and spread to the Israelites (1 Samuel 5:6–6:9); a mortal disease struck the Israelites during the reign of King David (2 Samuel 24:15–16); and a deadly contagion apparently afflicted the army of Sennacherib, king of Assyria, which besieged Jerusalem, saving the city from destruction.[2] That deliverance was attributed at the time to divine intervention (2 Kings 19:35–36).[3]

The city of Jerusalem had to be protected from the foul accumulation and putrefaction of waste that might breed mice and rats and other undesirable pathogenic agents. The garbage was periodically collected and carried to a dump outside the city. One place that apparently served as Jerusalem's dump was the vale of Hinnom, just south of the city. That was also where the outcasts of the city were relegated and where abominable rituals of child sacrifice were held. So despicable and abhorrent did the place become that its name, Gehenna, eventually came to connote "hell."

Clearly, the essential requirement for the life of a city was a dependable source of water, such as an accessible river or perennial spring. The original Jebusite city of Jerusalem depended on the spring of Gihon, which emerged just outside the city, on its southeastern edge. So important was that spring that it was chosen as the site for the coronation of Solomon, the first king crowned in Jerusalem.

---

1. It was at city gates that Abraham purchased the cave of Machpelah (Genesis 23:10), Boaz obtained the right to marry Ruth (Ruth 4:1–10), and Queen Jezebel contrived to falsely accuse and condemn Naboth the Jezreelite (1 Kings 21:8–18).

2. The dire conditions endured by the people of Jerusalem during the siege were expressed in the words of the Assyrian emissary Rabshakeh as he taunted the defenders of the city: "the men who are sitting on the wall, who will have to eat their dung and drink their own urine" (2 Kings 18:27).

3. An early use of antibiotics to cure a contagious skin disease is described in the Bible: "Hezekiah fell dangerously ill. . . . Then Isaiah said, 'Get a cake of [dried] figs.' And they got one and they applied it to the rash, and he recovered" (2 Kings 20:1, 7).

erately exaggerated and embellished.[16] Indeed, the united kingdom lasted for only about seventy-five years before splitting into the northern kingdom of Israel, based in Samaria, and the southern kingdom of Judah, which maintained Jerusalem as its capital.

But whether the grandeur of David and Solomon so vividly described and so highly exalted in the Bible was entirely real or perhaps partly imagined, and whatever the scale of construction and the level of wealth achieved during their reigns, that era clearly initiated a progression by which the city of Jerusalem became the focal point of Judaic monotheism. What was begun at the time of David and Solomon continued (albeit with many rises and declines) and was greatly reinforced by Josiah, and—much later—by the Hasmoneans and the Sanhedrin during the latter part of the Second Temple period (the last few centuries B.C.E. and the first century C.E.). So powerful was—and is—its mystique that Jerusalem has actually transcended the bounds of its own corporal existence to become a lasting beacon of faith not only for Judaism but for Christianity and Islam as well.

The ultimate vision of Jerusalem's destiny was for it to become the spiritual center of the entire world:

> In the days to come,
> the Mount of Yahweh's House
> shall stand firm above the mountains
> and tower above the hills;
> and all the nations shall gaze on it with joy.
> And the many peoples shall go and say:
> Come, let us go up to the Mount of Yahweh,
> to the house of Jacob's Elohim;
> that He may instruct us in His ways,
> and that we will walk in His paths.
> For instruction shall come forth from Zion,
> the word of Yahweh from Jerusalem. . . .
> They shall beat their swords into plowshares
> and their spears into pruning hooks;
> nation shall not take up sword against nation;
> they shall never again know war. (Isaiah 2:2–4)

10

THE EXILE DOMAIN

Expulsion, Survival, Revival, and Return

By the rivers of Babylon,
there we sat, sat and wept
as we remembered Zion. . . .
Our captors
asked us there for songs,
our tormentors for amusement:
"Sing us of the songs of Zion."
How can we sing a song of Yahweh
on alien soil?
If I forget you, O Jerusalem,
let my right hand wither;
Let my tongue stick to my palate
if I cease to think of you.

PSALM 137:1–6

As was the urban domain, but not the five ecological domains of the ancient Near East, the exile domain was a cultural and experiential realm, apart from nature, in which a nation expelled from what had been its homeland for centuries finally had to contend with the most fateful of all choices—in the most literal and dire sense—to be or not to be. Each of the other contemporary nations and cultures in the Near East, attached to and, indeed, defined by its own domain of origin, would lose its distinct identity and disappear from the stage of history once uprooted from its birthplace, from the source of its cultural sustenance and growth. Indeed, that was the rationale for the imperial policy (practiced by both the Assyrians and the Babylonians) of expelling and then intermingling entire populations: erase their traditional identities and assimilate them into an imposed imperial order.

For the Judeans, however, exile from the Land of Israel and, specifically, from its cultural heart, Jerusalem, was only a physical separation. It was not a detachment of the mind and the heart from what had long been the focus of their faith and hope. They remained rooted in spirit, if not in body. The exiled Judeans, or at least an important nucleus among them, carried with them memories of all their past experiences and yearned to return to the place and circumstance that they believed had been assigned them by the grace of Yahweh. In exile, even more devotedly than before, they continued and intensified their internal ideological and moral quest. They wrestled with their fate in a profound and painful effort to decipher the cause and ultimate meaning of the misfortune that had befallen them and to distill from it an essence of hope for renewal and restoration. That inner struggle could be resolved—indeed, had to be resolved—only in the context of an exile that was perceived to be a disciplinary act of God, urging the remnants of Israel to engage ever more deeply in introspection and rededication in order to be worthy of redemption.

IN THE PERIOD preceding the exile, the schism and internecine rivalry between the two half-nations, the northern kingdom of Israel and the southern kingdom of Judah, had weakened both. Each had to contend separately with hostile neighbors: Israel with the Aramaeans to the northeast and the Phoenicians to the northwest, and Judah with the Philistines to the west, the Amalekites to the south, and the Ammonites, Moabites, and Edomites to the east and southeast. And while contending with one another, all these small kingdoms, including Israel and Judah, lived under the threat of a possible attack by the overwhelmingly powerful great empires looming just beyond the horizon: Egypt to the southwest and Babylonia and Assyria to the east.

A city under siege: the assault by the Egyptian army on the Canaanite city of Ashkelon. The attackers break down the gates of the city and scale its walls, while the defenders beseech their gods to save them. A priest atop the wall (*right*) burns incense ceremonially, as children are sacrificed, apparently to placate the local god. This carving, on a wall in the temple of Amon at Karnak, in Upper Egypt, dates to the reign of Rameses II or Merneptah (thirteenth or twelfth century B.C.E.).

In the ninth and eighth centuries B.C.E., the Assyrians (so named after their first capital city, Ashur or Assur, on the Tigris River in northern Mesopotamia) gained the upper hand against the Babylonians of central and southern Mesopotamia. They then embarked on a campaign of conquest and established a mighty empire that, at its zenith, encompassed most of the Fertile Crescent.

In the second half of the eighth century B.C.E., King Tiglathpileser III, after having subdued Babylonia, turned his attention westward. King Ahaz of Judah then appealed to him for help against the northern kingdom of Israel and against the Philistines, who had risen once again to take advantage of Judah's weakness. The northern kingdom of Israel, in contrast, tried to resist Assyria by forming an alliance with its former enemy, the Aramaic kingdom of Damascus. However, the Assyrians vanquished Damascus (ca. 733 B.C.E.) and invaded Israel. They annexed Galilee and the Israelite territories to the east of the Jordan River and drove the Transjordanian tribes of Reuben, Gad, and Manasseh into captivity. Judah escaped that onslaught by submitting itself to the hegemony of Assyria.[1]

The practice of deporting defeated nations in order to break their resistance began in the second half of the eighth century B.C.E. and became the policy of the Assyrian Empire. That the Assyrians occasionally tried to entice their captives to submit to exile willingly is attested by Rabshakeh, speaking for King Sennacherib of Assyria, as he addressed the besieged defenders of Jerusalem:

> Make your peace with me and come out to me, so that you may all eat from your vines and your fig trees and drink water from your cisterns, until I come and take you away to a land like your own, a land of grain [fields] and vineyards, of bread and wine, of olive oil and honey, so that you may live and not die. Don't listen to Hezekiah, who misleads you by saying, "Yahweh will save us." Did any of the gods of other nations save his land from the king of Assyria? Where were the gods of Hamath and Arpad? Where were the gods of Sepharvaim, Hena, and Ivvah? [And] did they save Samaria from me? (2 Kings 18:31–34)

The commanders of the city, fearing that Rabshakeh might demoralize the defenders on the walls, implored him to speak Aramaic rather than Judean (Hebrew), but he persisted in his propaganda (2 Kings 18:26–30).

What remained of the kingdom of Israel became a vassal state and was required to pay a heavy tribute to Assyria. When Israel rebelled, under the delusion that Egypt might help it, King Shalmaneser V besieged Samaria, the capital of Israel. His successor, Sargon II, captured the city in 721 B.C.E. and then the entire kingdom. Assyrian records praise Sargon's victory and relate

The subjugation of the northern kingdom of Israel to Assyrian domination: "Jehu, the son of Omri" prostrating himself before King Shalmaneser III. This carving is on an obelisk found in 1846 at the site of the ancient city of Nimrud, in present-day Iraq.

that he exiled 27,290 residents from the capital. The following year, some of the remnant forces of Israel rebelled once again by joining with the Aramaic king of Hamath, but were ruthlessly crushed. Sargon then effectively abolished the kingdom of Israel. He rebuilt Samaria as the capital of an Assyrian satrapy, and populated it with captives from elsewhere.

This is how the terrible event is described in the Bible:

> The king of Assyria caught Hosea [king of Israel] in an act of treachery; he had sent envoys to King So of Egypt, and he had not paid the tribute to the king of Assyria as in previous years. The king of Assyria arrested him and put him in prison. Then the king of Assyria marched against the whole land; he came to Samaria and besieged it for three years. In the ninth year of Hosea, the king of Assyria captured Samaria. He deported the Israelites to Assyria and settled them in Halah, at the [River] Habor, at the River Gozan, and in the towns of Media. (2 Kings 17:4–6)

The biblical writer then explains the cause of the disaster:

> This happened because the Israelites sinned against Yahweh their Elohim who had freed them from the land of Egypt . . . worshipped other gods and followed the customs of the nations. . . . They worshipped fetishes. . . . They rejected all the commandments of Yahweh their Elohim; they made molten idols . . . two calves . . . and they worshipped Baʻal. They consigned their sons and daughters to the fire; they practiced augury and divination. . . . Yahweh was incensed at Israel and He banished them from His presence; none was left but the tribe of Judah alone. (2 Kings 17:7–18)

In place of the banished Israelites, the Assyrians settled strangers from disparate locations and gave them possession of the vacated cities and lands (2 Kings 17:24). The new inhabitants eventually merged with the remnants of the Israelites (mostly of the lower social and economic classes, living in villages) to form, in a process of syncretism that took place over the ensuing generations, a more or less distinct national entity that came to be known as the Samaritans.[2] The religious practices of the Samaritans were based on earlier Israelite modes of worship (complete with a temple on Mount Gerizim, in which animal sacrifices were offered by the priests), but they had been corrupted (in the opinion of the Judean purists) by pagan influences.

What happened to the exiled Israelites? The Bible tells us that they were resettled in various places (2 Kings 17:6, 18:11; 1 Chronicles 5:26). Some may have retained their national and religious identity and later joined the Judeans who were subsequently exiled to Babylonia. Most of the exiles from the northern kingdom, however, appear to have blended into the local populations of their new homes and to have lost their distinct nationality and culture. Thereafter, they were consigned to the realm of legend, the storied "Lost Tribes of Israel."[3]

Two decades after destroying Samaria, the Assyrians, this time under King Sennacherib, "marched against the fortified towns of Judah and seized them" (2 Kings 18:13).[4] Trying desperately to avoid the fate of Israel, Hezekiah, king of Judah, sought to placate Sennacherib, but the Assyrian king demanded an exorbitant payment of "three hundred talents of silver and thirty talents of gold" (2 Kings 18:14). So Hezekiah gave Sennacherib all the silver and all the gold he had, even including the gilded doorposts of the Temple of Yahweh.

To no avail. Sennacherib was not satisfied. He sent a strong contingent to besiege Jerusalem (ca. 701 B.C.E.). Sennacherib's emissary, Rabshakeh, admonished the Judeans in scathing terms: "On whom are you relying, that you have rebelled against me? You rely . . . on Egypt, that splintered reed of a staff, which enters and punctures the palm of anyone who leans on it! . . . And if you tell me that you are relying on 'Yahweh our Elohim' . . . Do you think I have marched against this land to destroy it without Yahweh? Yahweh Himself told me: Go up against that land and destroy it" (2 Kings 18:20–25). In so saying, Rabshakeh was—knowingly or otherwise—echoing the admonitions of the Hebrew prophets, who had repeatedly described the forces arrayed against the Israelites as merely Yahweh's instruments to punish them for their disobedience to his commandments. But this time, Rabshakeh's taunts were countered scornfully by the prophet Isaiah:

Fair Maiden Zion despises you, she mocks at you. . . .
Whom have you blasphemed and reviled? . . .

Against the Holy One of Israel! . . .
Because you thought, "Thanks to my chariotry,
it is I who have climbed the highest mountains
to the remotest parts of the Lebanon,
and have cut down its loftiest cedars. . . ."
Have you not heard?
Of old I [Yahweh] planned that very thing,
I designed it long ago . . .
laying waste fortified towns in desolate heaps,
their inhabitants helpless. . . .
They were but grass of the field . . .
grass of the roofs[5]
that is blasted before the standing grain. . . .
Because you have raged against Me . . .
I will place My hook in your nose
and My bit between your jaws,
and I will make you go back. (2 Kings 19:21–28)

The Assyrian assault against Judah, although it resulted in the ruin of many of its cities and the impoverishment of its treasury, failed to breach Jerusalem itself. For reasons not completely clear (perhaps because of a pestilence that struck the invading army, or because trouble elsewhere in their empire diverted the Assyrians from completing the destruction of Jerusalem), Sennacherib withdrew his forces and Jerusalem was saved. The narrowly averted disaster was celebrated at the time as a miraculous deliverance and as a vindication of Hezekiah's righteousness and loyalty to Yahweh. Alas, Judah's independence was not to last. Little more than a century after the fall of its sister kingdom, Judah suffered the same fate.

IN 621 B.C.E., the power of Assyria was supplanted by the resurgence of Babylonia, an empire no less cruel, which had adopted the same method of exiling recalcitrant nations within its sphere of dominance. When the Judeans resisted the Babylonians, in 598 B.C.E.:

King Nebuchadnezzar of Babylon advanced against the city [Jerusalem] while his troops were besieging it. Thereupon King Jehoiachin of Judah, along with his mother, and his courtiers, commanders, and officers, surrendered to the king of Babylon. The king of Babylon took him captive. . . . He carried off from Jerusalem all the treasures of the House of Yahweh and the treasures of the royal palace; he stripped off all the golden decorations in the Temple of Yahweh. . . . He exiled all of Jerusalem: all the commanders and all the warriors—ten thousand exiles—as well as all the craftsmen and smiths; only the poorest people in the land were left. (2 Kings 24:11–16)

Nebuchadnezzar then appointed Mattaniah, Jehoiachin's uncle, to be king in his stead, changing his name to Zedekiah.[6] But Zedekiah, too, ignoring warnings by the prophet Jeremiah, rebelled against the king of Babylon. Consequently,

> Nebuchadnezzar moved against Jerusalem with his whole army. He besieged it; and built an embankment against it all around. The city continued in a state of siege until the eleventh year of King Zedekiah. By the ninth day [of the fourth month] the famine had become acute . . . there was no food left for the common people. Then the wall of the city was breached. . . . On the seventh day of the fifth month . . . Nebuzaradan, the chief of the guards, officer of the king of Babylon, came to Jerusalem. He burned the House of Yahweh, the king's palace, and all the houses of Jerusalem; he burned down the house of every notable person. The entire Chaldean force . . . tore down the walls of Jerusalem on every side. The remnant of the people that was left in the city . . . were taken into exile. . . . Thus Judah was exiled from its land. (2 Kings 25:1–4, 8–11, 21)

The exile evidently took place in 587/586 B.C.E. The number of Judeans who were actually deported is unclear.[7] It appears that the expulsion was selective rather than total. While the members of the ruling urban elite were exiled, a substantial number of common folk ("some of the poorest in the land") were left in the small villages (2 Kings 25:12), disorganized and demoralized, but surviving nonetheless. Moreover, it seems that the Babylonians refrained from doing in Judea what the Assyrians had done in Samaria: bring in settlers from afar to replace the exiles. So the territory of Judah remained in a state of partial ruin for the two or three generations between the Babylonian expulsion and the Persian-sponsored return (Ezra 2).

The Babylonians settled the expelled Judeans in several centers along the River Chebar, in a territory whose earlier population may have been transferred elsewhere or decimated during the wars between the Babylonians and the Assyrians.[8] The exiles probably engaged in farming at first, but later may have also engaged in trade. In time, some of them seem to have become prosperous (according to statements about the precious gifts they later sent to Jerusalem [Ezra 1:6, 2:68–69]). Most remarkably, they were able to maintain and even develop their distinct community and national-religious identity. So when, two generations after they had been exiled, the opportunity arose for the Judeans to return to their homeland, a significant number (although only a minority) of them did so.

That opportunity occurred with the ascendance of the kingdom of Persia under the leadership of Cyrus the Great. In 540/539 B.C.E., he attacked and conquered Babylonia. Although fierce toward his enemies, Cyrus was in general a more enlightened ruler than his predecessors in his treatment of the peoples who lived in the various reaches of his kingdom. He seems to have realized that a large and multiethnic empire cannot maintain control over disparate nations in the long run just by coercion and intimidation, enforced by murdering leaders and exiling populations. Instead, he tried to win the allegiance of his subjects by returning exiled peoples to the territories of their origin, rebuilding temples, and treating differing cultures with tolerance and respect. In his benevolence toward the Judeans, he also may have been motivated by the practical prospect of establishing a buffer state between the Persian Empire and Egypt, which he had not conquered (although his successors did).[9]

In 538 B.C.E., Cyrus issued his historic proclamation: "Thus said King Cyrus of Persia: Yahweh the Elohim of Heaven has given me all the kingdoms of the earth, and has charged me with building Him a House in Jerusalem, which is in Judah. Anyone of you of all His people, Yahweh his Elohim be with him and let him go up" (2 Chronicles 36:23). The fuller version of the proclamation completes the last sentence and adds to it: "Anyone of you of all His people, may his Elohim be with him, and let him go up to Jerusalem that is in Judah, and build the House of Yahweh the Elohim of Israel, the Elohim that is in Jerusalem. And all who stay behind, wherever he may be living, let the people of his place assist him with silver, gold, goods, and livestock, besides the freewill offering to the House of the Elohim that is in Jerusalem" (Ezra 1:2–4).

Cyrus was hailed by the prophets of the period, especially Deutero-Isaiah,[10] as the one chosen by Yahweh to redeem Israel:

He is My shepherd,
he shall fulfill My purposes!
He shall say of Jerusalem, She shall be rebuilt,
and to the Temple: You shall be founded again. (Isaiah 44:28)

Thus said Yahweh to Cyrus, His anointed one [messiah]—
whose right hand He has grasped. (Isaiah 45:1)

Cyrus's proclamation was regarded as vindication of Yahweh's ultimate commitment to the redemption of his people Israel, provided only that they were willing to repent their sins and to adhere at last to his commandments.[11] His proclamation was indeed followed by the return of a significant number of

Judeans from their settlements in Babylonia, to rebuild Jerusalem and the Temple (as described in the Books of Ezra and Nehemiah).

The initial wave of returnees was led, in messianic fervor, by a scion of the House of David, Zerubbabel, who rebuilt the Temple in 520 to 516 B.C.E., in concert with a descendant of the high priests, Jeshu'a (whose very name means "redemption"). During the next hundred years, successive parties of Jews returned from Babylonia to help in the restoration of Jerusalem and Judea. Important among them were Ezra the Scribe and Nehemiah the Governor, who restored the walls of the city in 444 B.C.E. It appears, however, that a great number of the descendants of the Judean exiles chose to remain in the relative prosperity and security of Babylonia. Over the ensuing centuries, they constituted a large and thriving community, with a particularly creative cultural and religious life.[12]

How to explain the survival and resurgence of the exiled Judeans and their ability to overcome despair after the tragedy of destruction and expulsion,[13] in marked contrast to the demise and disappearance of their brethren, the exiled northern Israelites?

There were many differences. In the first place, the northern kingdom of Israel was composed of ten disparate and often squabbling tribes, each ensconced in a jealously held exclusive territory of its own, that never completely fused into a cohesive nation. Nor did the kingdom establish a central religious focus or a set of shared rituals that might have cemented the nation spiritually and religiously. The heterogeneity of the tribes and their proximity to their pagan neighbors caused them to be strongly influenced by the polytheists' popular and seductive rituals and faiths, including the worship of Ba'al and Ashera. (Especially strong was the impact of the Phoenicians, with whom some of the Israelite kings intermarried.) The failure of the northern kingdom to establish and maintain a single ruling dynasty also thwarted the attainment of national cohesion, especially as the crown was ever under threat of usurpation by rival aspirants. And when the Israelites were exiled by the Assyrians, they apparently were not resettled in one district but were scattered over distant and dissimilar areas, so their separated groups lacked the critical mass to constitute an autonomous cultural community. Their own land, in turn, had been repopulated with a mixture of alien peoples who were brought in from afar. Finally, their prolonged period of exile (having been driven from their land about 130 years, some five or six generations, earlier than the Judeans) made it difficult for them to maintain their distinct identity.

In comparison, the southern kingdom of Judah was a homogeneous nation that grew primarily out of the single and dominant tribe of Judah, which had completely absorbed the smaller tribe of Simeon as well as parts of the tribe of Levi (engaged in the service of the Temple) and of the decimated neighboring tribe of Benjamin. The Judean nation was ruled continuously by a single and indisputable dynasty, the tradition-honored and religiously sanctioned House of David. The royal court held sway in Jerusalem, which was also associated with the centralized religious life of the nation, focused in that city and, specifically, in the Holy Temple within it. Thus the secular and sacred authorities generally reinforced each other. The Babylonians evidently settled the exiled Judeans in several centers within contiguous districts, which allowed them to maintain and even develop their distinct cultural life, inspired and encouraged by the eloquent prophecies of Deutero-Isaiah and Ezekiel. The Judeans had been in exile for less than fifty years before they were offered the opportunity to return to their own country, so the elders among them could recollect and describe the beloved ancestral homeland, keeping the collective memory of it vivid. And their return from exile was feasible because the land had not, in the interim, been settled by strangers, but still supported a residual population of the exiles' national kin and coreligionists (the rural folk who had never been expelled).

In any case, the period of the Babylonian exile was not merely a passive hiatus. It was, in fact, a time of intense cultural ferment, collective soul-searching to understand the reason for the misfortune (which the prophets considered to be Yahweh's punishment for the people's misdeeds), and reaffirmation of identity and faith. Severance from the Temple and its sacrificial rites brought about greater emphasis on the spiritual and ethical aspects of the religion: on observing the Sabbath and the dietary laws, on praying and confessing (Daniel 9:3–19; Ezra 8:21, 10:1; Nehemiah 1:4–11), and on fasting to expiate the sins that had brought on the tragic expulsion (Zechariah 7:3, 5, 8:19). It required the faithful to rely on the written text as the depository and fount of the collective memory and aspiration. Thus the experience of exile provided the motivation to collect and select and correlate and interpret the old scrolls, to compose new ones, and to draw meaning from the assemblage.[14] All this led to the advent of a new form of communal worship: the synagogue.

In exile, the old scrolls were cherished and sanctified. The compendium of these scrolls (the Bible-in-the-making, which was, literally, a book of books) served, in effect, as a portable Temple. Reciting and studying and interpreting it became the major form of religious devotion. In it, especially in the accounts of earlier tribulations and redemptions, the exiles found solace and hope. It helped them to define and keep their identity, to reaffirm their sense of destiny. Thus was begun the process that was brought to closure only after the return to Jerusalem: the completion and canonization of the Hebrew Bible. That work of redaction was conducted very probably by, or under the guidance of, the foremost spiritual leader of the returnees: Ezra the Scribe.[15]

THE EVOLUTION OF HEBREW WRITING

Two different systems of writing were developed toward the end of the fourth millennium B.C.E. by the riverine civilizations at the two ends of the Fertile Crescent: Sumer in the east and Egypt in the west. The Sumerian system, later adopted by the Akkadians, Babylonians, and Assyrians, is called cuneiform, from the Latin word *cuneus* (wedge). Pressed into soft clay with a stylus, the original inscriptions were pictographs, which eventually were transformed into abstract symbols as they were turned on their side and written in horizontal lines. The symbols represented objects, words, concepts, and syllables. Some of the symbols are intelligible to even the uninitiated. For example, the rays of a star denoted "heaven" and, by extension, "deity"; a bowl meant "food." Some combinations of symbols are also readily understandable: the pictographs for "great" and "man" were combined to indicate "king."

The Egyptian system is called hieroglyphics, from the Greek words *hieros* (holy) and *gluphe* (carving). Generally written with ink on sheets of papyrus or carved on slabs (stelae) or walls of stone, hieroglyphs originally were pictographs, which, like cuneiform, evolved into symbols. Both cuneiform and hieroglyphics consisted of many hundreds of symbols. Hence the art of writing required long training and thus was an exclusive skill perfected and practiced by a relatively small number of professional scribes.

During the second millennium B.C.E., Mesopotamian cuneiform was adopted throughout the Fertile Crescent (except in Egypt) and was used as the medium of international communication. Even when Egypt had hegemony over Canaan (1450–1100 B.C.E.), Akkadian cuneiform script remained the medium of official communication between the city-states of Canaan and the royal court of Egypt.[1] Evidence of this is the collection of clay tablets discovered at Akhenaten's capital (now called Tel el 'Amarna).

Although Egyptian hieroglyphs were not adopted as such in Canaan, they figured significantly in the development of the Canaanite alphabet. This radically new form of writing apparently resulted from trade and other contacts between Canaanite-speaking Semites and Egyptians. Since the hieroglyphic symbols came to denote single or multiple consonants (as well as ideograms), the practical Canaanites could select from among them the few that were suitable for the phonetic representation of the consonants in their language.

The Canaanite alphabet thus constitutes a great simplification of the Egyptian hieroglyphs. In fact, the alphabet was so simple a form of writing that it could be learned by all—even by children—rather than remain the monopoly of an exclusive priesthood. Of the hundreds of hieroglyphs, just twenty-seven were selected at first, and these were later reduced to twenty-two, many of which still bear a resemblance to the original hieroglyphs from which they were taken. To cite a few examples, *aleph* (𐤀) depicts the head of a bull; *bet* (𐤁), a house; *gimel* (𐤂), a camel; and *dalet* (𐤃), a door.

A site in southern Sinai known as Serabit El Khadem, near the ancient Egyptian turquoise mines, is one place where the so-called proto-Sinaitic alphabet (dated to the eighteenth century B.C.E.) can be observed. Several inscriptions found in the area once known as Canaan represent the proto-Canaanite variant of the early alphabet (dated to the seventeenth century B.C.E.). By the beginning of the first millennium B.C.E., the alphabet was further refined by the Phoenicians, who disseminated it throughout their far-flung colonies. They also transmitted it to the Greeks, who modified it by adding symbols to represent vowels and by reversing the direction of writing: from left to right, rather than the right to left of Hebrew.

After the Babylonian exile, the original Canaanite (proto-Hebrew) script was replaced by the Aramaic (so-called square) script, which is used to this day.[2] Other changes took place as well. In the Hebrew writing of the first millennium B.C.E., no specific symbols were used to designate vowels. Instead, four of the regular letters (*aleph, heh, vav,* and *yud*,

continued

popularly called *ahoi*), known as *imot hakri'a* (Latin, *matres lectiones*), were considered all that was needed for the purpose. However, in post-exilic times, when the original Hebrew was no longer spoken and properly pronounced by all Jews, it became necessary to add special vowel signs to ensure correct pronunciation in the reading of the Torah, which had become an integral part of synagogue worship. Several systems of vocalization were tried until the standard "Masoretic," or "Tiberian,"[3] form was adopted after the ninth century C.E.[4] It is the one that is used in standard renditions of the Bible.

---

1. E. von Dassow, "What the Canaanite Cuneiformists Wrote: Review Article," *Israel Exploration Journal* 53 (2003): 196–217.

2. J. Naveh, *The Beginning of the Alphabet* (Jerusalem: Carta, 1989); A. Yardeni, *The Book of Hebrew Script* (Jerusalem: Carta, 1991); A. Saenz-Badillas, *A History of the Hebrew Language* (Cambridge: Cambridge University Press, 1993); N. Naaman, *The Past that Shapes the Present: The Creation of Biblical Historiography in the Late First Temple Period and After the Downfall* (Jerusalem: Hess, 2002).

3. Several groups of scholars in various locations were engaged in the effort and devised different techniques for phoneticizing the Bible and for standardizing the pronunciation of Hebrew in general. The most influential of these groups was centered in the city of Tiberias, in the Land of Israel. The notation devised by the Tiberian Masoretes (most prominent among whom were several generations of the Ben Asher family) was the most widely adopted and is still considered the standard. We are not certain, however, to what degree their rendition of Hebrew was an accurate interpretation of the Hebrew dialect (or dialects) spoken in biblical times. In modern times, different Jewish communities had their own versions of Hebrew pronunciation, the most important of which were the Sephardic and the Ashkenazic. Present-day Israeli Hebrew was intended to be based on the Sephardic pronunciation, but has evolved to be rather different from what its nineteenth-century initiators intended. Although strongly influenced by the language of the Bible, modern Israeli Hebrew probably would be incomprehensible to any of the characters in the Bible.

4. Still, some words in the Bible are of uncertain vocalization. One example is the second word of Genesis: ברא. It is generally read as *bara*, which casts the verb in the past tense (created)—as in the opening statement: "In the beginning Elohim created heaven and earth, and the earth was unformed and void." The alternative vocalization, which seems more plausible, is *bro*, which changes the sense of the statement to "When Elohim began to create heaven and earth, the earth was unformed and void."

The cultural development that took place in exile was not at all hermetically sealed. As had been their way throughout their many earlier wanderings, the Hebrews (or, rather, the remnant that came to be known as the Yehudim: Judeans or Jews) continued to absorb elements of culture and lore from their old-new Mesopotamian environment and to infuse additional components into their own special synthesis. Thus, for example, they adopted the Aramaic language (the region's lingua franca), which both supplemented and partly merged into their ancestral Hebrew,[16] and some of them adopted Mesopotamian names (such as Zerubbabel).[17] Earlier Mesopotamian influences (including myths and laws) were reinterpreted and selectively incorporated into Judaism's complex, ever-evolving religious culture, which became more elaborate and profound as it interacted repeatedly or continually with a panoply of other cultures over the generations.

The renewal and distillation of the Yahwistic religion that took place in exile assumed a kind of puritanical exclusivity. One manifestation of it after the return to Jerusalem was the rejection of the neighboring territory's "other" inhabitants (particularly the Samaritans) who offered to join in the reconstruction of the Temple (Ezra 4:1–3).[18] An even more extreme manifestation was the draconian repudiation of the foreign wives whom some returnees had married (Ezra 10:1–3, 10–12).[19]

THE PORTENT or premonition of exile seems to pervade the Bible, which is replete with actual or metaphorical allusions to it. The expulsion of Adam and Eve from Eden is the archetype. Other examples are the banishment of Cain and his consignment to vagabondage, the Great Flood, the scattering of the builders of the Tower of Babel, the eviction of Hagar and Ishmael, the escape of Jacob from the wrath of his brother, the "descent" of Jacob's clan into Egypt and its enslavement there, the extended wanderings in the desert of the Sinai Peninsula, and, finally, the deportations from Samaria and Jerusalem. In each case, exile was perceived as a divine punishment for misdeeds, generally of disobedience to God. But the fear of impending doom was always tempered by an equally powerful belief in the possibility of restoration and fulfillment.

Without the experience of the Babylonian exile and the return to Jerusalem, it is entirely possible that the scrolls written over the centuries would have remained separate from one another and perhaps been lost entirely. Preserving and reconciling the various scrolls and integrating them into what later came to be known as the Hebrew Bible was made necessary by the exile and was initiated during it. That task was finally consummated by religious leaders who had returned from the exile. Reciprocally, the Bible has contributed to the survival and vitality of the Jewish people through the subsequent exiles of their history and to the resurgent faith that redemption is possible. Disasters are only temporary setbacks: the metaphoric dry bones of Ezekiel's stunning vision can indeed rise from the valley of the shadow of death and spring to life again:

> Thus said Adonai Yahweh to these bones: "I will cause breath to enter you and you shall live again. . . . And you shall know that I am Yahweh." Then the breath entered them, and they came to life and stood up on their feet, a vast multitude. And He said to me, "O mortal, these bones are the whole House of Israel! They say, 'Our bones are dried up, our hope is gone, we are doomed.' O My people, I shall bring you to the Land of Israel and I will set you upon your own soil."
> (Ezekiel 37:5–14)

11

THE OVERARCHING UNITY

Culmination of Ethical Monotheism

He has told you, O man, what is good,
and what Yahweh requires of you:
only to do justice, and to love goodness,
and to walk modestly with your Elohim;
then will your name achieve wisdom.

MICAH 6:8

Tʜᴇ ʙɪʙʟᴇ proclaims that the unity of God was first vouchsafed to Abraham, was later reaffirmed to his son Isaac and to his grandson Jacob,
and was consequently revealed to Moses in Sinai, along with an elaborate set
of tenets and laws to guide the individual and communal life of the Israelite
people and, ultimately, of all peoples. The clear implication is that the belief
in a single all-powerful, omnipresent God was a singular revelation, a radical
departure from the beliefs of the polytheistic cultures of the ancient Near East,
and that it occurred at a particular place and time and to a particular, chosen
group of people with hardly a precedent.

Contrariwise, a case can be made that monotheism did not arise spontaneously and full fledged in the form of a sudden epiphany,[1] but evolved out of a
background of paganism. According to this view, the notion of a single deity
progressed in stages from polytheism through henotheism to monotheism:
from the deification of disparate physical forces and objects; through their
symbolic representation as idols, the notion of a chief god presiding over all
other gods, the realization of a single God, and the abstraction of the one God
as a disembodied pervasive presence; and, finally, to the identification of God
with the highest values of justice and compassion. Concurrently, an ideological transformation occurred from worshiping a private and exclusively tribal
chieftain to revering an all-embracing, universal, parent figure, and from fearing a fierce, jealous God of Hosts to loving a kind and just exemplar and judge.
That development evidently took place over extended space and time, and it
led step by step to the ultimate perception of God as a pure spirit suffusing the

whole world, manifesting and demanding adherence to standards of moral behavior from all humans. That is to say, the process culminated in the universal principle of ethical monotheism.

Many of the polytheistic elements originally present in the protobiblical scrolls were apparently eliminated by later redactors who gave the composited Bible (especially the Pentateuch) its final form. Those redactors did not necessarily reflect the popular beliefs and mores that prevailed at any particular time (or that changed progressively over time), but were dedicated to the advancement of their own ideology of pure monotheism. Yet numerous traces of the old polytheistic and henotheistic myths, metaphors, and beliefs can be discerned in the Writ as it has been transmitted to us.

Among the remnants of ancient myths preserved in the Bible are the references to animals with the power of speech, such as the serpent in the Garden of Eden (Genesis 3:1) and Balaam's ass (Numbers 22:30); to the birds that supplied Elijah with bread and meat (1 Kings 17:6); to Yahweh's fight with the sea dragons (Isaiah 27:1, 51:9); to Sheol, the kingdom of the dead (Isaiah 14:15; Psalms 49:15, 86:13); and to the morning star that tried to set its throne above that of God and was hurled into the depths (Isaiah 14:12). The Bible retains such allusions to vestigial pagan motifs, but gives them new meanings in the context of faith in the one God who governs all phenomena.

One of the most telltale signs of the evolution of monotheism is the number of names assigned to God in the Bible. That multiplicity is obscured somewhat in translation, but it stands out clearly in the original Hebrew. Why would a single God require so many different names? Might the contrasting names in the various sections of the Bible have referred originally to separate gods, who only later were merged into one? Such names as El, El-Elyon, El-Tsevaot, Eloha, Elohim, Shaddai, Yah, Yahweh, and Adonai may once have implied differences in nuance or in perceived manifestation of the divine that have long since been forgotten or deliberately obscured. Several of the most frequently mentioned names are cast grammatically in the plural: Elohim seems to mean, literally, "gods"; Adonai means, literally, "my lords"; while Shaddai, whose exact meaning is unknown, also seems to be in the plural mode. Some of those names (for example, El and Elohim) bear a distinct resemblance to names of pagan gods known from Mesopotamian and Canaanite records. Thus they likely were retained as the residual names of several distinct pagan gods who had been worshiped separately before a process of syncretism brought about their conceptual amalgamation, rendering them, in effect, synonymous.

Numerous passages in the Bible suggest the coexistence of an array of gods, all of whom, however, were portrayed as inferior to the Great God. In the first story of creation, Elohim stated: "Let us make an *adam* [earthling, human] in our image, after our likeness" (Genesis 1:26). After Adam and Eve partook

Scholars have long suggested that the various names of God originally represented different gods, who in the Bible were merged into a single entity. The name El—which means "god" in a generic sense, related to *elu*, the Akkadian word for "god"—may have been adopted from a Canaanite and Ugaritic deity of the same name. The word *el* has also been related to *ayil* (ram), implying "leader" or "chief," as well as to *eyal* (strength) (Psalms 22:20, 88:5). Some scholars have suggested that El was the original god of Israel, hence the name Isra-El.[1] In the Bible, the word *elim*, the plural form of *el*, designates the multiplicity of pagan gods. Eloha is a synonym for El. Its plural form, Elohim (gods), may have initially referred to a group of gods and then evolved to connote the head god of the pantheon[2] and, eventually, the composite of all gods: the one God. Another designation for God was 'Elyon, which means "most high" (Genesis 14:18–20; Psalm 82:6).

The name Shaddai first appears in the Bible when Yahweh says to Abraham: "I am El Shaddai" (Genesis 17:1). It is subsequently used on forty-eight other occasions. Of all the names assigned to God in the Bible, Shaddai is one of the most enigmatic. Its etymology is uncertain. It may imply "to overpower" and hence is rendered "Almighty" (Genesis 48:3; Exodus 6:3).[3] Some scholars compare it with the Akkadian word *shaddu*, which means "mountain" or "field," suggesting that Shaddai was a nature god. Still others connect it with *shad*, which means "breast" and connotes "provider, sustainer." Still others derive it from the Aramaic word *shda*, which implies "hurler of lightning and thunder," or from the Arabic term *sadda*, which signifies "heaper of benefits." In any case, it appears to be a name and a concept specific to the pastoral tradition.

The term El Tsevaot is not really a name, but a title, apparently derived from the word *tsava*, which means "assemblage" or "host." More specifically, it can refer to a military force, an army. Hence El Tsevaot (Lord of hosts) means, figuratively, "god of armies" or—by implication—"god of war."

Another name for God is Adon (master) or, more commonly, Adonai (my masters). It appears on numerous occasions, often in association with the explicit name Yahweh. (In reading the Bible aloud, Jews customarily say Adonai, Hashem [The Name], or Ado-shem in place of Yahweh, which is considered ineffable—too sacred to be pronounced.) Yahweh or Jehovah (vowels were absent in the original Scriptures, the name being written with the consonants YHWH or YHVH) is considered to be the ultimate name of the God of the Hebrews. The cult of Yahweh may have originated in the southern territories of Edom, Teman, and Sinai. In her triumphal ode, Deborah sang:

Yahweh, when you came forth from Seir,
advanced from the country of Edom,
the earth trembled,
the heavens dripped,
yea the clouds dripped water,
the mountains quaked, before Yahweh,
Him of Sinai, before Yahweh, the Elohim of Israel. (Judges 5:4–5)

Yahweh originally may have been a desert god, responsible for producing sudden bursts of rain and flash floods in an otherwise dry domain. Over time, the other gods were subsumed into his name.

Although capable of appearing in various forms, Yahweh was perceived to be a formless spirit. The letters YHWH or YHVH in Hebrew seem to suggest a composite of the three tenses of the third-person-singular verb "to be": *haya* (was), *hoveh* (is), and *yihyeh* (shall be). When Moses first encountered God in the burning bush, he asked for God's name and was given the answer: Eheyeh-Asher-Eheyeh (Exodus 3:13–14), which means "I am what I am" or "I shall be whatever I shall be." Implied in that enigmatic appellation is a pervasive presence that is not a person and hence needs no "personal" name.[4]

The names used for God have served as important clues in identifying the four major sources that make up the Pentateuch. The oldest of them is believed to be the Yahwist source, designated J (for Jahweh, the German spelling of Yahweh), which uses the name Yahweh.[5] It seems to have been written in the ninth century b.c.e. in the southern kingdom of Judah. The Elohist source, designated E, uses the name Elohim and seems to be concerned mainly with the northern tribes of the kingdom of Israel. It probably was composed somewhat later than the J source, and it presents material that parallels or supplements (but occasionally contradicts) the narrative of J. The Priestly source, designated P, uses the name El Shaddai and is replete with descriptions of ritual. It may date to the sixth century b.c.e. The fourth is the Deuteronomic source, designated D, which reprises, summarizes, and redefines essential material from the first four books of the Bible. It seems to have been composed during the later years of the kingdom of Judah, in the seventh or even the sixth century b.c.e. All these sources were collated, correlated, and edited by redactors between the sixth and fourth centuries b.c.e., partly in Babylonia and partly after the return of the exiles to Jerusalem.

1. M. S. Smith, *The Origins of Biblical Monotheism: Israel's Polytheistic Background and the Ugaritic Texts* (New York: Oxford University Press, 2001); K. Van der Toorn, B. Becking, and P. W. van der Horst, eds., *Dictionary of Deities and Demons in the Bible* (Leiden: Brill, 1999).

2. The Egyptian pantheon, for example, included a panoply of deities, each ruling over a specified function or realm: river, sky, Earth, moon, sun, resurrection. Some were superior to others in the hierarchy of the gods, but none (except perhaps the sun disk, Aten, worshiped by the dissident pharaoh, Akhenaten) excluded other gods. Hence the notion of a sole and ethereal god who was not one of the forces of nature but the one force of all nature, who chose an alien tribe of slaves and commanded them to follow him into the desert, would have been inconceivable to the Egyptians.

3. E. Klein, *A Comprehensive Etymological Dictionary of the Hebrew Language* (Jerusalem: Carta, 1987), 641.

4. J. M. Hoffman has postulated that the Tetragrammaton—God's most sacred name, called Ha-Shem ha-mefurash (explicit name)—was never meant to be pronounced (*In the Beginning: A Short History of the Hebrew Language* [New York: New York University Press, 2004]). That supposition is contradicted by the fact that the major part of the name (Yahu) was included in such men's names as Eli-yahu (my El is Yahu, transliterated rather poorly as Elijah), Yish'a-yahu (Yahu shall redeem, transliterated as Isaiah), and Yeho-shu'a (Yahu shall redeem, transliterated as Joshua, rather than Isaiah, because of the reversed word order). These and numerous other names that included an explicit reference to Yahweh were pronounced freely in everyday life.

5. J. Wellhausen, *Prolegomena to the History of Ancient Israel* (1885; reprint, New York: Meridian Books, 1957); R. E. Friedman, *Who Wrote the Bible?* (New York: Summit Books, 1987); T. Hiebert, *The Yahwist's Landscape: Nature and Religion in Early Israel* (New York: Oxford University Press, 1996); N. Naaman, *The Past that Shapes the Present: The Creation of Biblical Historiography in the Late First Temple Period and After the Downfall* (Jerusalem: Hess, 2002).

of the forbidden fruit of the tree of the knowledge of good and evil, Yahweh Elohim (note the change in God's name) continued in the same vein: "Now that the *adam* has become like one of us, knowing good and bad, what if he should stretch out his hand and take also from the tree of life and eat, and live forever!" (Genesis 3:22). Implied in these passages is the notion that Elohim or Yahweh Elohim was accompanied by a retinue of lesser gods, with whom he would speak or consult from time to time.

As though fearful of the humans' potential to assume God-like qualities, Yahweh Elohim banished them from the Garden of Eden, consigning them to till the soil out of which they had originally been taken. Then he "stationed east of the Garden of Eden the cherubim [plural of the Hebrew *kruv*, transliterated as "cherub"], and the fiery ever-turning sword, to guard the way to the tree of life" (Genesis 3:24). So to whom was Yahweh Elohim speaking when he created humans "in our image, after our likeness"? Alternatively, could he have been using the "royal we" merely as a figure of speech. But, then, mention of the cherubim clearly indicates that he had angels (or subservient lesser gods) at his service.

Among the numerous references, direct or oblique, to the presence of a number of lesser gods or of divine angels (messengers of Yahweh or Elohim, disguised as humans) are the angel who met Hagar in the desert (Genesis 16:7, 21:17–18), the three (who later, somehow, became two) angels who visited Abraham (Genesis 18) and then went down to smite the iniquitous cities of Sodom and Gomorrah (Genesis 19), the angels who encountered Jacob at Gilead (Genesis 32:3) and the one who struggled with him at the crossing of the Jordan (Genesis 32:25–31), the angel who appeared to Moses in a blazing fire in the wilderness (Exodus 3:2), the angel who confronted Balaam and his donkey (Numbers 22:22–31), the angel who met Joshua in Jericho (Joshua 5:13–15), the angel who recruited Gideon (Judges 6:11–23), the angel who foretold the birth of Samson (Judges 13:3–7), the angel who was sent to chastise David for having counted the people of Israel (2 Samuel 24:16–17), and the angel who summoned Elijah (2 Kings 1:3).

The imagined existence of a council of "sons of the Elohim" is portrayed in the Book of Job (which is a parable questioning the faith in God in a situation of undeserved misfortune): "One day the sons of the Elohim presented themselves before Yahweh, and the satan came along with them" (Job 1:6, 2:1). Obviously, the picture drawn in Job is merely metaphoric. But it seems to evoke an ancient tradition that accepted as reality the existence and the occasional appearance and interference of god-like figures (either benevolent or malevolent) in human life. That tradition, in turn, was a modified version of a still earlier one that had recognized the coexistence of many gods, all engaged in complex associations that could be either complementary or competitive. Coordinating all such gods and subjecting them to the authority of a supreme god—the first among equals—as children under the authority of a father, was an intermediate step between polytheism and monotheism.[2]

Other remnants of a more primitive depiction of God, endowing him with anthropomorphic traits, abound in the early books of the Bible. Adam and Eve heard the "sound of Yahweh Elohim moving about in the garden at the breezy time of the day" (Genesis 3:8) and hid from him, whereupon he "called out to the man and said to him, 'Where are you?'" (Genesis 3:9). After the Flood, Yahweh "smelled the pleasing odor" of Noah's burnt offering (of sacrificed animals) and was so pleased that he "said in his heart: 'Never again will I doom the earth because of the *adam*'" (Genesis 8:21). On another occasion, "Yahweh came down to look at the city and tower that the children of the *adam* had built," and he was displeased. He then said, again speaking in the plural to unspecified listeners: "Let us, then, go down and confound their speech there, so that they shall not understand one another's speech" (Genesis 11:7). Even more curious is the episode in which Moses, on his way to Egypt, approached a lodging place, and there "Yahweh met him, and sought to kill him" (Exodus 4:24). Then there is the incident of Yahweh revealing himself bodily to Moses, but only partially: "He [Moses] said: 'Oh, let me behold Your Presence!' . . . And He said, 'You cannot see My face, for man may not see Me and live . . . [but] you will see My back'" (Exodus 33:18, 20–23). At other times, Yahweh became furious and acted impetuously, and Moses implored him to be more compassionate.

An even more startling passage reveals that the Elohim actually had an unspecified number of sons, who had distinctly human lusts: "When men began to increase on earth and daughters were born to them, the sons of the Elohim saw how beautiful the daughters of men were and took wives from among those that pleased them. . . . It was then, and later too, that the Nephilim appeared on earth—when the sons of the Elohim cohabited with the daughters of men, who bore them offspring. They were the heroes of old, the men of renown" (Genesis 6:1–4). The notion that gods could mate with humans to produce demigods with superhuman qualities was prevalent in pagan societies. Such semidivine origin was claimed by some of the rulers in Mesopotamia and Egypt (and, later, some of the emperors in Rome). Demigods abound in Greek mythology, and the lingering belief in them may well have preconditioned people in the Hellenized Roman Empire during the early centuries of the common era to accept the divinity of Jesus.

The passage about the sons of God raises an interesting question: How did the Elohim beget sons? Of course, he could have ordered them into being by verbal command, as he did with the *adam* in the first story of creation (Genesis 1:26). Alternatively, he could have fashioned them from clay, as he did with the *adam* in the second story of creation (Genesis 2:7). But there is a third possibility: he may have had a female consort![3] Although the very thought seems preposterous or even blasphemous, there is evidence for it. Some years ago, the Israeli archaeologist Zeev Meshel discovered an inscription at a site in eastern Sinai that may have been an outpost of the northern kingdom of Israel.[4] The inscription explicitly mentions, and even illustrates, "Yahweh of Samaria and his Ashera." The notion of Yahweh's possible liaison with a female mate was banished from the Hebrew Bible, which grew out of the southern, Judean, offshoot of the Israelite tradition. The goddess Ashera, as the long-time counterpart of the god Ba'al in Canaanite tradition, was regarded as an abomination by the prophets of Yahweh.[5]

The persistent lure of paganism was due to the tenuous circumstances of life in Canaan. To many of the early Israelites, the formless Yahweh may have seemed too stern, formidable, and distant, as well as practically impossible to please. Thus it was natural for people to try other types of religious practice. The rites of the indigenous cults, designed to propitiate relatively tangible gods, offered an additional or alternative way, an often more gratifying way, to appeal to those mysterious forces that were believed to control events. There was a sensuality and joy attending the pagan rites that evidently was lacking in the austerely authoritarian practice of Yahweh worship. But when either Yahweh rites or pagan rites (or both) failed to improve their lives, the people felt doubly punished—physically by the drought or pestilence or sword that was threatening them, and spiritually by the guilt cast on them by their prophets for having caused the trouble by their failure to follow the true path of Yahweh. Only gradually did the Israelites come to accept Yahweh alone.

Clearly, the notion of God, initially anthropomorphic or theriomorphic (possessing animal form), underwent a gradual change, corresponding to the needs and perceptions of his adherents. Because they needed rain, God had to be a provider of rain. Because they needed protection against powerful enemies, God had to be a powerful warrior. Beyond material needs, however, a nation in a constant state of anxiety had spiritual needs for solace, emotional assurance, faith in the future—a sense of meaning and mission and destiny. Such needs eventually brought about the sublimation of the very notion of God, the ultimate result of which was the projection of an entirely abstract or ethereal spirit, without form or material substance, but with total power over nature and the lives of humans.

God was perceived as having a dual nature: compassionate, forgiving, and loving, on the one hand, but strict, harsh, and punitive, on the other: "Yahweh, Yahweh, El merciful and gracious, patient and abundant in goodness and truth; keeping mercy unto the thousandth generation, forgiving iniquity and transgression and sin, that will by no means clear the guilty; visiting the iniquity of the fathers upon the children and upon the children's children, unto the third generations" (Exodus 34:6–7).

Even the modes of worship, initially predicated on the physical presence of a god who savored burned flesh, were transformed. In time, the prophets railed against the ghastly practices of slaughtering animals and sprinkling their blood performed by the hereditary priests. These rituals had become an abomination because they had been used as a substitute for, and an evasion of, the truly important duties of upholding justice and practicing charity. Justice and charity were abstract concepts, to be sure, but consistent with the ideological evolution toward faith in a transcendent God of righteousness and mercy—that is, toward ethical and compassionate monotheism.[6]

The concept of one God did not take place in a vacuum. It seems unlikely that an individual could have conceived of monotheism in a sudden flash of enlightenment without precedent. Nor does it seem likely that a council of wise elders could have conjured it up in closed session and then foisted it on a reluctant nation. Rather, it must have arisen gradually out of the cumulative experience of generations who lived and struggled throughout an ecologically and culturally varied region and had to adjust—physically, ideologically, and spiritually—to changing circumstances in different domains. Only from that background of initially inchoate notions could any person or persons of special insight have been able to extract an essential realization and to articulate it in a way that convincingly expressed the shared perceptions and aspirations of the nation.

The Israelites whose culture is reflected in the Bible did not live in isolation, but in constant interaction—both synergistic and antagonistic—with the cultures of their neighbors. They absorbed all the cultural strands of the period from all the ecological domains of the ancient Near East that they experienced and assimilated them into their own culture. The tendency to abstraction is a characteristic of a society in flux, not bound to a stable regimen or rigid custom. The originality of the Israelites lay not in their creation of cultural elements, but in their recomposition of them to synthesize and distill a new culture, a new faith, elevated to ever-higher ideological, ethical, and humane levels. Central to their culture was the concept of a single and singular God, whom they had invoked in their minds out of necessity rather than out of choice, since no existing pagan god provided them with the intimacy and the sense of power that could begin to answer their emotional needs and their intellectual comprehensions. Hence they conceived of a God so complete and perfect and devoted that they could depend on him to provide for them[7]—on the condition, of course, that they would obey his commandments. If he did not provide, that could only be due to their own inadequacy, which they had to correct.

Polytheistic societies were, in general, rather tolerant of one another's beliefs. Since they all worshiped multiple gods in their own domains, they were in principle willing to accept the possible existence of other gods in other domains. Indeed, the gods of one polytheistic society often resembled those of another, notwithstanding the differences in names, functions, or relative rankings of the similar gods in the hierarchies of the societies. Societies often accepted the gods of their neighbors into their own pantheons, albeit at a level lower than that of their indigenous gods. Not so was the attitude of the monotheistic purists among the Israelites. Having committed their faith to the one and only Yahweh, they rejected absolutely the existence of any other god or gods. Hence the Israelite monotheists, especially the biblical prophets, were fundamentally and totally intolerant of other religions and of those among their coreligionists who tended to hedge their commitment to Yahweh by retaining vestiges of polytheistic beliefs and practices. The prophet Isaiah, for example, heaped scorn and ridicule on the makers and worshipers of idols:

> Those who squander gold from the purse
> and weigh out silver on the balance,
> they hire a metalsmith to make it into a god,
> to which they bow down and prostrate themselves.
> They must carry it on their backs and transport it;
> when they put it down, it stands,
> it does not budge from its place.
> If they cry out to it, it does not answer.
> It cannot save them from their distress. (Isaiah 46:6–7)

In time, even some of the polytheistic religions underwent a process of syncretism, by which various gods were integrated in the direction of monolatry: the worship of one god while recognizing the possible or actual existence of other gods.

OTHERS BEFORE the Israelites had had inklings of the unity of nature and its creator, but none had elevated the one God to such pure abstraction and ethical standards as had the Israelite monotheists. And none had so tightly and exclusively committed their faith and life to that unique God, whom they called Yahweh, but whose very name they came to regard with such reverence and awe as to constrain them from even pronouncing it aloud.

Again and again, the fundamental paradox posed itself: God made humans as they are. Humans are imperfect, prone to transgress, "for the devisings of man's mind are evil from his youth" (Genesis 8:21). The Israelites (or their heirs, the Judeans or Jews) were assigned—or took on themselves—the major burden of accepting and implementing God's commandments. Repeatedly, it seems, they fell short of that goal and were punished for their weaknesses. What was it, then, in their condition and mind-set that induced them to continue, even to intensify, their belief in their intangible God and their own special destiny as his chosen people?[8]

The basic reason for their troubles lay in the environment where, by historical coincidence, the Israelites staked out their life as a nation. It is a land of unstable climate, at the edge of the desert, with some years or succession of years blessedly rainy and others accursedly dry. Why would a land so humid, lush, and fertile turn suddenly so sere and barren? Surely, the reversal could not be random. The very thought that phenomena may be random contradicts the fundamental premise that all objects and events are controlled, even caused, by a single supreme being. All things emanate from him, except the evil impulse, which comes from humans. They are like impulsive children, foolishly and brazenly disobeying their parent. Hence they must be scourged.

The land became a sort of moral seismograph, an indicator of the nation's collective behavior. Its manifestations were to be watched at all times for telltale signs of the return of the desert. The fluctuating polarity of the environment imposed a kind of dialectic mind-set on its inhabitants. To the Israelites, it was not merely the land that might revert, but—in a symbolic way—they themselves. Returning to the desert meant retrogressing to the utter destitution and homelessness that had been their condition before entering the promised land, even before Moses granted them the Law and declared them to be God's chosen people. In a way, the title to the land that had been given to Abraham and reaffirmed to Moses and Joshua placed the Israelites in a permanent state of probation, requiring them to be morally sensitive, as well as physically insecure, at all times. The labile nature of the land amplified itself in the consciousness of the people living on it.

Being at the crossroads of continents, the land could never be isolated from the swirling events of the region. The enormous difficulty of maintaining independence and of adhering to a unique identity and culture in the midst (and in defiance) of the nations that surrounded them or the militarily dominant powers that periodically encroached on them understandably tempted some of the Israelites to forgo their separateness.

Those who assimilated—over the centuries, they were numerous—were not those who wrote and sanctified and adhered to the Hebrew Bible. Only those willing to defy expedience and to affirm the uniqueness of their national-religious identity and destiny remained faithful to the Bible and its tenets. The wonderment is that enough of the people elected to remain within the fold, despite the vicissitudes, to keep the national and religious identity alive through all the changes of environment and culture that occurred over the centuries.

The real situation of the Israelites as a nation in Canaan, and certainly in exile, seemed at times to be desperate. The easy way to escape that predicament would be to give up their national and religious distinctness, to submit to whatever temporal power held sway at any time, and thus to lose their identity (as eventually did most of the region's peoples). The only means that remained for avoiding such cultural self-negation was to detach mentally from the difficulties of the moment and pin hopes on a miraculous, "end of days" redemption (stated with the greatest eloquence in Isaiah 2, 9, and 11), in which perfection would reign at last. So luminous and inspiring was this vision that a core of believers among the Jews could continue to live by its light for many centuries, even though they were cut off from their land and subject to cruel persecution. With this attitude, the situation of the people, although repeatedly disastrous, could never be hopeless. The desire to believe can be so strong as to banish doubt. The faithful believed that deliverance was always attainable and that it depended on their own acts. All they had to do was—in the ritual words intoned on their most solemn holiday, the Day of Atonement—practice repentance, charity, and prayer. Thus, paradoxically, the greater their troubles, the stronger was the Israelites' inducement to adhere, ever more tenaciously, to the faith. In dire circumstances, only absolute faith could provide hope, which, in turn, gave the faithful the strength to endure.

If Yahweh is mighty and just, and if the Israelites were his chosen people, whom he had redeemed from slavery and led into the promised land, why should they have continued to suffer so much misfortune, deprivation, humiliation, and subjugation? The very existence of evil in a world created by a perfect God is baffling, and it is even harder to understand why it should have befallen the very people who had accepted him as the only God.

There can be but two possible explanations. One is that the very concept of God is wrong. That is to say, God either does not exist as perceived or is not all-powerful, compassionate and just, or concerned about the Israelites. To accept any part of that proposition would be to lose hope, purpose, and direction—indeed, to be left alone and helpless in a cruel world. The other explanation may seem difficult to accept and even more difficult to apply, yet is more hopeful. Misfortune cannot be random, since God directs all events. Nor was it an inevitable consequence of the Israelites' situation as a small nation clinging to a location and situation that were inherently precarious. Rather, the troubles that befell them must have been God's doing, either to punish them for their transgressions[9] or to test their faith. Being God's chosen entailed an awesome responsibility and a moral burden:

> You alone have I singled out
> of all the families of the earth,
> that is why I will call you to account
> for all your iniquities. (Amos 3:12)

If God punished the Israelites, it must have been because he cared for them and wished them to mend their ways. That reasoning may seem paradoxical, but it made perfect sense to the religiously committed; the self-assumption of guilt requiring repentance became a source of hope: if wrongful behavior caused misfortune, then, surely, rightful behavior will bring good fortune.[10]

The second explanation was first promoted by the major Hebrew prophets, who castigated the people for having committed all manner of sins—in short, for "doing what was evil in the sight of Yahweh." Although rejected at first by most members of the ruling class (most especially by the priests, who strove to maintain their own monopoly over the Temple-based cultic practices) as the ravings of madmen, the teachings of those prophets also held a positive message of conditional redemption. Indeed, they were eventually accepted by the compilers of the biblical text and included in the canonized version as divine revelations.[11]

The idea that people, knowing right from wrong (once having partaken, symbolically, of the fruit of the tree of the knowledge of good and evil), are responsible for their fate and can affect it by their behavior in the light of Yahweh's commandments vested the believers with a sense of power. Instead of wallowing in self-pitying victimhood or surrendering in the face of seemingly overwhelming odds, people of faith always had something they could do—try ever harder to please their God. That attitude granted to the believers a measure of free choice, including the possibility of resisting the dictates of arbitrary authority. God can and will redeem those who make the right choice, just as he once redeemed the Israelites from bondage in Egypt. That seminal liberation, as the Bible reminds the faithful repeatedly, was not achieved by the people directly, but was accorded to them by Yahweh conditionally, provided that they obey—and continue to obey—his commandments. Adherence to Yahweh could not remain a merely abstract belief, but must be actualized through performance.

THE TRAUMATIC EVENT that brought the evolving faith of the Judeans into sharp focus was the destruction of the First Temple and the exile from Jerusalem in the sixth century B.C.E. That enormous misfortune might have led to the total abandonment of the faith in Yahweh and therefore to the dissolution of Judaic culture and nationhood (as, indeed, happened to the Israelites from the northern kingdom, who had been expelled from Samaria by the Assyrians more than a century earlier). Instead, after the destruction of its sister kingdom of Israel, the kingdom of Judah, under the leadership of Josiah, rededicated itself to the worship of Yahweh even more strictly. Nevertheless, Josiah was killed in a battle with the Egyptians, and, eventually, all of Jerusalem was destroyed by the Babylonians.

A haunting question posed itself inescapably to the surviving victims of that debacle: Why was the life of their nation so cruelly disrupted? Might the entire basis of their national culture, as it was intimately associated with their national God, Yahweh, have been wrong? The righteousness of Yahweh must be beyond question! So why had Yahweh abandoned them? It must be that they—despite their efforts—still failed him. What more could they have done, or could still do, to win the approval of their demanding, yet potentially beneficent, God?

The Bible was compiled, edited, and canonized in the wake of the Babylonian exile. It was an act of national and cultural self-preservation. Even after the Babylonians were conquered by the Persians, the exiled Judeans could have assimilated into the culture of the Zoroastrian Persians, whose dualistic belief included a force of evil that is in constant struggle with the force of good. The existence of an evil force could have provided a convenient explanation for the Israelites' misfortune and absolved them of self-blame for it. Instead, the exiles were exhorted by their spiritual leaders to adhere to the faith in Yahweh as the single Almighty God, who had punished them for their own iniquity and would be prepared to forgive and redeem them—if only they would mend their ways. And, as they always thought of themselves as a nation descended from a single tribe, even a single family, they could not imagine salvation in any but a collective, national context. The Bible thus connected the Jews with their past and directed them toward their destined future. It served, in effect, as a bridge of parchment that linked the generations.

Although created in God's image, people are repeatedly tempted to do evil and must therefore be exhorted to pursue its opposite, which is compassionate justice. The prophets urged the well-to-do to care for the disadvantaged members of society, including the strangers in their midst, the orphans, and the widows. Farmers were instructed to leave the corners of their fields uncut and to allow the poor to glean in their fields after the harvest (Leviticus 23:22). Paying the wages of a day laborer was not to be delayed (Deuteronomy 24:14–15). Lending to the poor was to be done unhesitatingly (Deuteronomy 15:7–11) and without interest (Exodus 22:24; Deuteronomy 23:20–21).

Loss of faith in the veracity, or even the existence, of Yahweh and his special relationship to the people of Israel might have weakened the sense of national destiny, even of national identity. A rationale had to be found for the failure, so far, to fulfill the utopian promises, presumed to have been made long before to the Patriarchs and repeated ever since to the prophets, of a secure life in a land flowing with milk and honey, of becoming a light unto the nations, of a future with perfect justice and harmony. The fault could not be that of Yahweh, who was perfect by definition. Rather, the fault must lie within the people themselves.

What, then, might they have done wrong? The question was not *if* they had done wrong, for that was axiomatic from the very fact that they were in trouble. That quandary led to a tendency toward constant introspection.

But there was also a positive side: if they had not been good enough in the past to deserve the material bounty and spiritual comfort that omnipotent Yahweh alone could provide, then they could make themselves more deserving in the present and the future. How? By being, and doing, better. At first—long ago—the traditional way to placate the Almighty had been through the pagan ritual of offering animal, or even child, sacrifices, administered by priests. But in time, that gory practice seemed less and less efficacious, as the image of the Deity became increasingly ethereal. The prophets railed against sacrifice as being worse than irrelevant, a hypocritical evasion of the real task, which they perceived to be the pursuit of justice and charity. In the words of Isaiah:

"What need have I of all your sacrifices?" says Yahweh.
"I am sated with burnt offerings of rams
and suet of fatlings,
and blood of bulls;
and I have no delight in lambs and he-goats. . . .
Trample My courts no more. . . .
Your hands are stained with crime—
wash yourselves clean. . . .
Cease to do evil.
Learn to do good.
Devote yourselves to justice.
Aid the wronged.
Uphold the rights of the orphan.
Defend the cause of the widow." (Isaiah 1:11–17)

Jeremiah, himself a priest, also denounced forms of pagan idolatry that had been insinuated into the practice of Judaism. Other prophets extended and reinforced that theme. Typical is the statement by Micah:

What Yahweh requires of you:
only to do justice, and to love goodness,
and to walk modestly with your Elohim. (Micah 6:8)

The only alternative to self-negation and loss of identity was a resolute adherence to faith in the God-ordained destiny of the nation. If God seemed to abandon his support, then something could and must be done to restore it. That something was sought within, through self-examination, acceptance of individual and communal responsibility, admission of faults, atonement, and rectification of wrongs. Thus did monotheism acquire its ethical dimension and become democratized.

SOME SECULAR CRITICS may regard the religious fervor of the Hebrew prophets as a delusion, an escape from harsh reality, a yearning for a magical dream world of perfect justice and harmony. As reality grew more threatening, the refusal to face it intensified. In some individuals, the mental withdrawal became extreme, to the point of aberration. Bands of ecstatic, God-intoxicated prophets couched their visions in sublimely poetic language, which they attributed to God himself and by which they attempted to captivate and enrapture their audiences. In their extreme manifestations, the prophets gave reason to the expression *evil hanavi, meshuga ish haruah* ("The prophet is a fool, / the man of the spirit is mad!" [Hosea 9:7]). To the extent that the religiously absorbed could not ignore reality, they interpreted it to fit their basic premises, which were absolute and inviolable. Yet in the way they offered hope to the hopeless, dignity to the downtrodden, and an ideal of beauty and perfection to an imperfect world, they inspired people to strive for their own betterment.

The prophets were not entirely divorced from the here and now. Some of them were exceedingly wise and insightful. From time to time, however, they would engage in flights of fancy, disengage from the issues of the day, to attain a more encompassing view of the present, past, and future and of universal as well as local concerns.

In time, the very notion of God was distilled and internalized, transformed from a formidable and demanding power to be feared and blindly obeyed into an inspiring human ideal of perfect harmony, equanimity, certitude, justice, and mercy to be consciously and willingly emulated. The physical taskmaster without became the spiritual guide within, a source of reassurance and solace. In the words of the prophet:

> Strengthen the hands that are slack,
> make firm the tottering knees!
> Say to the anxious of heart,
> Be strong, fear not;
> behold your Elohim!
> Requital is coming,
> the recompense of Elohim—
> He himself is coming to redeem you. (Isaiah 35:3–4)

In the process, however, Judaism in the Diaspora became increasingly separated from its ecological roots, its attachment to the natural environment. Restoring the precious balance between the spiritual and the earthly is now an essential task for Judaism, as it is for other religions.

EPILOGUE

The Lasting Relevance of Early Ecological Influences

Remember the days of old,
consider the years of ages past.
DEUTERONOMY 32:7

L ITERALLY FROM the words "Go forth" (Genesis 12:1), spoken by a mysterious voice commanding Abram to leave his native land and wander toward an unspecified promised land, the early history of the Israelites, as portrayed in the Bible, is of a nation of seekers that was never confined entirely or for long to a single, distinct environment or realm, but ranged over and spanned all the natural, cultural, and political domains of the ancient Near East. Practically every other nation in the region derived its identity from its particular locale and would have lost its character if detached physically from its place of origin and frame of reference. Not so the Israelites, who—seemingly alone among their contemporaries—established a kind of portable identity. Initially it was the identity of a nomadic tribe rather than of an organized society with a stable mode of life.

The national ethos of the Israelites was imbued with a strong sense of history, ethnicity, and religion. If that ethos was rooted in myth, it was a myth so thoroughly self-inculcated as to be accepted over the generations as consensual reality: the sense of shared ancestry, background of nomadic pastoralism in the semiarid domain, slavery in the riverine domain of Egypt and liberation from it, sojourn and spiritual revelation in the desert domain of Sinai, settlement in the rainfed domain of Canaan, interactions with the cultures of the maritime domain of the Mediterranean, coalescence of tribes into a centrally governed nation-state in the urban domain of Jerusalem, repeated episodes of drought and of invasions by superior military powers, defeat and loss of homeland, survival in exile and subsequent return, and revival of nationhood and faith. The tradition of inconstancy and mobility, misfortune and eventual recovery, found its strongest expression in terms of a creed enshrined and sanctified for all time in a written text. Their unique experience and the creed that arose from it set the Israelites apart from all other nations in the region. In a cognitive sense, their ancient and lasting alienation from established societies made the history of the Israelites and their descendants a tale of more than 3500 years of solitude.

The formative ecological experiences of the ancient Israelites affected their outlook and behavior not only for a time but, indeed, for all time. What began as serial adaptations to shifting circumstances and contrasting environments eventually became an integrated, internalized characteristic, so deeply embedded in their lore as to acquire cultural permanence. Attitudes formed long before were infused into their character as a nation and etched into their collective psyche. Those early influences and lingering memories conditioned, consciously or otherwise, the amazingly persistent beliefs, habits of thought, and behavioral patterns of the Israelites—and of their heirs, the Jews—ever since.

Just as a mighty river is a confluence of many tributaries that cease being separate when they converge and mix their waters, so the Bible is a compilation of many sources that have been brought together to express a singular motif. And just as a tree reaching to the sky continues to draw its sustenance from the soil in which it began to grow as a frail seedling, so the Jews—being heirs to the Israelites who wrote the Bible—continue to draw values and mores from the environmental circumstances that conditioned their culture at its inception. The Israelites have always been a remembering people. The collective experience is at the core of their culture. In their Scriptures and rituals and prayers, they remind themselves endlessly to teach their children about the past, and their children to teach their children in turn: an unbroken chain of remembrance.

One of the most profound and lasting of the historical memories was of the nomadic life led by the early Israelites, the beginnings of which are attributed in the Bible to the Patriarchs. Because the semiarid or arid pastoral domain offered only meager and ephemeral resources of grass and water at any one place, it dictated a lifestyle based on small clans, or tribes, moving from place to place, often in competition with other migrant clans. Since individuals on their own could hardly survive in the semidesert, they needed mutual support and protection. The security of every person therefore depended on affiliation with a tightly knit family group. Necessity fostered a fierce loyalty to family and cohesion within the clan.

Having to move repeatedly across barren land, and—in times of dire necessity—to intrude into occupied greener land, made the pastoralists (in strong contrast to the sedentary farmers of the rainfed domain) strangers and so-journers everywhere they went. It also induced an attitude of estrangement from and distrust of other peoples—an attitude that was reciprocated by those "others"—and a constant, gnawing anxiety born of insecurity. Mutual estrangement and distrust were reinforced as the Israelites evolved a unique religion that set them still farther apart. Alienation and insecurity caused the Israelites much suffering, but also spurred their extraordinary adaptability and cultural creativity.

The need to adjust to ever-changing circumstances in a labile environment required the Israelites to be resourceful and flexible, to improvise one way or another, to avail themselves of shifting opportunities, and to avoid pitfalls. Their nationhood marked its historical origin by a revolt against the authoritarian regime of Egypt. Their mental tendency was therefore to question and resist any fixed authority—except that of their all-knowing God. The Israelites' faith in Yahweh alone and in their role as his chosen people was a source of spiritual comfort for them, but also served to further isolate them and to intensify their confrontations with other nations, as those conflicts became religious as well as territorial. They paid a high price in those struggles to preserve their defiantly unique identity.

The meaning of the concept "God's chosen people" has been argued endlessly. What is incontrovertible is that the Israelites themselves had chosen their God, that their choice made them distinctive, that they suffered greatly in consequence, and that they persevered largely as a result of their steadfast adherence to their distinctiveness.

By the time the Israelites settled in the rainfed domain of Canaan and adapted to farming, and even to life in the urban domain of Jerusalem, their sense of distinct identity and separateness had been fixed, and they were disinclined to blend with the peoples with whom they shared the region. They saw themselves as a nation apart, a people unto themselves—even as they always lived among other nations. (Surely this was not true for all Israelites, many of whom over the centuries left the fold and assimilated into other cultures; however, it was true for enough of them to keep and nurture the national identity.)

Did such tendencies or traits precondition the Jews to adapt to even a life in exile and to endure and survive its enormous hardships? It has long been a puzzle, even a reason for astonishment, that the Jews, uniquely, have managed to persist and to retain their cultural bonds so long after having been driven out of their homeland and scattered around the globe. (Indeed, they have lived far longer in exile—or, rather, in the Diaspora—than in their own homeland) How these heirs to the biblical tradition have defied all historical patterns has baffled numerous historians, none of whose theories can explain the Jews' continuing existence and—more remarkable—enduring cultural vitality and ongoing contributions to civilization. Is there some clue in the ancient history of the people that can help solve this mystery? It would be simplistic to suggest that the nomadic background of their ancestors was *the* major factor that preconditioned the Jews to prevail and to maintain, even to develop, their ancient culture in the hundred or so lands in which they have lived. But it seems reasonable that it was *a* factor, and an important one.

There are hardly any Jewish pastoralists today and not very many farmers. (Even in modern Israel, their numbers are few.) Most of the Jews in the world now live, and have lived for many centuries, in environments far removed from those of their distant progenitors. But the nomadic tradition has persisted, albeit in a very different sense. The stereotype of the Wandering Jew is not entirely fictional. Vestiges of the past, traceable to the formative stages in the development of Israelite culture, are still discernible. Archetypal images pervade Jewish prayers and literature and outlook. A peripatetic people that began its course in a quest for permanence—and, indeed, seemed to find permanence for a time, only to be deprived of it repeatedly—is a people in a state of constant, restless yearning, with a unique talent to resurrect itself after each calamity. A people ridden with self-consciousness and self-doubt, at once skeptical yet hopeful, disillusioned yet idealistic, keeps longing for its promised land while yet able to fit in everywhere (although hardly ever with complete security). Through all the suffering, the cultural heirs of the ancient Israelites have remained attuned to the awe-inspiring yet comforting ancient voice that has repeatedly restored their faith despite all misfortunes and doubts. Their sense of mission and destiny has helped to compensate for the adversity and humiliation that they have had to endure. The very fact of their insecurity has made it impossible for them to sink into complacency and has provided the impetus for their creativity and their determination not merely to survive but, indeed, to overcome and excel.

A characteristic trait with deep roots in the past is the tendency to introspection, to self-examination, leading to ethical judgment of self and others and to a propensity to accept as well as ascribe guilt. The soul-searching was aimed at finding where the people had erred and why they deserved the punishment they had sustained as a small nation in a series of ever-vulnerable domains. They were constantly concerned with the need to please a stern and demanding God who alone could protect from dire threats (or so they desperately wanted to believe, else there would seem to be no hope). The fear of God was exceeded only by the fear of having to live without God. The tendency appears to have had its origin in the tenuous lives of the ancient Israelites at the juncture of continents and on the fringes of the rainfed domain, always beset by uncertainty and danger. The dangers of living in Diaspora in the midst of

Much in the lore of a nation is myth derived from idealized fantasy. Trying to actualize the fantasy may lead to disappointment. Such may have been the fantasy of the promised land, which in actuality turned out to be something less than "a land flowing with milk and honey." Such may have been the Temple on the lofty mount, where the presence of God was presumed to be centered, but that in actuality was turned into an abattoir where bleating animals were slashed, their blood splattered and fat sizzled on the smoking altar. No wonder the prophets were repelled by the ritual and sought to spiritualize the religion. In so doing, though, they fancied so perfect a kingdom of God that no reality on Earth could possibly resemble it. Perhaps the religion would not have survived if the nation had not been banished from the grim reality of that disappointment. Perhaps it survived precisely because it was turned into something surreal, something so idealized as to enter the realm of longing and striving.

The phenomenon of the survival and continuing cultural vitality of the Jews defies simple explanation. It is an enigma that has baffled and eluded observers (both sympathetic and otherwise) for many ages, from the Alexandrian Greek writer Apion[1] in the first century C.E. to the Russian novelist Aleksandr Solzhenitsyn[2] in the twentieth. Prominent among the writers who have grappled with that enigma was the British historian Arnold J. Toynbee, who characterized Jewish culture as a "fossilized" relic of ancient history.[3] Far from being a mere anachronism, however, Jewish culture has been ever dynamic, vital, self-renewing, and influential—imbued and impelled and sustained by a seminal and powerful Bible-derived idea, both particular and universal: that the people of Israel were the earliest recipients and continuing conveyors of the idea and ideal of ethical monotheism. In even the darkest of times, that ideal has offered a promise and posed a challenge to each individual and to the community as a whole.

As outsiders seeking to become insiders, the Jews have tried to adjust to every new environment while retaining their history-rooted identity and special sense of purpose. Living with constant change, having to bear and overcome repeated misfortunes and dislocations, kept them from stagnation or complacency and forced them to continue striving, to draw inspiration and hope and purpose even out of recurrent tragedy. Over the centuries, an untold number of Jews chose or were forced to leave the fold. Some may have remained Jewish out of sheer inertia or lack of choice. However, a sufficient number elected to affirm their Jewish identity and to sustain their communities through all the changes of culture and environment they had undergone. In other words, the survival of the Jews is not a mysterious miracle or destiny. It is, in the final analysis, a matter of positive choice—although never an easy one—to continue the quest. The literary expression of that choice and that quest has been, and continues to be, the monumental work of generations long past—the Bible.

---

1. Apion was a noted writer in his day. He composed the extensive *History of Egypt*, in which he included the derogatory tract "Against the Jews," claiming that the Jews were the descendants of undesirable lepers and misfits who had been expelled from Egypt. In response, the Jewish historian Josephus wrote *Contra Apionem* (*Against Apion*). See *Encyclopaedia Judaica* (Jerusalem: Keter, 1971), s.v. "Apion."

2. In 1992, Solzhenitsyn published a book in Russia on the complex relationship between the Russians and the Jews, in which he pondered the issue of Jewish survival throughout history. See R. Pipes, "Solzhenitsyn and the Jews: Review of *Alone Together*," *New Republic*, 25 November 2002.

3. Toynbee's twelve-volume *Study of History* makes numerous references to the history of the Jews. A more accessible version is that edited by D. C. Somervell: *A Study of History*, 2 vols. (New York: Oxford University Press, 1956).

estranged and potentially hostile societies could only reinforce that ancient sensitivity and wariness.

As implied in Lamentations, written in grief after the destruction of Solomon's Temple and the exile to Babylonia, "we were punished for our sins" (Lamentations 1:8, 14, 3:41, 4:13, 5:7, 5:16). To feel both guilty and vulnerable is surely not an exclusive Jewish trait, but it has long been a characteristic one. The Jews begin their new year not with merriment, but with a ten-day ritual of atonement for their own sins. Beyond guilt, however, beckons hope: "Let us search and examine our ways, / and turn back to Yahweh" (Lamentations 3:40), then "O Adonai, You have redeemed my life" (Lamentations 3:58), and, finally, "Your iniquity, Fair Zion, is expiated; / He will exile you no longer" (Lamentations 4:22). The repeated plea is "Take us back, O Yahweh, to Yourself . . . / Renew our days as of old" (Lamentations 5:21).

The fierce loyalty to a tribe that feels alienated and beleaguered may lead to a pattern of thought and behavior that is self-isolating. Such may have been the reason for the rejection by the returnees from Babylonia of the local folk who wished to join in rebuilding Jerusalem, and Ezra's refusal to accept the foreign wives (Ezra 4, 9). While such exclusiveness may seem self-defeating, it may have been crucial to self-perpetuation as a distinct culture.

The ensuing centuries of Jewish history were fraught with physical and spiritual struggles, much suffering, and repeated expulsions. Scattered over the entire world, the Jews became the most adaptable of people, even while retaining their sense of otherness. If they did not adapt, they would perish; if they relinquished their otherness, they would lose their soul. The more wretched was their lot, the more sublime became their dream of redemption.

An even greater paradox is that, along with a sense of alienation from the nations of the world, the Israelites perceived themselves to be the appointed carriers of a universal message. Ultimately, the conception of universal humanity emerged as an inclusive extension of the tribal identity. This was expressed as a belief that a time would come when Israel would no longer be an exclusive entity, but a conduit through which God's message is delivered to all the nations. The narrow view of Israel as a nation apart was countered by the wider view of Israel as God's prototype for humanity at large and of the experience and eventual redemption of Israel as a stage in the process of achieving his ultimate purpose of redeeming the whole world. That universalist view is supported by the story of Jonah, which suggests that God cares as much for Nineveh as for Israel. The notion of a mission to the gentiles was expressed most eloquently by the prophet Isaiah: "In that day there will be a highway from Egypt to Assyria. The Assyrians shall join with the Egyptians . . . and then the Egyptians together with the Assyrians shall serve [Yahweh]. In that day Israel shall be a third partner with Egypt and Assyria as a blessing on earth; for Yahweh Tsevaot will bless them, saying, 'Blessed be My people Egypt, My handiwork Assyria, and My very own Israel'" (Isaiah 19:23–25).

As expressed in the Bible, salvation is not achieved by faith alone, but by righteous and generous deeds. The universal practice of justice and charity is an expression of the love of and loyalty to God, the way to be worthy of his approval and beneficence. That is the true culmination of ethical monotheism. Sectarian concerns are regarded to be too narrow and parochial, hence not commensurate with God's transcendent greatness. God is the exemplar of a higher morality, to be obeyed and emulated by all.

The prophecy of Joel is especially noteworthy in the way it invokes nature to describe the bliss of being at peace with Yahweh and living securely on the land:

In that day,
the mountains shall drip with wine,
the hills shall flow with milk,
and all the watercourses of Judah
shall flow with water.
A spring shall issue from the House of Yahweh
and shall water the vale of the Shittim. (Joel 3:18)[1]

Joel mixed the tribal with the universal, as he invoked God's power and proclaimed the final judgment of the multitudes to be assembled in the valley. His prophesy is no less relevant today than when it was first sounded. For we have only lately begun to awaken from our anthropocentric delusion of mastery to realize at last that the final judgment may be visited on us all, on humanity as a species, as a consequence of our own deeds. As we devour the Earth's resources and despoil the environment, we are driving ourselves into a metaphoric vale of tears. The biblical threat and promise loom before us at present more poignantly than ever.

THE ISRAELITES were not noted for their grandiose architecture, colossal statues, or awesome pyramids or ziggurats. Theirs was the legacy of the mind and the spirit, encapsulated in the eloquent passages and ringing messages of their written legacy, the book that gave Western civilization the concepts of ethical monotheism and of law based on justice. It is not merely a remnant of bygone times. We are never through with it, and it never ceases to speak to us. As an expression of an ongoing spiritual quest, the Bible retains its compelling force, disturbing and comforting, posing the most profound ageless questions and suggesting answers that are sometimes pregnant with meaning and other times fraught with ambiguity. Those of us who do not hear God's voice directly may yet hear subliminal echoes of it emanating from inspired sections of the ancient Writ, which continues to mystify, to intrigue, to fascinate, to disturb, to challenge. What have we really discovered, and what have we missed? Is there a secret passage, a revealing sentence, or a clue somewhere that can open the long-sealed gate of understanding and thus help us fathom the age-old yet ever-fresh power of this tome? Is there a pathway through turmoil and tribulation toward greater harmony? Has the time come for all descendants of Adam and Eve to truly savor the finally ripened fruit from Eden's tree of the knowledge of good and evil?

APPENDIX 1

ON THE HISTORICAL VALIDITY
OF THE BIBLE

A S FAR AS we know, the Bible is the first attempt to describe a people's history as a continuous progression of events, circumstances, personalities, and cultural development. But it is not an objective, disinterested composition. It encapsulates the way the Israelite nation chose to tell its story to itself and its progeny and to extract meaning from its distilled experience. It is the narrative of a troubled nation, restless, vulnerable, insecure, haunted by a primal existential anxiety—yet constantly aspiring and ever hopeful. It is a very human story of imperfect, error-prone personalities struggling in austere—at times wretched, humiliating, and dangerous—circumstances. In that sense, it is entirely realistic. In another sense, however, it is a narrative imbued with the immanence of the divine, ever capable of intervening miraculously. Perhaps the greatest of the miracles is the very improbable—indeed, triumphal—fact of this small nation's survival as a distinctive culture, alone among all its neighbors in the ancient Near East, and its ultimate influence on Western civilization.

The Bible expresses and promotes a decided worldview—that of a profoundly religious, Jerusalem-centered group of prophets, priests, scholars, and scribes, committed to the concept of a single omnipresent, all-powerful, and just God (Yahweh). Their commitment was a matter not merely of accepting a set of principles in theory, but also of fulfilling a comprehensive set of commandments in practice, based on ethics.

Throughout the greater part of the twentieth century, most scholars—including historians and archaeologists who studied the strip of land between the Jordan River and the Mediterranean Sea—were convinced that the biblical account of the history of the Israelites, from their entry into Canaan in the twelfth century B.C.E. to their exile from it in the sixth century B.C.E., is essentially true. In recent years, however, some archaeologists have found evidence that in some cases refutes the assumptions or interpretations of their "classical" predecessors. These "revisionist" archaeologists have argued that David's and Solomon's capital of Jerusalem was merely a small and poor hillside village, not the glorious capital depicted in the Books of Samuel, Kings, and Chronicles.[1] The views of some of these scholars have been widely publicized, through numerous popular articles and a few books. The controversy is not merely an academic dispute about ancient history, as it seems to carry far-reaching national and international implications regarding the cultural and actual roots and rights of modern Israelis to the land that they have long regarded as the homeland of the Jews (who claim descent from the ancient Israelites, particularly from the tribes of Judah and Levi).

So now the question is posed more starkly than ever before: How valid are the sections of the Bible that pertain to the history of the Israelites and of the ancient Near East as a whole? The question has aroused much controversy among scholars. At one end of the spectrum of opinion are the confirmers, who insist on the complete veracity of the biblical account and purport to find corroboration for all the described events, including even the supernatural miracles.[2] At the other end of the spectrum are the refuters, who find no objective evidence for (but some evidence against) many of the biblical accounts, particularly those of the period predating the establishment of the separate kingdoms of Israel and Judea, in the late tenth or early ninth century B.C.E.[3]

Most scholars who have joined the fray have taken an intermediate position. Their balanced proposal is that many (albeit far from all) of the events related in the Bible have historical roots, even though they are typically described from a particular point of view that the narrative tends to emphasize or even advocate.[4] Since the Bible is not a monolith, some accounts in it may be factually based, while others may be largely mythical. In each case, the question comes down to where the burden of proof lies: whether a particular version of history is accepted unless proved otherwise by independent historical or archaeological evidence, or whether it is accepted only if independent evidence is found to verify it.

The controversy is not easy, and perhaps not possible, to resolve entirely. Although the quantity of archaeological evidence is great, it is still fragmentary, with significant gaps in its chronological sequence and spatial coverage. Moreover, the meaning of many of the findings is ambiguous and lends itself to differing interpretations.[5] Some may claim that the material remains in the field provide the only objective evidence, but, in reality, many archaeologists are no less subjective than other scholars whose expertise lies in the study of the biblical text. In mathematics, a hypothesis can be proved rigorously; in classical science, it can be tested experimentally or established statistically. In archaeology, however, the evidence is generally incomplete, so a proposition can be weighed only on a scale of seeming reasonableness. Moreover, in science an experiment can generally be repeated, by the original investigator as well as by others. In archaeology, unfortunately, each site is typically destroyed as it is being studied, leaving only the evidence selected and interpreted by the person who conducted the excavation.

Part of the problem lies in the mistaken assumption of strict "scientific objectivity" when applied to human affairs. Humans by nature are driven by perceptions and emotions that are deeply subjective (although seldom acknowledged explicitly). So were the protagonists whose lives and deeds are depicted in the Bible, so were the scribes who wrote about them, so were the compilers and redactors who selected and correlated the written records to produce the final Scriptures, so were the translators and interpreters who explained the Scriptures, and so are the archaeologists and textual analysts who try to assess the original meaning and current relevance of the Bible.

Skeptics often point out that the biblical narrative is partisan, ethnocentric, contrived, and committed to the advocacy of a particular theological doctrine—that it is primarily a monotheistic manifesto addressed to the Israelite nation rather than a dispassionate historical treatise. Of course, many other records found in the Middle East (inscriptions on stelae, paintings on walls, cuneiform symbols impressed into clay tablets, or ink writings on shards and papyri) are at least equally partial in support of or in opposition to a deity (whether the Egyptian Amon, the Babylonian Marduk, the Assyrian Ashur, the Hittite Tartuna, the Aramid Hada, or the Moabite Chemosh) or a ruler in whose service the writing was undertaken. No historical text exists that is completely unbiased and independent of the ideology of its author.

An often repeated contention is that no external (that is, nonbiblical) written records exist that parallel and hence confirm biblical accounts of events that occurred before the end of the tenth century B.C.E. Therefore, some archaeologists ascribe little credence to the biblical narratives that describe the earlier periods (the lives of the Patriarchs, the sojourn in Egypt, the Exodus, the wanderings in Sinai, the conquest of Canaan by Joshua, and the deeds of the Judges), believing them to be fictitious compositions written long after their supposed times. However, those who deny any historical basis for a biblical account simply because they have no evidence to support it may themselves be ignoring the fundamental scientific principle that the absence of proof does not in itself constitute proof of absence. That is to say, an event may have occurred even though archaeologists have not (yet?) found evidence for it on the ground where they have searched.

To counter that criticism, archaeologists working in Israel may claim that they have searched so intensively (Israel having been excavated more thoroughly than any other country except, perhaps, Greece) that the probability of anything major escaping their painstaking scrutiny is very low. Nonetheless, they acknowledge that many of the remnants of the older periods were obscured or destroyed by those of subsequent eras or have long since become inaccessible. For example, the remains of the Temple of Solomon (presumably built in the early tenth century B.C.E.) probably are buried under those of the Second Temple (fifth century B.C.E.), whose remnants lie under the Herodian Temple (first century B.C.E.), which, in turn, is below the Muslim mosques built well over a thousand years ago on the same Temple Mount (known in Arabic as Haram esh-Sharif [Noble Sanctuary of the Muslims]). For these reasons, the failure (to date) of archaeologists to find vestiges of a particular object or event that is mentioned in the Bible does not in itself prove that it never existed.

The chancy nature of archaeological findings is illustrated by the lack, until very recently, of any direct evidence of the existence of King David, even though the Bible represents him as a very important king, who greatly enhanced and extended the kingdom of Israel and established the Judean dynasty. Indeed, the Bible devotes more space to David, his life and deeds, than to any other individual. Because of the absence of archaeological evidence, we might have been tempted to consider David as a minor historical character or an entirely fictitious one. But in 1993, excavations at Tel Dan, at the far northern tip of the state of Israel, uncovered an inscription with an explicit mention of the "House of David."[6] This single finding, dated to the ninth century B.C.E., confirms beyond doubt the historical verity not only of David himself, but also of his role as the founder of an important dynasty.

What does raise many doubts in the mind of an impartial reader of the Bible is the apparent lapse of time between the occurrence of an event and its later (sometimes much later) formal documentation, and the additional time between the composition of that document and its editing and selective insertion into the Bible. Many changes occur as an event becomes an experience and then a memory, which is transmitted in the form of an oral story, which, in turn, is committed to paper or parchment, which, in its turn, is compiled with other recorded stories and redacted. Further changes take place as the written document is translated and interpreted in a place and time far removed from the original event. (These problems are not unique to the Bible, of course. One example among many is Homer's classical account of the Trojan War, written centuries after the event it supposedly described, leaving us to wonder how much of it, if any of it at all, is "true.")

Fragment of an inscription boasting of the victory of the Aramaeans against the Israelites of the northern kingdom, in which the "House of David" is mentioned. The inscription is on a stela, found at Tel Dan in northern Israel, that dates to the ninth century B.C.E.

The period between the occurrence of an event and its formal documentation may extend over many generations. In the interim, the transmission of the memory is subject to preferential inclusions or deletions, as well as distortions and embellishments, in the interest of national or individual justification or glorification. This problem is greatly compounded by the work of redactors who may have their own agendas, influencing their choice of accounts to include and emphasize or to exclude and discount. In the form in which it was finally canonized, the Bible was obviously intended to convey a grand message, expressing the supremacy of the one God, his laws, his relationship with Israel, and the course of the history of the Israelites as defined by that relationship.

Long before the advent of systematic textual analysis of the Bible, initiated in Germany at the end of the nineteenth century by Julius Wellhausen, some scholars began to doubt the orthodox view of the Bible as an internally consistent monolith.[7] Baruch (Benedict) Spinoza, who may have been the first to express his skepticism openly, opined as early as 1670 that the historiography of the Bible from the Pentateuch through the Book of Kings was composed by Ezra the Scribe, who used earlier sources to sequence the chronology of the events.[8]

Today's scholars are fully aware that the various books of the Bible integrate disparate sources, composed at various times during and after the period of the First Temple. The process by which the diverse materials were amalgamated involved a great deal of juxtaposition, rearrangement, and efforts at reconciliation of apparent contradictions.[9] However, the redactors who, toward the end of the first millennium B.C.E., compiled and canonized the Bible, more or less in the form in which it has been handed down to subsequent generations, were not altogether successful in eliminating all the disparities.

Careful analysis has revealed, and in some cases disentangled, the strands that were woven together by those redactors. Diverse parts of the Bible, varying in time of composition and in authorship, differ in point of view as well as in vocabulary and style. (This discrepancy is partially obscured by the translation of the Bible into uniform English, as well as into numerous other languages, especially as some of these translations were not made from the primary Hebrew, but from an earlier translation into Greek.)

THE EFFORT to compile and correlate the historical books of the Bible in order to form a comprehensive continuum evidently began during the later years of the kingdom of Judah (seventh century B.C.E.), most likely under the influence of the religious reformation during the reign of King Josiah.[10] The work continued and intensified during the Babylonian exile (sixth century B.C.E.), when it became crucial to preserve the written legacy in order to ensure the survival of the faith despite the separation from the Land of Israel. The task was more or less completed after the return from exile, between the fifth and third centuries B.C.E. The compilers must have made use of numerous scrolls and fragments of earlier writings, including stories of heroic deeds, tales of wars and tribulations, works of poetry, odes of adulation, prophesies, laws, liturgical tracts, preachings, and chronologies of the royal houses. The records of more recent events were naturally more historically accurate, whereas the tales of more ancient times were likely to be shrouded in myth and lore.

Some writings extant at the time were probably not considered appropriate for inclusion in the canon and thus were excluded by the compilers, or they may have been lost entirely. A few of the missing sources are even mentioned in the Bible explicitly (for example, the Books of Yashar, Sayings of Solomon, and Chronicles of the Kings of Israel and of Judah). Consequently, the biblical chronology is rather uneven: some periods are described in great detail, whereas others, perhaps no less eventful, are mentioned only cursorily. This may be due either to gaps in the available records or to the judgment of the writers or compilers about what period or event was worthy of inclusion or emphasis and what was less important, from their particular viewpoint.

The era of the Patriarchs is especially problematic. The descriptions of that time in Genesis do not accord very well with what is known about the historical reality that prevailed during the early to mid-second millennium B.C.E., the presumed period of the Patriarchs. Some archaeologists date the composition of the Patriarchal stories to the time of the Judges (eleventh century B.C.E.), and some to the much later era of the kingdom (eighth or even seventh century B.C.E.).[11]

Some scholars have gone so far as to deny that the time of the Patriarchs ever existed in anything like the way it is portrayed in the Bible. They contend that the nomadic or seminomadic mode of life ascribed to the Patriarchs would not have been possible before the domestication of the camel, which is supposed to have occurred only at the end of the second millennium B.C.E.[12] They reason that because present-day Bedouin depend on the camel as a beast of burden to transport their possessions as they migrate, the ancient pastoralists, without camels, could not have followed the same mode of life. But the camel may have been domesticated earlier than is supposed, and, even if not, pastoralists in the ancient Near East used donkeys (which were domesticated

---

APPROXIMATE CHRONOLOGY OF BIBLICAL PERIODS

Sixteenth to fifteenth centuries B.C.E.	Pastoral Patriarchs
Fourteenth to thirteenth centuries B.C.E.	Sojourn in riverine Egypt
Late thirteenth century B.C.E.	Exodus and wandering in desert Sinai
Twelfth to eleventh centuries B.C.E.	Settlement in rainfed and pastoral Canaan
Tenth century B.C.E.	United monarchy

Northern Kingdom (Israel)

Late tenth to late eighth centuries B.C.E.	720 B.C.E. Sargon II destroys Samaria
	Selective deportation to Assyria
	Foreigners brought in to mix with remnants of local populace
	Nation becomes known as the Samaritans

Southern Kingdom (Judah)

Late tenth to early sixth centuries B.C.E.	586 B.C.E. Nebuchadnezzar destroys Jerusalem
	Selective deportation to Babylonia and exile there
	538 B.C.E. Cyrus the Great allows Judeans to return to Jerusalem
Late sixth to early fifth centuries B.C.E.	Reconstruction of city walls and Temple
Fifth and fourth centuries B.C.E.	Bible completed

much earlier than camels) to carry their goods, as is depicted in Egyptian tomb paintings of Asiatics traveling to Egypt, dated to about 1890 B.C.E.

The biblical stories seem to convey a flavor of authenticity in the way they portray the characters and lifestyle of nomadic pastoralists. Even though they may contain anachronistic elements, these accounts also seem to include essential aspects of what was once a reality.[13] The inauthenticity may not inhere in the essential fact that there was a period of nomadic pastoralism, but in the specific details of how it was later recalled[14] and in the period of time presumably ascribed to it.

Another controversial issue is the Israelites' sojourn in Egypt and exodus from it, long considered the definitive events in the history of the Israelites. The humiliation of slavery, the exhilaration of freedom, the passage through the desert, and the opportunity and challenge of self-determination constitute the core of the Israelites' national and religious ethos. The biblical stories of these events are so dramatized as to seem surreal. Yet these narratives have inspired Jews and non-Jews alike for many generations. If they were entirely contrived, they could hardly have had such lasting power. Underlying the surrealism, there appears to be a believable core of authenticity. Moreover, many of the details reveal an intimate familiarity with Egypt, including the proliferation of frogs and gnats (typical of the riverine domain), the effects of occasionally anomalous weather conditions on typical Egyptian crops (such as flax, barley, wheat, and emmer [Exodus 9:31–32]), and the reedy banks of the Nile, to mention just a few.

The descriptions of Egypt in the Bible are very vivid and very realistic. The method of making bricks by mixing wet mud with straw is followed in Egypt to this very day. Another instance is the nostalgia expressed for the crops grown in Egypt's irrigated gardens, but generally not cultivated in Canaan's rainfed fields: "We remember the fish that we used to eat free in Egypt, the cucumbers, the melons, the leeks, the onions, and the garlic" (Numbers 11:5).

Forced laborers (perhaps enslaved Semitic immigrants) in ancient Egypt preparing a mud slurry, molding it into bricks, and then carrying the bricks to a construction site. This painting is on a wall of the tomb of Rekhmire, near Thebes.

THE REALITY OF MOSES ACCORDING TO AHAD HAAM

In a brilliant essay published a century ago, the philosopher Ahad Haam[1] answered the early doubters of the Bible's factuality and, indeed, anticipated some of the contentions of the present-day refuters of the Bible. Although the specific topic of his essay was the personhood of Moses—whether his existence and character were grounded in fact or fancy—the larger implications of the essay pertained to the whole issue of the historical reality of the Bible.[2]

Ahad Haam's reasoned conclusion was that the exact facts pertaining to the lifetime of the original Moses (obscured as they are by the passage of millennia) matter much less than does the image of Moses projected by tradition and retained in the minds of countless people through many generations. The person Moses is long dead, while an essential quality of his personality remains alive and influential. The real hero of history, the one whose effect on the course of human civilization never wanes and may indeed continue to grow, is the image or the character or the message that is attributed to the hero and that is etched into the collective memory and faith of the people.

The character of the leader retained (and perhaps amplified beyond the original reality) by a nation's tradition is the character designed (consciously or otherwise) to answer the people's needs and aspirations, which are indeed real, and to embody their values and ideals, which inform and inspire their view of the past and their anticipation of the future. Accordingly, the strict archaeological "truth" (even if it were completely determinable, which it seldom is) does not necessarily express the relevant historical truth. Even a myth can play a powerful role in history and constitute a reality all its own. What matters ultimately is what people wish (or feel the need or are led) to believe, and how strongly they commit themselves to what they do believe, individually and collectively.

---

1. Ahad Haam (one of the people) was the pen name of Asher Ginzberg (1856–1927), one of the foremost Jewish thinkers and writers of the late nineteenth and early twentieth centuries.

2. Ahad Haam, "Moses," *Hashelah* 13 (1904): 342–347.

Moreover, some of the names ascribed to the Israelite religious leaders at the time of the Exodus are distinctly Egyptian in origin, including Pinhas the priest (grandson of Aaron), and Moses himself. Embedded in Moses's name is the designation *mose*, meaning "born to" or "child of," commonly applied to Egyptian kings, such as Rameses (Ra-Mose) and Thut-Mose.

No direct evidence yet has been found of the Israelites' presence in Egypt or their dramatic departure from the eastern fringes of the Nile delta and subsequent wandering in the desert of the Sinai Peninsula. However, the historical circumstances that prevailed in Egypt during the New Kingdom accord more or less with the biblical description of groups of nomadic pastoralists entering the eastern delta district during periods of severe drought that affected their traditional (semiarid) grazing grounds.[15] Egyptian records from several periods describe sizable populations of "Asiatic" slaves.[16] If, indeed, the exodus of the Israelites did occur, it may have been the sort of minor episode or embarrassing incident that the Egyptians would have preferred to ignore rather than record.

The inscription of Merneptah (ca. 1220 B.C.E.),
which mentions "Israel" (*enlarged*).

Egyptian documents dated to the fourteenth century B.C.E. refer to people called the Habiru, who apparently were landless bands present in Canaan. Some scholars have speculated that the Habiru may have been the early Hebrews.[17]

The inscription of Merneptah mentions the name Israel explicitly, in listing the victories of this pharaoh (the son of Rameses II) over many cities and peoples during his military foray into Canaan: "Israel is laid waste, his seed is ended."[18] The location where Israel was vanquished is unstated, which suggests that the Israelites still were wandering in Sinai, before their settlement in Canaan. But at the very least, the inscription appears to corroborate the existence of a people called Israel, apparently one of the tribes that inhabited Canaan or its environs, at about the time of (or shortly after) the Exodus.

Another argument in support of the sojourn in Egypt in the history of the Israelites is the unlikelihood that a nation would ascribe to itself so humble and humiliating a national beginning as slavery, unless it had some basis in truth.[19]

Critics of biblical historiography are especially dismissive of the events described in the Book of Joshua concerning the Israelites' conquest of Canaan and the settlement of the twelve tribes. They find no evidence that the major cities in the country were sacked during a concerted campaign at the supposed time of the biblically claimed invasion by the Israelites.[20] But much evidence has been found of extensive, probably gradual, settlement activity that took place in the thirteenth and twelfth centuries B.C.E. in the district of Samaria (in the central highlands)[21] and, to a lesser extent, in the highlands of Judea (in the south) and Galilee (in the north). What was the origin of those settlers? Were they invaders from across the Jordan River, as described in the

Bible, or indigenous Canaanites, from either the cities or the pastoral hinter-land? Could the new settlers in the heartland where the Israelite nation was forged[22] have been a mixture of these disparate socioethnic groups? It seems that an amorphous collection of peoples would not have evolved into a cohesive nation unless there was among them a strong ethnic and religious core that could generate and inspire such a distinctive national ethos. In any case, Israel's occupation of the land seems to have been a gradual process of infiltration and settlement by pastoralists alongside the sedentary populations of the Late Bronze Age cities of Canaan rather than a conquest following a single organized military campaign.

Controversy also surrounds the biblical account of the united kingdom, founded by Saul and consolidated and enlarged by David, who established the capital in Jerusalem, and his son Solomon, who built the Temple there. The reign of King Solomon is described in glorious terms in 1 Kings 3–10. It was supposed to mark the zenith of ancient Israel's power, prestige, wealth, and territorial extent. However, the skeptics among present-day archaeologists point out that the evidence on the ground fails to support those lavish descriptions. The relatively few remains found in Jerusalem that date to that period (tenth century B.C.E.) lack any of the characteristics indicative of a mature kingdom, such as an enlarged urban center with monumental structures and written records. Other archaeologists, more prone to affirm the biblical account, counter that vestiges of some of the monumental structures may lie inaccessible under the Temple Mount. Even if the city of Jerusalem was then

The inscription on the stele of Mesha, king of Moab (written in a language nearly identical with ancient Hebrew and dated to the ninth century B.C.E.), claiming that he, not the Israelites, won the war described in 2 Kings 3:4–27. This is one example of how the writing of history by one side or another may be biased by the writer's point of view or motive.

small, it still could have served as the formative center of the Israelite, and later the Judean, kingdom that established itself there.[23] The very possibility that the kingdom of Judah was only a weak and marginal entity makes it all the more remarkable that it could have given rise to such a monumental cultural contribution as the Bible.

Much less controversy attends the historical accuracy of the biblical account of the period of the divided kingdom,[24] especially of the southern kingdom of Judah, which survived after the final destruction of the northern kingdom of Israel at the end of the eighth century B.C.E. Confirmation of the historical events during that time is abundant in extra-biblical records (including those of Egypt, Assyria, Babylonia, Aramaea, and Moab, as well as a large number of Hebrew inscriptions found throughout the country).[25] The Bible, however, offers the only surviving consecutive record of a nation's purported history from any area of the ancient Near East.[26]

THE ACADEMIC DISPUTES over specific issues cannot change the paramount fact that the Bible as a written document—notwithstanding its shortcomings as objective history, its internal inconsistencies, and its points of variance from the archaeological record—has rooted itself so deeply in the consciousness of the Israelite people and of their descendants, the Jews, as to be accepted forevermore as their collective legacy. In the end, what matters most is the Israelites' perception of their own experience and the culture and faith that arose out of that perception. Here lies the chief power and importance of the Bible: its influence on the nation, and—ultimately—on the entire development of Western civilization. That is a historical fact that cannot be denied. Although it arose initially out of a sectarian context, the greater message of the Bible has become transcendent and universal. Whether or not the Exodus occurred as described, its message of the divine right of humans to attain freedom from bondage is the earliest manifest against tyranny: "Proclaim liberty throughout the land, and to all the inhabitants thereof!" (Leviticus 25:10).[27] Similarly, it was the preaching of the Hebrew prophets, inspired by their faith in a just God, that laid the ideological and ethical foundations for the evolution of a society dedicated to equality and justice. Even if the Bible were a totally contrived invention without a shred of provable fact, it is still a work of monumental genius, pregnant with profound meaning, an inspiring message to future generations. Not the least of its powers is that it is rooted in the authentic environment of a region that is indeed the cradle of Western civilization. The wonderment is that such a rich and complex composition as the Bible, imbued with such loftiness of spirit, could have arisen out of such humble and constrained circumstances.

APPENDIX 2

PERCEPTIONS OF HUMANITY'S
ROLE ON GOD'S EARTH

T HE FIRST two chapters of Genesis reveal much about the Israelites' perceptions of the natural world. They provide profoundly symbolic accounts of the momentous act of creation, the beginning of life on Earth, and the origin and role of humanity. Much misunderstanding attends these stories. One view that was popularized during the early stages of the environmental movement and still is advocated by some environmentalists blames our society's abuse of the environment on the Judeo-Christian perception of humanity's role in nature, as derived from the biblical story of creation.

Here is the version of that story as summarized in an oft-quoted article by Lynn White:

> By gradual stages a loving and all-powerful God had created light and darkness, the heavenly bodies, the earth and all its plants, animals, birds, and fishes. Finally, God had created Adam and, as an afterthought, Eve to keep man from being lonely. Man named all the animals, thus establishing his dominance over them. God planned all this explicitly for man's benefit and rule: no item in the physical creation had any purpose save to serve man's purpose. And, though man's body is made of clay, he is not simply part of nature: he is made in God's image.[1]

Numerous articles and books have since been written both to support and to counter that simplistic and muddled rendition. Yet the notion persists and in some circles has become accepted as a truism. Many environmental courses still include White's often quoted article in the list of required reading. My purpose is not to defend the Bible in its entirety, but to reveal its complexity and to explore the differing views expressed by the authors who originally wrote the creation stories.

Completely ignored in White's interpretation of the biblical story of creation is the important fact that Genesis presents not one but two widely differing accounts. Of the many disparities between the two versions, perhaps the most important to us is the role assigned to humanity in the scheme of life on Earth.

The first story is generally attributed to the P (Priestly) source. It describes how, after summoning up radiant energy with the command "Let there be light!" the creator (who, although singular, is called by the plural name Elohim) imposed form and order on the primeval fluid chaos by separating land from water, and Earth from sky. The sea and the land were then made to spawn a multitude of living species.

Then Elohim decided to create humans. He announced this to an unspecified entourage (presumably, a retinue of lesser gods or of angels), speaking in the plural: "Let us make [an] *adam* [earthling, human] in our image, after our likeness. They shall rule the fish of the sea, the birds of the sky, the cattle, the whole earth, and all the creeping things that creep on earth" (Genesis 1:26). So Elohim created the *adam*, apparently by command: "male and female created He them" (Genesis 1:27). And Elohim blessed them, saying: "Be fruitful, multiply, and fill the earth and subdue it, and dominate the fish of the sea and the fowl of the air, and every animal that tramples upon the earth" (Genesis 1:28). And Elohim said: "Here, I have given you every herb yielding seed, that is upon the face of the all the earth, and every tree, in which is the fruit of a tree yielding seed—to you it shall be for food" (Genesis 1:29). This version can indeed be construed as a divine ordination of humans to dominate the Earth and to use every nonliving or living thing on it for their own purposes, without reservation or restraint.

The entire act of God and the injunction to humanity are described quite differently in the second version of the creation story, which is attributed to the J (Yahwist) source. In these passages, the specific name of God, Yahweh (commonly rendered Jehovah in English), is used explicitly.

In this account of creation, the birth of the human species is described thus: "Yahweh Elohim formed the *adam* out of the material of the soil and blew into its nostrils the breath of life, and the earthling became a living soul" (Genesis 2:7). Next comes the crucial statement: "Yahweh Elohim planted a garden in Eden in the east, and placed the earthling in the Garden of Eden to serve and preserve it" (Genesis 2:8).[2] Thus humanity was not given license to rule over the environment and use it for its own purposes alone, but—quite the contrary—was charged with the responsibility to nurture and to protect

God's creation, which in its pristine state was indeed a veritable *gan eden* (garden of delight).

Thus we have two diametrically opposed perceptions of humanity's role on God's Earth. The first, in Genesis 1, is anthropocentric: humans were put in charge of nature. They were endowed by God with the power and the right to dominate all other creatures, without explicit limitations: "The heavens belong to Yahweh, / but the earth He gave over to man" (Psalm 115:16). (The same notion was expressed in the fifth century B.C.E. by the Greek philosopher Protagoras: "Man is the measure of all things.")

The second perception of humanity's role, in Genesis 2, is much more modest. The human creature was made of soil and given a "living soul," but is not said to be made "in the image of God." Humans were not given the power to rule over other creatures. What power they do have is constrained by duty and responsibility. Humanity's appointment was not an ordination, but an assignment. Humans are neither owners nor masters of the Earth. Rather, they are its custodians, entrusted with the stewardship of God's garden: "For the land is Mine; you are but sojourners resident with Me" (Leviticus 25:23) and "The earth is Yahweh's and all that it holds" (Psalm 24:1). This view of humanity's role accords with the modern ecological principle that the life of every species is rooted not in separateness from other forms of life in nature, but in integration with the entire living community.

The verb used in Genesis 2:7 for Yahweh's forming of the earth-creature, *vayitzer,* is the specific verb for pottery making (Isaiah 41:25; Jeremiah 18:4, 6; Lamentations 4:2). As such, it is very different from the verb *bara,* which was used for Elohim's acts of creation in Genesis 1. Rather than creation by command, the forming of the *adam* in Genesis 2:7 is described as an active manual deed. On the ceiling of the Sistine Chapel, Michelangelo missed the point in his portrayal of God hovering over Adam and touching his hand delicately. Shaping him out of clay and blowing the breath of life into him would have been more in the spirit of the second account of creation. The originally formed earth-being, *ha-adam* ("the" *adam,* implying "the earthling") was not specifically male. Only later was that creature differentiated into *ish* (man) and *ishah* (woman), with the respective names Adam and Hava (Eve).

Readers of the Bible in translation are deprived of the imagery and poetry of the evocative verbal associations in the Biblical Hebrew. The indissoluble link between humanity and the Earth is manifest in the word *adam* and the proper name Adam, which derives from *adamah*—a Hebrew word of feminine gender that means "soil" or "earth." What's in a name? In Adam's, an encapsulation of man's material origin and fate: his livelihood derived from the Earth,[3] to which he is rooted throughout his life and to which he is destined to return at the end of his days. In similar vein, the name of Adam's mate, Hava, rendered Eve in transliteration, means, quite literally, "living." In the words of the Bible: "Man called his wife Eve because she was the mother of all living [humans]" (Genesis 3:20). Together, therefore, Adam and Eve signify "Earth [soil] and Life."[4]

As the story unfolds in Genesis 3, the first humans soon disobeyed God's trust. One interpretation is that they turned greedy, succumbed to unrestrained temptation, and consumed beyond their needs. Perhaps in so doing they banished themselves from the Garden of Delight by despoiling it, so that it was no longer Eden but, indeed, a "cursed land."

Another interpretation of the story is that humans are driven by an innate curiosity to learn all things and try all things. It is an impulse fraught with risk and filled with promise. It is, indeed, the drive that has given rise to science. Significantly, when faced with the choice, the archetypal humans partook of the tree of knowledge first, rather than of the tree of life, even though the latter was not explicitly forbidden (Genesis 2:16–17).

Yet another interpretation is that, out of an unbounded instinct to meddle, they committed the cardinal sin of hubris, fancying themselves to be free of all constraints, to be practically divine. But, as the story suggests, there are two attributes to divinity—knowledge and immortality: "Yahweh Elohim said: 'Now that the man has become like one of us, knowing good and evil, what if he should stretch out his hand and take also from the tree of life, and eat, and live forever!' Hence Yahweh Elohim banished him from the Garden of Eden to till the soil from which he was taken" (Genesis 3:22–23).

Perhaps the most interesting aspect of the story is that the knowledge acquired by humans cannot be neutral or value-free. It carries a moral charge. It is, explicitly, *knowledge of good and evil*, not merely the possession of facts but the ability to discern between right and wrong, the very foundation of conscious responsibility. The important lesson is that, after the initial acquisition of moral knowledge by humans, the remainder of the Bible becomes, in essence, the story of their struggle to apply that knowledge in varying circumstances. Alas, it is hard enough to know good and evil in theory, yet much harder to apply that knowledge in practicing good deeds and avoiding evil ones.

Whatever the reason for or meaning of their action, Adam and Eve were indeed expelled from the Garden of Delight. Thenceforth, they and their descendants were condemned to suffer the consequences and sentenced to a tenuous life of endless toil:

> Cursed is the ground because of you;
> by toil you shall eat of it. . . .
> Until you return to the earth,

for from it you were taken,

for *afar* [soil] you are,

and to *afar* [soil] you shall return. (Genesis 3:17–19)

The expulsion from the Garden of Eden is a folk memory of the beginning of agriculture. With that transition, humans no longer dwelled idyllically in a parkland, feeding on wild fruits or animals, but had begun the toilsome cultivation of cereals.[5]

Curiously, the account of creation in Genesis 2 appears to be older than that in Genesis 1, so the order of the chapters really should have been reversed. However, the redactor who assembled the disparate sources and decided how to arrange them chose to place the P account first. That sequence may have seemed more appropriate because the J version leads directly to the stories related in the following chapters, which also derive from the J source. To arrange them otherwise, the redactor would have had to interrupt the continuity of the J narrative to insert the P story of creation between the J story of creation and the J story of the banishment from the Garden of Eden, thereby making the narrative all the more incongruous. Although the redactor's decision may have made sense on the grounds of textual logic, the placement of the P account ahead of the J account gave the former more emphasis than the latter. In hindsight, that decision may seem unfortunate.

Over the generations, the arrogant and narcissistic attitude toward nature implied by the first creation story has for too long prevailed over the more responsible role assigned to humanity in the second creation story. Taken out of context and represented as the primary biblical message, the injunction to dominate nature has been used for generations as a religious rationale for humans' unbridled exploitation of the environment—even to the point of justifying the destruction of ecosystems and the extinction of species. And that attitude has persisted despite its obvious self-contradiction: the power to control must not carry the right to destroy God's creation, for such a right would imply not only humanity's equality with God but, indeed, humanity's superiority to him. That is a notion that no religion based on faith in God and respect for God's creation can possibly sanction, inasmuch as it suggests that the creator has abdicated responsibility or care for his own creation. It would seem, therefore, that any religious concern for God's creation should deny humanity's right to destroy any part of it.

Although a partial and hence misleading reading of the biblical message has been used as an excuse to exploit rather than to protect the environment, the primary impulse to do so does not come from religion. In the rich industrialized societies, it often comes from sheer greed, while in poverty-ridden societies, it comes from sheer necessity in the seeming absence of an alternative. In fact, human interference with and modification of the environment began very early in the prehistory of our species, long before the Bible was conceived. That exploitation has greatly increased since the advent of the Industrial Revolution, even as Western societies have become increasingly secular and materialistic, and hence less concerned about obeying messages in the Bible.

Using the Bible to justify or even to explain the abuse of nature is an abuse of the Bible. The Bible is not a monolithic or monochromatic document, with a single unequivocal point of view. It contains many passages that carry different meanings and that can be interpreted variously. Attempts to select one passage while ignoring others that express an alternative view are unbalanced and hence deceiving. A fuller understanding of the range of attitudes toward nature expressed in the Bible requires a more discerning and subtle approach, based on recognition of its complex environmental, cultural, and historical context.

APPENDIX 3

SELECTED PASSAGES
REGARDING THE SEVEN DOMAINS

T HE BIBLE abounds with references to nature in its multiple and complex manifestations and to the interactions of humans with each of the various domains in which they lived. Although the natural environment seldom serves as a subject in its own right, it is reflected indirectly as the backdrop to described events. Of the great profusion of references to natural phenomena (for example, landscapes, weather, plant and animal life, and the milieus of farming and grazing), many occur as figurative expressions, metaphors, poetic similes, aphorisms, parables, and allegories.

The task of revealing the implicit role of nature in the Bible consist of gleaning and interpreting those fragmentary references in an effort to understand the environmental context within which the biblical drama was enacted. In so doing, one is repeatedly struck and captivated by the vividness and evocative quality of the depictions of nature. Quite obviously, the various writers of the Bible, although living in different periods and domains, were intimate with nature and closely familiar with its variable manifestations. They evidently knew much about each of the domains, especially about the lives of shepherds and of farmers. Moreover, they were keen observers of plants, animals, ecological habitats, and weather patterns. Altogether, the Bible mentions as many as 150 species of animals and about 100 species of plants.[1]

Of particular interest are the prophets, who expressed the sense of their times with deep insight and high eloquence. Isaiah, for instance, was an urbanite who nonetheless could evoke natural scenes most colorfully and in profuse detail, with an amazing precision and economy of language. Jeremiah, who was born in Anatoth, near the edge of the Judean Desert, was imbued with a sense of the desolate wilderness. His powerful warnings of an impending drought (which he attributed to the transgressions of the people of Judah) describe the shepherds' desperate cry for water to quench the thirst of their flocks as they seek the puddles remaining in pits along the dry streambeds, and the farmers in the villages as they helplessly watch their withered crops while awaiting the reluctant rain (Jeremiah 14). He even described the suffering of wild animals, wandering about like shadows in a parched land. Ezekiel was an artist of the surreal, whose imaginative scenery seems drawn from the realm of dreams and fantastic realism, yet is rooted in the natural world. The Book of Psalms contains poems of nature with such compelling realism as to transport the reader into the vast silent desert, the flames of a burning forest, or the waves of a raging sea. The Book of Job provides an expansive view of the starry cosmos and a dizzying sensation of a sweeping whirlwind. The peak of lyrical poetry is achieved in the enchanting Song of Songs, where youthful abandon and delirious love are manifested in earthy detail, amid the flowering of trees and vines in springtime.

THE RIVERINE DOMAIN

A River issues from Eden to water the garden, and then it divides and becomes four branches. (Genesis 2:10)

Lot looked about him and saw how well watered was the whole plain of the Jordan, all of it . . . like the garden of Yahweh, like the land of Egypt. So Lot chose for himself the whole vale of the Jordan. (Genesis 13:10–11)

Hold out your arm with the rod over the rivers, the canals, and the ponds, and bring up the frogs on the land of Egypt. (Exodus 8:1)

As that people has spurned the gently flowing waters of Siloam, . . . Adonai will bring up against them the mighty, massive waters of the Euphrates, the king of Assyria and his multitude. It shall rise above its channels and overflow its beds and swirl through Judah like a flash flood reaching up to the neck. (Isaiah 8:5–8)

Yahweh . . . will raise His hand over the Euphrates
with the might of His wind
and break it into seven streams,
so that it can be trodden dry-shod. (Isaiah 11:15)

Water shall fail from the ponds [of Egypt],
rivers dry up and be parched,
channels turn foul as they ebb,
and Egypt's canals run dry.
Reed and rush shall decay,
and the Nile papyrus by the Nile-side
and everything sown by the Nile shall wither,
blow away, and vanish.
The fishermen shall lament,
all who cast lines in the Nile shall mourn,
and those who spread nets on the water shall languish.
The flax workers, too, shall be dismayed,
both carders and weavers chagrined.
Her foundations shall be crushed,
and all who make dams shall be despondent. (Isaiah 19:5–10)

Thus said Yahweh:
I will extend to her peace [well-being] like a river,
the wealth of nations like a flowing stream,
and you shall drink of it. (Isaiah 66:12)

He whose trust is in Yahweh . . .
shall be like a tree planted by waters,
sending forth its roots by a stream;
it does not sense the coming of heat,
its leaves are ever fresh;
it has no care in a year of drought,
it does not cease to yield fruit. (Jeremiah 17:7–8)

A fount of living waters, Yahweh. (Jeremiah 17:13)

Who is this that rises like the Nile,
like streams whose waters surge?
It is Egypt that rises like the Nile,
like streams whose waters surge. (Jeremiah 46:7–8)

Waters nourished it,
the deep made it grow tall,
washing with its streams
the place where it was planted,
making its channels
well up to all the trees of the field. (Ezekiel 31:4)

It is Adonai Yahweh Tsevaot
at whose touch the earth trembles . . .
and all of it swells like the Nile
and subsides like the Nile of Egypt. (Amos 9:5)

Like a tree
planted beside streams of water,
which yields its fruit in season,
whose foliage never fades,
and whatever it produces thrives. (Psalm 1:3)

He who lets out water begins strife. (Proverbs 17:14)

I went down to the nut grove,
to see the buddings along the stream;
to see if the vines had blossomed,
if the pomegranates were in bloom. (Song of Songs 6:11)

THE PASTORAL DOMAIN

Abram went up into the Negev with his wife and all he possessed, together with Lot. Now Abram was very rich with livestock. . . . Lot, who went with Abram, also had flocks and herds and tents, so the land could not support them staying together. . . . So Abram said to Lot, "Let there be no strife between you and me, between my herdsmen and yours, for we are kinsmen. Is not the whole land before you? Let us separate: if you go left I will go right; and if you go right I will go left." (Genesis 13:1, 5–6, 8–9)

Now Jacob became incensed and took up his grievance with Laban. . . . These twenty years I have spent in your service, your ewes and she-goats never miscarried, nor did I feast on rams from your flock. That which was torn by beasts I never brought to you; I myself made good the loss. . . . Often, scorching heat savaged me by day and frost by night; and sleep fled from my eyes. (Genesis 31:36–40)

If you obey the commandments that I enjoin upon you . . . I will provide grass in the fields for your livestock, and thus you shall eat and be satisfied. (Deuteronomy 11:13, 15)

You shall shepherd my people Israel. (2 Samuel 5:2)[2]

David spoke to Yahweh when he saw the angel that smote the people, and said: "Lo, I have sinned . . . but these sheep, what have they done?" (2 Samuel 24:17)

All flesh is grass,
all its goodness like flowers of the field;
grass withers, flowers fade
when the breath of Yahweh blows on them.
Indeed, people are but grass:
Grass withers, flowers fade—
but the word of our Elohim is always fulfilled. (Isaiah 40:6–8)[3]

Like the shepherd He pastures His flock;
He gathers the lambs in His arms
and carries them in His bosom;
gently He drives the mother sheep. (Isaiah 40:11)

Lo, I will send serpents against you,
adders that have no whisper,
and they shall bite you. (Jeremiah 8:17)[4]

Ah, shepherds who let the flock of My pasture stray and scatter!—declares
Yahweh . . . concerning the shepherds who should tend to My people. It is you
who let My flock scatter and go astray. . . . I Myself will gather the remnant of
My flock from all the lands to which I have banished them, and I will bring
them back to their pasture, where they shall be fertile and increase. And I will
appoint over them shepherds who will tend them. (Jeremiah 23:1–4)

Thus said Adonai Yahweh: ". . . As a shepherd seeks out his flock when some
[animals] in his flock have gotten separated, so I will seek out My flock . . . and
I will bring them to their own land, and will pasture them on the mountains of
Israel, by the watercourses. . . . I will feed them in good grazing land, and the
lofty hills of Israel shall be their pasture. There, in the hills of Israel, they shall
lie down in a good pasture and shall feed on rich grazing land. . . . I will look
for the lost, and I will bring back the strayed; I will bandage the injured, and I
will sustain the weak. . . . I will banish vicious beasts from their land, and they
shall live secure in the wasteland, they shall even sleep in the woodland. . . . For
you, My flock, the flock that I tend, are men; and I am your Elohim—declares
Adonai Yahweh." (Ezekiel 34:12–14, 25, 31)

I will let you dwell in your tents again
as in the days of old. (Hosea 12:10)[5]

The words of Amos, a sheepbreeder from Tekoa, who prophesied concerning Israel in the reigns of Kings Uzziah of Judah and Jeroboam son of Joash of
Israel, two years before the earthquake. (Amos 1:1)[6]

The seacoast . . .
shall become an abode for shepherds
and folds for flocks
and it shall be a portion
for the remnant of the House of Judah.
On these [pastures] they shall graze [their flocks]. (Zephaniah 2:6)[7]

Woe to the idol-worshipping shepherd
who abandons the flock! (Zechariah 11:17)[8]

Yahweh is my shepherd, I lack nothing.
He lays me down in green pastures,
He leads me along still waters.
He enlivens my spirit.
He guides me in right paths,
as befits His name.
Though I walk through a valley of the shadow of death,
I fear no harm,
for You are with me.
Your rod and Your staff—
they comfort me. (Psalm 23:1–4)[9]

Go follow the tracks of the sheep,
and graze your kids
by the tents of the shepherds. (Song of Songs 1:8)

Your hair is like a flock of goats
streaming down from Mount Gilead.
Your teeth are like a flock of ewes
climbing up from the washing pool. (Song of Songs 4:1–2)

THE DESERT DOMAIN

Early in the morning Abraham took some bread and a skin of water, and gave them to Hagar. He placed them over her shoulder, together with the child [Ishmael], and sent her away. And she wandered about in the wilderness of Beer-sheba. When the water was gone from the skin, she left the child under one of the bushes, and went and sat at a distance, a bowshot away, for she thought, "Let me not look on as the child dies." And sitting thus, she burst into tears. Elohim heard the cry of the boy. . . . Then Elohim opened her eyes and she saw a well of water. She went and filled the skin with water, and let the boy drink. Elohim was with the boy and he grew up; he dwelt in the wilderness and became a bowman. (Genesis 21:14–20)

Take care lest ... your heart grow haughty and you forget Yahweh your Elohim ... who led you through the great and terrible wilderness with its *seraph* serpents and scorpions, a parched land with no water in it, who brought forth water from the flinty rock, who fed you in the wilderness with manna ... in order to test you by hardships only to benefit you in the end. (Deuteronomy 8:11–16)

Like the gales [whirlwinds] that race through the Negev,
it comes from the desert,
the terrible land. (Isaiah 21:1)[10]

The arid desert shall be glad,
the wilderness shall rejoice and blossom like a rose.
It shall blossom abundantly,
it shall also exult and sing. (Isaiah 35:1–2)

He who turns his thoughts from Yahweh
shall be like a bush in the desert,
which does not sense the coming of good;
it is set in the scorched places of the wilderness,
in a saline land without inhabitants. (Jeremiah 17:5–6)

The watercourses are dried up
and fire has consumed
the pastures of the desert. (Joel 1:20)

Fear not, O beasts of the field,
for the pastures in the wilderness are clothed with grass. (Joel 2:22)

The voice of Yahweh convulses the wilderness,
the voice of Yahweh convulses the Wilderness of Kadesh. (Psalm 29:8)

He turns the wilderness into pools,
parched land into springs of water.
There He settles the hungry;
they build a place to settle in,
they sow fields and plant vineyards
that yield a fruitful harvest....
He pours contempt on great men

and makes them lose their way in trackless deserts,
but the needy He secures from suffering
and increases their families like flocks. (Psalm 107:35–41)

Restore our fortunes, O Yahweh,
like watercourses in the Negev. (Psalm 126:4)[11]

THE RAINFED DOMAIN

When you enter the land that I assign you, the land shall observe a sabbath to Yahweh. Six years you may sow your field and six years you may prune your vineyard and gather in the yield. But in the seventh year the land shall have a sabbath of complete rest. . . . But you may eat whatever the land during its sabbath will produce. (Leviticus 25:2–6)

Yahweh your Elohim is bringing you into a good land, a land with streams and springs issuing from plain and hill; a land of wheat and barley, of vines, figs, and pomegranates, a land of olive trees and honey; a land where you may eat food without stint, where you will lack nothing. . . . When you eat your fill, give thanks to Yahweh your Elohim for the good land which He has given you. (Deuteronomy 8:7–10)

If you obey the commandments that I enjoin upon you. . . . I will grant the rain for your land in season, the early rain and the late, so you shall gather your new grain and wine and oil. (Deuteronomy 11:13–14)

May my discourse come down as the rain, my speech distil as the dew, like showers on young growth, like droplets on the grass. (Deuteronomy 32:2)

All the days of Solomon, Judah and Israel from Dan to Beersheba dwelt in safety, everyone under his own vine and under his own fig tree. (1 Kings 5:5)

Let us revere Yahweh our Elohim,
Who gives . . . the early and late rains in season;
Who keeps for our benefit
the weeks appointed for harvest. (Jeremiah 5:24)[12]

When He makes His voice heard, there is a rumbling of water in the skies.
He makes clouds rise from the end of the earth.
He makes lightning for the rain,
and brings forth wind from His treasuries. (Jeremiah 10:13)[13]

Yahweh named you verdant olive tree, fair, with choice fruit. But with a great roaring sound He has set it on fire, and its boughs are broken. Yahweh Tsevaot, who planted you, has decreed disaster for you, because of the evil wrought to the House of Israel and the House of Judah, who angered Me by sacrificing to Ba'al. (Jeremiah 11:16–17)[14]

He shall come unto us as the rain,
as the latter rain that waters the earth. (Hosea 6:3)[15]

Your goodness is as a morning cloud
and as the dew that early passes away. (Hosea 6:4)[16]

Ephraim is a trained heifer
but preferred to thresh.
I placed a yoke on her sleek neck.
I will make Ephraim do first plowing;
Judah shall do [main] plowing!
Jacob shall harrow.
Sow righteousness for yourselves;
reap the fruits of goodness,
break for yourselves fresh ground of seeking Yahweh. (Hosea 10:11–12)[17]

O children of Zion, be glad,
rejoice in Yahweh your Elohim,
for He has given you the early rains in [His] kindness.
Now He makes the rain fall [as] formerly—
the early rain and the late.
And threshing floors shall be piled with grain,
and vats shall overflow with new wine and oil.
I will repay you for the years
consumed by swarms and hoppers,
by grubs and locusts,
the great army that I loosed against you.
And you shall eat your fill,
and praise the name of Yahweh your Elohim,
who dealt so wondrously with you. (Joel 2:23–26)

I therefore withheld the rain from you
three months before harvesttime.
I would make it rain on one town
and not on another.

One field would be rained upon
while another on which it did not rain would wither. (Amos 4:7)[18]

A time is coming—declares Yahweh—
when the plowman shall meet the reaper,
and the treader of grapes him who holds the [bag of] seed;
when the mountains shall drip wine
and all the hills shall wave [with grain].
I will restore My people Israel.
They shall rebuild ruined cities and inhabit them.
They shall plant vineyards and drink their wine.
They shall till gardens and eat their fruits.
And I will plant them upon their soil,
nevermore to be uprooted from the soil
I have given them—
said Yahweh your Elohim. (Amos 9:13–15)

The remnant of Jacob shall be
in the midst of many peoples,
as dew from Yahweh,
as showers upon the grass. (Micah 5:6)

Put me to the test, said Yahweh Tsevaot. I will surely open the floodgates of the
sky for you and pour down blessings on you; and I will banish the locusts from
you, so that they will not destroy the yield of your soil and the vines in the field
shall no longer miscarry. (Malachi 3:10–11)

You take care of the earth and irrigate it;
You enrich it greatly,
with the channel of Elohim full of water;
You provide grain for men;
for so do You prepare it,
saturating its furrows,
leveling its ridges,
You soften it with showers,
You bless its growth;
You crown the year with Your bounty. . . .
The meadows are clothed with flocks,
the valleys mantled with grain. (Psalm 65:10–14)

They who sow in tears
shall reap with songs of joy.

Though he goes along weeping
carrying the seed-bag,
he shall come back with songs of joy
carrying his sheaves. (Psalm 126:5–6)

The winter is past,
the rains are over and gone.
The blossoms have appeared in the land,
the time of pruning has come.
The song of the turtledove
is heard in our land.
The green figs form on the fig tree,
the vines in blossom
give off fragrance (Song of Songs 2:11–13)

Nard and saffron, fragrant reed and cinnamon,
with all aromatic woods,

AN ODE TO THE LOVE OF NATURE
AND TO LOVE IN NATURE

The Song of Songs depicts a romance between a young man and a young woman in the glorious season of spring. It is a delicious and delirious fantasy. The scenery is beautifully evoked, with charming descriptions of the country's plant and animal life. The frolicking lovers are compared to gazelles and hinds of the field, "leaping upon the mountains, / skipping upon the hills" (Song of Songs 2:8). But love is not to be rushed:

I adjure you, O daughters of Jerusalem . . .
that ye awaken not, nor stir up love,
until it is desired. (Song of Songs 2:7)

Soon enough, it is desired: "My love is mine, and I am his, / that feeds among the lilies" (Song of Songs 2:16).

As in numerous other parts of the Bible, the motifs of nature and of agriculture are invoked, though this time not teaching a direct moral lesson but alluding to a frankly sexual courtship. The young man admires the beauty of his beloved, whose breasts are

like two fawns
that are twins of a gazelle
that feed among the lilies.
Until the day expires,
and the shadows flee away,
I will get to the mountain of myrrh,
to the hill of frankincense. (Song of Songs 4:5–6)

The love-struck pair joins, parts, and reunites. The maiden then ceases to "guard her own vineyard" as she literally falls in love with her lover, whose "left arm is under my head / and his right embraces me" (Song of Songs 2:6).

AGRICULTURAL METAPHORS

The epitome of happiness is that of a farmer at the ingathering of his crop: "They have rejoiced before You / as they rejoice at reaping time" (Isaiah 9:2). Freedom for people is compared to breaking the bonds of a draft animal:

The yoke that they bore
and the stick on their back—
the rod of their taskmaster
You have broken. (Isaiah 9:3)

The destruction wrought by war on farmland is lamented thus:

Ended are the shouts over your fig and grain harvests.
Rejoicing and gladness are gone from the farm land.
In the vineyards no shouting or cheering is heard.
No more does the treader tread wine in the presses. (Isaiah 16:9–10)

The acts of harvesting grain with a sickle and gathering olives by beating the branches are described succinctly:

Like the standing grain harvested by the reaper,
who reaps ears by the armful. . . .
As when one beats an olive tree:
Two berries or three
on the topmost branch,
four or five
on the boughs of the crown. (Isaiah 17:5–6)

The labor of a farmer in the vineyard comes to naught when the pruning is done at the wrong time:

Before the harvest, yet after the budding,
when the blossom has hardened into berries,
He will trim away the twigs with pruning hooks,
and lop off the trailing branches.
They shall all be left
to the kites of the hills
and to the beasts of the earth. (Isaiah 18:5–6)

A rhetorical question lists the operations of a farmer preparing his field for sowing common field crops, including grains and spices:

Does he who plows to sow plow all the time,
breaking up and furrowing his land?
When he has smoothed its surface,
does he not rather broadcast black cumin and scatter cumin,
or set wheat in a row,
barley in a strip,
and emmer in a patch? (Isaiah 28:24–25)

Evidently, draft animals were fed a sort of silage, which may have been made of hay that was compressed in a pit and stored for several months, during which it underwent partial fermentation. The compressed feed had to be loosened (winnowed) before being fed to the animals: "Your livestock . . . shall graze in broad pastures. As for the cattle and the asses that till the soil, they shall partake of soured fodder that has been winnowed with shovel and fan" (Isaiah 30:23–24).

Grass on rooftops, such as might germinate in a thatched roof following a rainstorm, may flourish for a short while, but tends to wilt very quickly as the thin layer in which it germinated dries out:

Fortified towns lie waste in desolate heaps,
their inhabitants helpless.
Dismayed and shamed,
they were but grass of the field and . . .
grass of the roofs
that is blasted by the east wind. (Isaiah 37:26–27)

God will change the direction of the invading Assyrians as a farmer controls his beasts of burden:

Because you have raged against Me,
and your tumult has reached my ears,
I will place My hook in your nose
and My bit between your jaws,
and I will make you go back by the road
by which you came. (Isaiah 37:29)

Isaiah makes poetic use of an implement still used today by *fellahin* (traditional farmers) in the Middle East: a wooden platform whose underside is studded with sharp stones. A beast of burden draws the platform, with a person standing on it, around and around over a layer of harvested wheat. The abrasion removes the chaff from the grain, after which the mix is heaved into the air so that the wind blows away the chaff:

I will make of you a threshing board,
a new thresher, with many spikes.
You shall thresh mountains to dust,
and make hills like chaff.
You shall winnow them and the wind shall carry them off.
The whirlwind shall scatter them.
And you shall rejoice in Yahweh. (Isaiah 41:15–16)

Similarly picturesque metaphors were conceived by the other prophets. To cite but one example:

Thus said Yahweh to the men of Judah and to Jerusalem:
"Break up [plow] the untilled ground,
do not sow among thorns." (Jeremiah 4:3)

The yoke is a symbol of bondage, which Yahweh commanded Jeremiah to place on himself: "Make for yourself thongs and bars and put them on your neck" (Jeremiah 27:2). The symbol was meant to be a warning to the king of Judah not to ally himself with the kings of Edom, Moab, Ammon, Tyre, and Sidon in the rebellion against Babylon, lest he bring on himself and the people of Judah the full wrath of what was then the mightiest empire. When Hananiah, the official court prophet who advocated the rebellion, accosted Jeremiah and broke his yoke, Jeremiah replied: "You broke bars of wood, but you shall have bars of iron instead. I [Yahweh] have put a yoke of iron upon the neck of all those nations, that they may serve Nebuchadnezzar, King of Babylon" (Jeremiah 28:13–14).

Horns symbolize masculinity and ascendancy, personifying the powerful and fierce bull that gores its enemies. Zedekiah, a would-be prophet who wished to ingratiate himself with King Ahab, fashioned a pair of iron horns to symbolize the certainty that Ahab would be victorious over the powerful Aramaeans: "With these shall you gore the Aramaeans, till you make an end of them" (1 Kings 22:11).

continued

The Book of Proverbs is a veritable treasure trove of metaphors, similes, allusions, allegories, and tropes of various sorts, many of which pertain to nature, the pastoral life, and agriculture:

Drink water from your own cistern,
running water from your own well. . . .
Your springs will gush forth. . . .
Let your fountain be blessed. (Proverbs 5:15–18)

Go to the ant, sluggard,
study her ways and learn. (Proverbs 6:6)

These aphorisms refer to the custom of farmers to plant thorny shrubs along the edges of their fields and plantations; trying to walk through such hedges can be painful:

The way of a lazy man is like a hedge of thorns,
but the path of the upright is paved. (Proverbs 15:19)

I passed by the field of a lazy man. . . .
It was all overgrown with thorns. (Proverbs 24:30–31)

This metaphor refers to the importance of the end-of-season rains, which help fill the grain and ensure a bountiful harvest: "His favor is as a cloud / of the latter rain" (Proverbs 16:15). This "much talk but no substance" aphorism conjures up the desperation of drought-afflicted farmers looking up at every whiff of cloud and praying vainly for rain: "Like clouds, wind—but no rain— / is one who boasts of gifts not given" (Proverbs 25:14).

The Book of Job, which describes the physical and mental anguish of a just man who undergoes a sudden reversal of fortune for no apparent reason, is also filled with allusions to natural phenomena and to farming: "Those who plow evil / and sow iniquity reap the same" (Job 4:8). Conversely, those who toil honestly will be deservedly rewarded: "He who tends a fig tree will enjoy its fruit" (Proverbs 27:18).

myrrh and aloes—
all the choice perfumes. . . .
Awake, O north wind;
come, O south wind!
Blow upon my garden
that its perfume may spread. (Song of Songs 4:14, 16)

THE MARITIME DOMAIN

Zebulun shall dwell by the seashore.
He shall be a haven for ships,
and his flank shall rest on Sidon. (Genesis 49:13)

Dan—why did he linger [dwell] by the ships? (Judges 5:17)

They shall roar against them in that day
like the roaring of the sea;

and if one look to the land,
behold darkness and distress,
and the light is darkened in the skies. (Isaiah 5:30)[19]

O you who dwell at the gateway to the sea, who trade with the peoples on many
coastlands, thus said Adonai Yahweh:

"O Tyre . . .
Your frontiers were on the high seas,
your builders perfected your beauty.
From cypress trees of Senir
they fashioned your planks.
They took a cedar from Lebanon
to make a mast for you.
From oak trees of Bashan
they made your oars.
Of boxwood from the isles of Kittim, inlaid in ivory,
they made your decks.
Embroidered linen from Egypt
was the cloth that served you for sails.
Of blue and purple from the coasts of Elishah
were your awnings.
The inhabitants of Sidon and Arvad
were your rowers.
Your skilled men, O Tyre, were within you,
they were your pilots.
Gebal's elders and craftsmen were within you,
making your repairs.
All the ships of the sea, with their crews, were in your harbor
to traffic in your wares.

"Men of Paras, Lud, and Put were in your army, your fighting men. They
hung shields and helmets in your midst, they lent splendor to you. Men of
Arvad and Helech manned your walls all around. And men of Gammad were
stationed in your towers. . . .

"Tarshish traded with you because of your wealth of all kinds of goods. They
bartered silver, iron, tin, and lead for your wares. Iavan, Tubal, and Meshech—they
were your merchants. They trafficked with you in human beings and copper utensils. From Beth-togarmah they bartered horses, horsemen, and mules for your
wares. . . . Many coastlands traded under your rule and rendered you tribute in
ivory tusks and ebony. Aram traded with you . . . in turquoise, purple stuff, embroi-
dery, fine linen, coral, and agate. Judah and the land of Israel . . . trafficked with you
in wheat of Minnith and pannag, honey, oil, and balm. . . . Damascus traded with

you in Helbon wine and white wool. Vedan and Iavan from Uzal . . . trafficked with
you in polished iron, cassia, and calamus. Dedan was your merchant in saddle-
cloths for riding. Arabia and all Kedar's chiefs . . . traded with you in lambs, rams,
and goats. The merchants of Sheba and Raamah . . . bartered for your wares all the
finest spices, all kinds of precious stones, and gold. Haran, Caneh, and Eden . . .
were your merchants in choice fabrics, embroidered cloaks of blue, and many-col-
ored carpets tied up with cords and preserved with cedar. . . .

"So you were full and richly laden
on the high seas.
Your oarsmen brought you out
into the mighty waters.
The tempest wrecked you
on the high seas.
Your wealth, your wares, your merchandise,
your sailors and your pilots,
the men who made your repairs, those who carried on your traffic,
and all the fighting men within you . . .
shall go down into the depths of the sea
on the day of your downfall." (Ezekiel 27:3–27)

Jonah prayed to Yahweh from the belly of the fish. He said:

"In my trouble I called to Yahweh,
and He answered me. . . .
You had cast me into the depths,
into the heart of the sea.
The floods engulfed me,
all Your breakers and billows
swept over me.
I thought I was driven away
out of Your sight. . . .
The waters closed in over me,
the deep engulfed me. . . .
Yet You brought my life up from the pit,
O Yahweh my Elohim." (Jonah 2:3–7)

The ocean sounds, O Yahweh,
the ocean sounds its thunder,
the ocean sounds its pounding.
Above the thunder of the mighty waters,
more majestic than the breakers of the sea,
is Yahweh, majestic on high. (Psalm 93:3–4)

There is the sea, vast and wide,
with its creatures beyond number,
living things, small and great.
There go the ships,
and Leviathan that You formed to sport with.
All of them look to You. . . .
Hide Your face, they are terrified; . . .
send back Your breath, they are created,
and You renew the face of the earth. (Psalm 104:25–30)

Those who go down to the sea in ships,
who ply their trade in the mighty waters,
they have seen the works of Yahweh
and His wonders in the deep.
By His word He raised a storm wind
that made the waves surge.
Mounting up to the heaven, plunging down to the depths,
disgorging in their misery,
they reeled and staggered like a drunken man,
all their skill to no avail.
In their adversity they cried to Yahweh,
and He saved them from their troubles.
He reduced the storm to a whisper;
the waves were stilled.
They rejoiced when all was quiet,
and He brought them to the port they desired. (Psalm 107:23–30)

By His power He stilled the sea;
by His skill He struck down Rahab [the sea-monster].
By His wind the heavens were calmed.
His hand pierced the Elusive Serpent. . . .
Who can absorb the thunder of His mighty deeds? (Job 26:12–14)

THE URBAN DOMAIN

Alas, she has become a harlot,
the faithful city that was filled with justice,
where righteousness dwelt—
now murderers. . . .
[But] I will restore your magistrates as of old,
and your counselors as of yore.

After that you shall be called [again] City of Righteousness,
Faithful City. (Isaiah 1:21, 26)

For instruction shall come forth from Zion,
the word of Yahweh from Jerusalem. (Isaiah 2:3)

What can have happened to you that you have gone,
all of you, up on the roofs,
O you who were full of tumult,
you clamorous town, you city so gay? . . .

You took note of the many breaches in the City of David. And you collected the water of the Lower Pool; and you counted the houses of Jerusalem and
pulled houses down to fortify the wall; and you constructed a basin between
the two walls for the water of the old pool. (Isaiah 22:1–11)

For the sake of Zion I will not be silent . . .
till her victory emerge. . . .
Nations shall see your victory,
and every king your majesty. . . .
You shall be a glorious crown in the hand of Yahweh,
and a royal diadem. . . .
Upon your walls, O Jerusalem,
I have set watchmen,
who shall never be silent. . . .
Until He establish Jerusalem
and make her renowned on earth. . . .
A city not forsaken. (Isaiah 62:1–12)

Desolation, devastation, and destruction!
Spirits sink, knees buckle,
all loins tremble,
all faces turn ashen. (Nahum 2:11)[20]

Draw water for the siege,
strengthen your forts;
tread the clay,
trample the mud;
grasp the brick mold! (Nahum 3:14)[21]

Fair-crested, joy of all the earth,
Mount Zion . . . city of the great king. . . .

Let Mount Zion rejoice!
Let the towns of Judah exult because of your judgments.
Walk around Zion, circle it,
count its towers,
take note of its ramparts;
go through its citadels,
that you may recount it to a future age. (Psalm 48:3, 12–14)

Our feet stood inside your gates, O Jerusalem,
Jerusalem built up,
a city knit together,
to which tribes would make pilgrimage,
the tribes of Yahweh. . . .
There the thrones of judgment stood,
thrones of the house of David.
Pray for the well-being of Jerusalem;
may those who love you be at peace.
May there be well-being within your ramparts,
peace in your citadels. (Psalm 122:2–7)

THE EXILE DOMAIN

When they [the people of Israel] sin against You—for there is no man who does not sin—and You are angry with them and deliver them to the enemy, and their captors carry them off to an enemy land, near or far; and then they . . . repent and make supplication to You in the land of their captors, saying: "We have sinned . ." and they turn back to You with all their heart and soul . . . and they pray to You in the direction of their land which You gave to their fathers, of the city which You have chosen, and of the House which I [Solomon] have built to our name—oh, give heed in Your heavenly abode to their prayer . . . and pardon Your people. . . . Grant them mercy in the sight of their captors that they may be merciful to them. . . . For You, O Adonai Yahweh, have set them apart for Yourself from all the peoples of the earth as Your very own. (1 Kings 8:46–53)

Thus said Adonai Yahweh: When I have cleansed you of all your iniquities, I will people your settlements, and the ruined places shall be rebuilt, and the desolate land, after lying waste in the sight of every passerby, shall again be tilled. And men shall say, "That land, once desolate, has become like the garden of Eden; and the cities, once ruined, desolate, and ravaged, are now populated and fortified." (Ezekiel 36:33–35)

The hand of Yahweh came upon me . . . and set me down in the valley. It was full of bones . . . there were very many of them . . . and they were very dry. He said to me, "O mortal, can these bones live again?" I replied, "O Adonai Yahweh, only You know." And He said to me, "Prophesy over these bones and say to them: O dry bones, hear the word of Yahweh! . . . You shall live again. And you shall know that I am Yahweh!" I prophesied as I had been commanded . . . and suddenly there was a sound of rattling, and the bones came together, bone to matching bone . . . and flesh had grown, and skin had formed over them. . . . Then breath entered them, and they came to life and stood up on their feet, a vast multitude. And He said to me, "O mortal, these bones are the whole House of Israel. . . . I will set you upon your own soil. Then you shall know that I Yahweh have spoken and have acted." (Ezekiel 37:1–14)

Assuredly, thus said Adonai Yahweh: "I will now restore the fortunes of Jacob and take the whole House of Israel back in love . . . from among the peoples and gathered them out of the lands of their enemies and have manifested My holiness through them in the sight of many nations." (Ezekiel 39:25–27)

Give ear, O shepherd of Israel. . . .
O Yahweh Elohim Tsevaot [Lord of hosts] . . .
You plucked up a vine from Egypt;
You expelled nations and planted it.
You cleared a place for it;
it took deep root and filled the land. . . .
Its branches reached the sea,
its shoots, the river.
Why did you breach its wall
so that every passerby plucks its fruit,
wild boars gnaw at it,
and creatures of the field feed on it? . . .
O Elohim Tsevaot turn again . . .
take not of that vine,
the stock planted by Your right hand. . . .
O Yahweh Elohim Tsevaot restore us,
show Your favor that we may be delivered. (Psalm 80:2–20)

Turn our captivity, O Yahweh,
as streams in a dry land. (Psalm 126:4)

By the rivers of Babylon,
there we sat, sat and wept,

as we remembered Zion.
On the poplars
we hung up our lyres,
for our captors
asked us there for songs,
our tormentors, for amusement:
"Sing us of the songs of Zion."
How can we sing a song of Yahweh
on alien soil?
If I forget you, O Jerusalem,
let my right hand wither;
let my tongue stick to my palate
if I cease to think of you,
if I do not keep Jerusalem in memory
even at my happiest hour. (Psalm 137:1–7)

Alas, lonely sits the city
once great with people!
She that was great among nations
is become like a widow. . . .
Judah has gone into exile because of misery
and harsh oppression. . . .
Zion's roads are in mourning,
empty of festival pilgrims;
all her gates are deserted. . . .
She is utterly disconsolate! . . .
Gone from Fair Zion
are all that were her glory. . . .
Jerusalem has greatly sinned,
therefore she is become a mockery. (Lamentations 1:1–8)

All who pass your way
wring their hands at you,
they hiss and nod their head
at fair Jerusalem.
Is this the city that was called
perfect in beauty,
joy of all the earth? (Lamentations 2:15)

Even jackals offer the breast
and suckle their young;

but my poor people has turned cruel,
like ostriches of the desert. (Lamentations 4:3)

The kings of the earth did not believe,
nor any inhabitants of the world,
that foe or adversary could enter
the gates of Jerusalem.
It was for the sins of her prophets,
the iniquity of her priests,
who had shed in her midst
the blood of the just. (Lamentations 4:11–13)

Because of this our hearts are sick,
because of these our eyes are dimmed;
because of Mount Zion, which lies desolate;
jackals prowl over it. . . .
Take us back, O Yahweh, to Yourself, and let us come back;
renew our days as of old! (Lamentations 5:17–21)

ENVIRONMENTAL CONSCIOUSNESS

So long as the earth endures, seedtime and harvest, cold and heat, summer and winter, day and night shall not cease. (Genesis 8:22)

The land must not be sold beyond reclaim, for the land is Mine; you are but strangers [sojourners] resident with Me. Throughout the land that you hold, you must provide for the redemption [salvation] of the land. (Leviticus 25:23–24)

When in war against a city you have to besiege it a long time in order to capture it, you must not destroy its trees, wielding the ax against them. You may eat of them, but you must not cut them down. Are the trees of the field human to withdraw before you into the besieged city? Only trees that you know do not yield food may be destroyed; you may cut them down for constructing siegeworks against the city that is waging war on you, until it has been reduced. (Deuteronomy 20:19–20)

If, along the road, you chance upon a bird's nest, in any tree or on the ground, with fledglings or eggs and the mother sitting over the fledglings or on the eggs, do not take the mother together with her young. Let the mother go. (Deuteronomy 22:6–7)

When you go out in [military] camp against your enemies . . . there shall be an area for you outside the camp, where you may relieve yourself. With your gear you shall have a spike, and when you have squatted you shall dig a hole with it and cover up your excrement. (Deuteronomy 23:10, 13–14)

The Creator . . .
who formed the earth . . .
did not create it a waste,
but formed it for habitation. (Isaiah 45:18)

Even the stork in the sky
knows her seasons;
and the turtledove, swift, and crane
keep the time of their coming. (Jeremiah 8:7)[22]

Fear not, O soil,
rejoice and be glad;
for Yahweh has wrought great deeds.
Fear not, O beasts of the field,
for the pastures in the wilderness are clothed with grass.
The trees have borne their fruit. (Joel 2:21–22)

Yahweh will roar from Zion
and shout from Jerusalem. . . .
"And in that day,
the mountains shall drip with wine,
the hills shall flow with milk,
and all the watercourses of Judah
shall flow with water." (Joel 4:16, 18)

Who made the Pleiades and Orion,
who turns deep darkness into dawn
and darkens day into night,
who summons the waters of the sea
and pours them out upon the earth—
His name is Yahweh. (Amos 5:8)

The mountains shall melt under Him
and the valleys burst open
like wax before fire,
like water cascading down a slope. (Micah 1:4)[23]

PARABLES OF PLANTS AND PEOPLE

Nature-based parables are frequently used in the Bible to score subtle (and, occasionally, not so subtle) points pertaining to people in various situations. All the parables and metaphors, although couched in poetic language, are linked to the environment within which the writers and their intended readers or listeners lived.

Trees (and other plants) are especially endowed with symbolic meanings:[1] witness the trees of knowledge and of life in the Garden of Eden; Abraham's shady oak and tamarisk trees; Jacob's blessing that his son may be like a fruitful vine; the burning bush in which God revealed himself to Moses; the gnarled acacia wood from which the Holy Ark was fashioned and the mighty cedarwood that served in the construction of Solomon's Temple; the flimsy broom tree under which the prophet Elijah sought shelter in the desert; the castor tree that sprang up overnight to shelter Jonah, only to wilt as fast as it appeared; the fruit-bearing trees that must not be felled during a siege, for "are they like a man?"; the comparison of the downtrodden people of Israel to the bare terebinth and oak trees that shall yet reawaken; and the promise that "the righteous man shall rise like a date tree, / shall flourish like a cedar of Lebanon" (Psalm 92:13).

The parable of Jotham is a particular example. After Gideon's victory over the marauding nomads (Judges 7–8), the people of Israel asked that he and his progeny rule over them, but he refused (Judges 8:22–23). After his death, however, Abimelech—a son born to Gideon by a concubine in Shechem—conspired ruthlessly to kill seventy of Gideon's other sons and then have the denizens of Shechem declare him king. Only the youngest of the brothers, Jotham, managed to escape. He ascended to the top of Mount Gerizim and shouted to the people of the city below:

> Once the trees went to anoint a king over themselves. They said to the olive tree, "Reign over us." But the olive replied, "Have I, through whom Elohim and men are honored, stopped yielding my rich oil, that I should go and wave above the trees?" So the trees said to the fig tree, "You come and reign over us." But the fig replied, "Have I stopped yielding my sweetness, my delicious fruit, that I should go and wave above the trees?" So the trees said to the vine, "You come and reign over us." But the vine replied, "Have I stopped yielding my new wine, which gladdens Elohim and men, that I should go and wave above the trees?" Then all the trees said to the thornbush, "You come and reign over us." And the thornbush said to the trees, "If you are truly anointing me king over you, come and take shelter in my shade; but if not, may fire issue from the thornbush and consume the cedars of Lebanon." (Judges 9:8–15)

Then Jotham made his bitter admonition explicit regarding the unworthy and dangerous (thornbush-like) character of Abimelech:

> If you acted truly and sincerely in making Abimelech king, if you have done right by Jerubba'al [Gideon] and his progeny . . . according to his deserts since my father fought for you and saved you from the Midianites at the risk of his life, and now you have turned on my father's progeny and killed his sons, seventy men on one stone, and set up Abimelech, the son of his handmaid, as king over the citizens of Shechem . . . then have joy in Abimelech. . . . But if not, may fire issue from Abimelech and consume the citizens of Shechem and consume Abimelech! (Judges 9:16–20)

Not long afterward, the people of the town became disenchanted with Abimelech and rebelled against him. The conflict ended badly for them, as well as for Abimelech, who was killed by a woman at Thebez (Judges 9:50–55).

A shorter but no less pungent parable was told by Joash, monarch of the kingdom of Israel. When challenged to a fight by Amaziah, king of Judah, he replied contemptuously: "The thistle in Lebanon sent this message to the cedar in Lebanon, 'Give your daughter to my son in marriage.' But a wild beast in Lebanon went by and trampled down the thistle. . . . You [Amaziah] have become arrogant. Stay home . . . rather than provoke disaster" (2 Kings 14:9–14). An insult indeed, but also a fair warning. Foolishly, Amaziah did not heed the warning, and—insisting on a fight—he was defeated and his capital, Jerusalem, ransacked.

Interestingly, the point of Joash's parable contradicts the similar parable of Jotham, many generations earlier. In Jotham's parable, it is the thorny shrub that threatens the mighty cedar with fire. Lowly as it is, the shrub has the capacity to initiate a conflagration that can indeed consume the mightiest of trees. But in Joash's parable, the insolent shrub is trampled ignominiously by a wild animal.

Ezekiel, in Babylonian exile, also used a lofty cedar and an inferior vine as metaphor:

> The word of Yahweh came to me: O mortal, propound a riddle and relate an allegory to the House of Israel. . . . The great eagle . . . came to Lebanon and seized the top of a cedar. He plucked off its topmost bough and carried it off . . . and planted it in a fertile field. . . . It grew but became [only] a spreading vine of low stature. . . . But there was another great eagle . . . and this vine now bent its roots in his direction and sent out its twigs toward him. . . . Will it thrive? . . . It shall wither. . . . And suppose it is transplanted, will it thrive? When the east wind strikes it, it shall wither. (Ezekiel 17:1–10)

Then comes the explanation:

> Do you not know what these things mean? The king of Babylon came to Jerusalem and carried away its king and its officers and brought them back with him to Babylon. He took one of the seed royal and made a covenant with him. . . . But [that prince] rebelled. . . . Will he succeed? . . . Right here in Babylon he shall die. . . . Thus said Adonai Yahweh: Then I in turn will take and set [in the ground a slip] from the lofty top of the cedar . . . and I will plant it in Israel's lofty highlands, and it shall bring forth boughs and produce branches and grow into a noble cedar. Every bird shall take shelter under it, shelter in the shade of its boughs. Then shall all the trees of the field know that it is I Yahweh who have abased the lofty tree and exalted the lowly tree. (Ezekiel 17:12–24)

In a more lighthearted vein is the statement: "As the crackling of [burning] thorns under a pot, / so is the laughter of a fool" (Ecclesiastes 7:6).

Isaiah compared human fear to the swaying of treetops: "His heart was shaken, and the heart of his people, as the trees of the forest are shaken with the wind" (Isaiah 7:2). The image of trees quivering in the wind tells us something about the type of forests that were present in Israel. Trees sheltered inside a dense forest hardly move in the wind. Only spatially separate trees, such as those that grow in an open savanna-like grove (typical of some Mediterranean scrub forests), are affected directly by the wind. Deciduous trees in dormancy are taken to illustrate the forlorn condition of those who disobey Yahweh and are "ravaged like the terebinth and the oak / that are in a leafless state" (Isaiah 6:13). Trees also served Isaiah in his warning against complacency:

> Ephraim and the inhabitants of Samaria,
> in arrogance and haughtiness. . . .
> Sycamores have been felled—
> we'll grow cedars instead![2]
> So Yahweh let the enemies . . . triumph. (Isaiah 9:8–10)

The self-destructiveness of evildoing is illustrated thus:

> Wickedness has blazed forth like a fire
> devouring thorn and thistle.
> It has kindled the thickets of the wood,
> which have turned into billowing smoke. (Isaiah 9:17)

continued

The tragic condition of humanity, compared with that of a tree, is stated most movingly by Job:

There is hope for a tree;
if it is cut down it will renew itself;
its shoots will not cease.
If its roots are old in the earth,
and its stump dies in the ground,
at the scent of water it will bud
and produce branches like a sapling.
But mortals languish and die;
man expires, where is he? (Job 14:7–10)

The opening of the Book of Psalms contrasts the fate of the righteous with that of the wicked in botanical and agricultural terms:

Happy is the man
who has not followed the counsel of the wicked. . . .
Rather, the teaching of Yahweh is his delight. . . .
He is like a tree
planted beside streams of water,
which yields its fruit in season,
whose foliage never fades,
and whatever it produces thrives.
Not so the wicked.
Rather, they are like chaff that wind blows away. (Psalm 1:1–4)

Finally, there is a universal message of hope for all humans and all living beings, and it, too, is couched in botanical terms:

A shoot shall grow out of the [dormant] stump of Jesse [the father of David],
a twig shall sprout from his stock.
The spirit of Yahweh shall alight upon him:
A spirit of wisdom and insight,
a spirit of counsel and valor. . . .
Justice shall be the girdle of his waist. . . .
The wolf shall dwell with the lamb,
the leopard lie down with the kid . . .
with a little boy to herd them. . . .
For the land shall be filled with devotion to Yahweh
as water covers the sea.

In that day the stock of Jesse that has remained standing shall become a standard to peoples—nations shall seek his counsel and his abode shall be honored. (Isaiah 11:1–10)

---

1. The relatively short prophesy of Hosea, for instance, referred to no fewer than eighteen plants: vine, fig, wheat, olive, barley, flax, oak, terebinth, styrax (*livneh*), juniper (*brosh*), lily (*shoshanah*), hemlock (*rosh*), aquatic plants (*ahu*), and five species of thorns and thistles: *kotz, dardar, kimosh, hoah*, and *sirim*. The contexts of Hosea's prophesy were the desert, the forest, the vineyard, and various agricultural activities.

2. Sycamores were locally grown trees that yielded timber of mediocre quality, whereas cedars imported from Lebanon provided the most desirable wood

When I behold your heavens, the work of Your fingers,
the moon and the stars that You set in place,
what is man that You have been mindful of him,
mortal man that You have taken note of him;
that You have made him little less than divine,
and adorned him with glory and majesty.
You have made him master over Your handiwork,
laying the world at his feet. (Psalm 8:4–7)

O Yahweh, You are very great. . . .
You spread the heavens like a tent cloth. . . .
He makes the clouds His chariots,
moves on the wings of the wind. . . .
Mountains rising, valleys sinking—
to the place You established for them. . . .
You make springs gush forth in torrents;
they make their way between the hills,
giving drink to all the wild beasts. . . .
The birds of the sky dwell beside them
and sing among the foliage. . . .
You make the grass grow for the cattle
and herbage for man's labor. (Psalm 104:1–3, 8, 10–14)

How many are the things You have made, O Yahweh;
You have made them all with wisdom;
the earth is full of Your creations. (Psalm 104:24)

He turns the rivers into a wilderness,
springs of water into thirsty land,
fruitful land into a salt marsh,
because of the wickedness of its inhabitants.
He turns the wilderness into pools,
parched land into springs of water.
He feeds the hungry;
they build places to settle.
They sow fields and plant vineyards
yielding fruitful harvests. (Psalm 107:33–37)

Four are among the tiniest on earth,
yet they are the wisest of the wise:
Ants are a folk without power,
yet they prepare food for themselves in summer;

the badger is a folk without strength,
yet it makes its home in the rock;
the locusts have no king,
yet they all march forth in formation;
you can catch the gecko in your hand,
yet it is found in royal palaces. (Proverbs 30:24–28)

All streams flow into the sea,
yet the sea is not full.
To the place [from] which they flow,
the streams return. (Ecclesiastes 1:7)

ETHICAL CONSCIOUSNESS

When you reap the harvest of your land, you shall not reap all the way to the edges of your field, or gather the gleanings of your harvest. You shall not pick your vineyard bare, or gather the fallen fruit of your vineyard; you shall leave them for the poor and the stranger. (Leviticus 19:9–10)

Love your neighbor as yourself. (Leviticus 19:18)

Yahweh your Elohim is the Elohim of the Elohim and the Adonim of the Adonim [master of masters], the great, the mighty, and the awesome El, who shows no favor and takes no bribe, but upholds the cause of the fatherless and the widow, and befriends the stranger, providing him with food and clothing. You must love the stranger, for you were strangers in the land of Egypt. (Deuteronomy 10:17–19)

Justice, justice shall you pursue, that you may thrive and occupy the land that Yahweh your Elohim is giving you. (Deuteronomy 16:20)

You shall not abuse a needy and destitute laborer, whether a fellow countryman or a stranger. . . . You must pay him his wages on the same day, before the sun sets, for he is needy and urgently depends on it; else he will cry unto Yahweh against you, and you will incur guilt. (Deuteronomy 15:14–15)

I Yahweh love justice,
I hate robbery with a burnt offering. . . .
For as the earth brings forth her growth,

and a garden makes the seed sprout,
so Adonai Yahweh will make charity and renown
shoot up in the presence of all the nations. (Isaiah 8–11)

Spare me the sounds of your hymns
And let me not hear the music of your lutes.
But let justice well up like water,
Righteousness like an unfailing stream. (Amos 5:22–24)

Thus said Yahweh Tsevaot: Execute true justice, deal with one another loyally
and compassionately. Do not defraud the widow, the orphan, the stranger, and
the poor; and do not plot evil against one another. (Zechariah 7:9–10)

Have we not one father? Did not one El create us? Why do we break faith with
one another, profaning the covenant of our ancestors? (Malachi 2:10)

Who may sojourn in Your tent,
who may dwell on Your holy mountain?
He who lives without blame, does what is right,
and in his heart acknowledges the truth;
whose tongue is not given to evil;
who has never done harm to his fellow,
or borne reproach for [his acts toward] his neighbor;
for whom a contemptible man is abhorrent,
but who honors those who fear Yahweh;
who keeps his oath even to his hurt;
who has not lent money at interest,
or accepted a bribe against the innocent.
The man who acts thus shall never be shaken. (Psalm 15:1–5)

Who may ascend the mountain of Yahweh?
Who may stand in His holy place?
He who has clean hands and a pure heart,
who has not taken a false oath
or sworn deceitfully. (Psalm 24:3–4)

My comrades are fickle,
like a dry streambed that once flowed [with water]. (Job 6:15)[24]

DIVINE PUNISHMENT AND FORGIVENESS

Should Your people Israel be routed by an enemy for having sinned against You, and then turn back to You and acknowledge Your name, and they offer prayer and supplication to You in this House, oh, hear in heaven and pardon the sin of Your people Israel, and restore them to the land that You gave to their fathers. Should the heavens be shut up and there be no rain, because they have sinned against You, and then they pray toward this place and acknowledge Your name and repent of their sins, when You answer them, oh, hear in heaven and pardon the sin of Your servants, Your people Israel, after You have shown them the proper way in which they are to walk; and send down rain upon the land which You gave to Your people as their heritage. So, too, if there is a famine in the land, if there is pestilence, blight, mildew, locusts or caterpillars, or if an enemy oppresses them in any of the settlements of the land. (1 Kings 8:33–37)

NOTES

PROLOGUE

1. *The Wilderness and the Sown* is the original title of a book by my late professor
   A. Reifenberg of the Hebrew University in Jerusalem, based on his pioneering
   research in the 1930s and 1940s.
2. After the wistful first sentence cited by Ben-Gurion, the prophet Jeremiah continues his bitter lament: "that I might leave my people. . . . / For they are all adul-
   terers, / a band of rogues. . . . / Beware every man of his friend! . . . / One man
   cheats the other, / they will not speak truth" (Jeremiah 9:2–5).
3. Ben-Gurion insisted, despite his age (then sixty-eight), to work physically for several hours each day, and it somehow became my task at first to set and supervise
   his daily assignment. Each evening, I would go to his hut, ostensibly to describe
   the work to be done on the next day, but really to converse with him. The conver-
   sation, however, was mainly one-sided. As he was nearing the end of his stormy
   political career, Ben-Gurion was in a contemplative mood and tended to rem-
   inisce and to discourse at length about the history of Israel (both ancient and
   recent). And I was an avid listener—until his protective wife, Paula, would interrupt these sessions at midnight to shoo me away so the "Old Man" could get his
   sleep.
4. Throughout his long career, Ben-Gurion pursued an active interest in the Bible.
   Busy as he was, he held regular sessions at his home at which biblical passages
   were read and discussed with scholars. On at least one occasion, he caused a
   political stir when he expressed his opinion publicly that the biblical account
   regarding the number of Israelites who had left Egypt at the time of the Exodus was highly exaggerated. Members of the Orthodox parties in the Knesset
   were aghast at this brazen denial of the authenticity of the Bible and threatened
   to topple the government. Ben-Gurion survived that small crisis (as, indeed, he

had many small and large crises throughout his career) and even published his unorthodox views in a book: *Biblical Reflections* (Tel Aviv: Am Oved, 1976). Sdeh Boqer, incidentally, is now a thriving village and a regional educational center. It is also a branch campus of Ben-Gurion University of the Negev, which is based in Beersheba.

5. K. W. Butzer, *Environment and Archaeology: An Ecological Approach to Prehistory* (Chicago: Aldine Atherton, 1971); D. Hillel, *Negev: Land, Water, and Life in a Desert Environment* (New York: Praeger, 1982); *Out of the Earth: Civilization and the Life of the Soil* (Berkeley: University of California Press, 1992); and *Rivers of Eden: The Struggle for Water and the Quest for Peace in the Middle East* (New York: Oxford University Press, 1994).

6. N. Glueck, *The River Jordan: Being a Pictorial Account of the Earth's Most Storied River* (Philadelphia: Westminster Press, 1946).

7. The dangers of translation are exemplified in the King James Version of 1 Kings 13:27, in which the wrong pronoun prompts unwitting comedy: "And he spake to his sons, saying: Saddle me the ass, and they saddled *him.*" Modern translations changed "him" to "it" or "and they did so." The excessive use of "and" at the beginning of sentences is also misleading. In Hebrew, the prefix *vav* (*waw*) is not necessarily the conjunction "and." It often is employed as a grammatical device to convert a verb from future tense to past tense (and, less often, from past to future). Hence it is called the "conversive *vav*" by grammarians.

8. In the words of William Blake: "Both read the Bible day and night, / But thou read'st black where I read white" ("The Everlasting Gospel").

9. A midrash is a traditional Hebrew term for a rabbinical exposition regarding a biblical text. Commonly translated as "an exegesis," it is generally more than mere explanation. A midrash can extend and even transcend the literal meaning of a passage, to draw deeper philosophical, ethical, and legal inferences. Because some midrashim can be very far-fetched (occasionally even fanciful), I have tried to avoid referring to them in my pursuit of a more realistic understanding of the biblical texts, insofar as possible in connection with their original environmental contexts. Hence my approach is closer to what the Mishnah called *pshat*—implying fealty to the direct meaning of a passage rather than speculating about its possible hidden theological implication.

10. Compared with those in the Bible, the stories in the mythology of pagan societies, however filled with symbolism and insights into the human condition, are generally unguided by any sense of morality. In them, humans are thrust into tragic situations, predestined by fate or manipulated by gods who are capricious and, more often than not, morally blind.

11. Ben Bag Bag, Avot 5:25, Mishnah.

1. ENVIRONMENT AND CULTURE

1. By definition, the word "ecology" refers to the interrelationship between living communities and their habitats. Human ecology deals with the ways human societies are affected by and, in turn, affect their physical and biological environments (including the landscape, soil, water, natural fauna and flora, and climate). The term "cultural ecology" describes the mutual influences of the environment and the culture of a society, including its perceptions and practices related to

nature. The manner in which a society adapts to and treats its environment can determine its eventual fate. See, for example, K. W. Butzer, *Environment and Archaeology: An Ecological Approach to Prehistory* (Chicago: Aldine Atherton, 1971); C. Ponting, *A Green History of the World: The Environment and the Collapse of Great Civilizations* (New York: Penguin Books, 1993); and D. Hillel, *Out of the Earth: Civilization and the Life of the Soil* (Berkeley: University of California Press, 1992).

2. A systematic interpretation of the biblical attitudes toward nature is offered in R. A. Simkins, *Creator and Creation: Nature in the Worldview of Ancient Israel* (Peabody, Mass.: Hendrickson, 1994).

3. Among the notable studies on how the environment of Canaan affected the political and religious life of the early Israelites are D. C. Hopkins, *The Highlands of Canaan: Agricultural Life in the Early Iron Age* (Sheffield, Eng.: JSOT Press, 1985), and "Life on the Land: The Subsistence Struggles of Early Israel," *Biblical Archaeologist* 50 (1987): 178–191; F. S. Frick, *The Formation of the State in Ancient Israel*, Social World of Biblical Antiquity, no. 4 (Sheffield, Eng.: JSOT Press, 1985); and T. Hiebert, *The Yahwist's Landscape: Nature and Religion in Early Israel* (New York: Oxford University Press, 1996).

4. The word "canon" derives from the Hebrew *qaneh*, which means "reed" or "measuring stick." By extension, it means "fixed standard" or "norm."

5. The association of each locale with its ruling god is illustrated in the words of Ruth, the Moabite woman who chose to leave her homeland and join her mother-in-law, Naomi, to live in Israel: "Wherever you go, I will go . . . your people shall be my people, and your Elohim [God] my Elohim" (Ruth 1:16).

6. J. Bronowski offered the following definition of creative originality: taking two or more facts or phenomena, perceiving their essential connectivity, and by their combination (as in the case of sexual reproduction) giving birth to new inferences (*A Sense of the Future: Essays in Natural Philosophy* [Cambridge, Mass.: MIT Press, 1977]).

7. Only in the twentieth century was scientific determinism tempered by the realization that natural phenomena may exhibit an inherent randomness and uncertainty. See, for example, J. Gleick, *Chaos: Making a New Science* (New York: Viking Books, 1987).

8. The practice of animal sacrifice continued, despite the protestations of the prophets, as long as the Temple rituals persisted. And so did the dichotomy between the religion's traditional ritualistic practices, on the one hand, and its spiritual-ethical manifestations, on the other hand. That dichotomy was analyzed brilliantly in a classic essay by Ahad Haam: "Prophet and Priest." See, for example, *Encyclopaedia Hebraica*, vol. 2 (Jerusalem: Encyclopaedia, 1955), s.v. "Ahad Haam."

9. See, for example, P. Kurtz, ed., *Science and Religion: Are They Compatible?* (Amherst, N.Y.: Prometheus Books, 2003); A. Einstein, *Ideas and Opinions* (New York: Crown, Wings Books, 1954), which includes four articles on science and religion; and S. J. Gould, *Rocks of Ages: Science and Religion in the Fullness of Life* (New York: Ballantine Books, 1999).

10. The concept of progressive time contradicts the concept of cyclic time, as it is observed in the annual alternation of seasons and river flow, the waxing and waning of the moon, the menstruation of women, and the succession of human generations. For discussions of time's arrow versus time's cycle, see, for exam-

ple, S. J. Gould, *Time's Arrow, Time's Cycle: Myth and Metaphor in the Discovery of Geological Time* (Cambridge, Mass.: Harvard University Press, 1987), and E. Schwartz, "Judaism and Nature: Theological and Moral Issues," in A. Waskow, ed., *Torah of the Earth: Exploring 4,000 Years of Ecology in Jewish Thought*, vol. 2, *Zionism and Eco-Judaism* (Woodstock, Vt.: Jewish Lights, 2000). See also T. Cahill, *The Gifts of the Jews: How a Tribe of Desert Nomads Changed the Way Everyone Thinks and Feels* (New York: Random House, 1998).

11. The notion of "climatic determinism" posited that cultures formed in an optimally temperate climate are inherently superior to cultures formed in either a warmer or a colder climate. By implication, this notion purported to confirm the "natural" superiority of the peoples of Europe over those of other regions. See, for example, E. Huntington, *Civilization and Climate* (New Haven, Conn.: Yale University Press, 1915), and E. Huntington and S. W. Cushing, *Principles of Human Geography*, 5th ed. (New York: Wiley, 1940).

2. THE ECOLOGICAL CONTEXT

1. K. W. Butzer, *Environment and Archaeology: An Ecological Approach to Prehistory* (Chicago: Aldine Atherton, 1971); J. D. Hughes, *Ecology in Ancient Civilizations* (Albuquerque: University of New Mexico Press, 1975); D. Hillel, *Out of the Earth: Civilization and the Life of the Soil* (Berkeley: University of California Press, 1992).

2. The end of the last ice age is the demarcation between the Pleistocene epoch and the Holocene epoch, in which we live.

3. For accounts of plant domestication, see D. Zohary and M. Hopf, *Domestication of Plants in the Old World* (Oxford: Clarendon Press, 1988), and J. R. Harlan, *The Living Fields: Our Agricultural Heritage* (Cambridge: Cambridge University Press, 1995). For accounts of animal domestication, see J. A. Clutton-Brock, *Natural History of Domesticated Mammals* (Cambridge: Cambridge University Press, 1999), and R. W. Bulliet, *The Camel and the Wheel* (New York: Columbia University Press, 1990).

4. See, for example, C. C. Lamberg-Karlovsky and J. A. Sabloff, *Ancient Civilizations: The Near East and Mesopotamia* (Prospect Heights, Ill.: Waveland Press, 1995).

5. M. Zohary, *Introduction to the Geobotany of Palestine* (Hebrew) (Tel Aviv: Sifriat Poalim, 1994); A. Danin, "Vegetation Units of Israel in Satellite Imaging at a Scale of 1:750,000" (Hebrew), in B. Z. Kedar and A. Danin, eds., *Remote-Sensing Aerial Photography and Satellite Imaging as Investigative Tools in Israel Studies* (Jerusalem: Yad Ben-Zvi, 2000), 20–30; J. Albert, M. Bernhardsson, and R. Kenna, eds., *Transformation of Middle Eastern Natural Environments: Legacies and Lessons* (New Haven, Conn.: Yale University School of Forestry and Environmental Studies, 1998).

6. C. L. Redman, *The Rise of Civilization: From Early Farmers to Urban Society in the Ancient Near East* (San Francisco: Freeman, 1978).

7. D. E. Vasey, *Ecological History of Agriculture, 10,000 B.C.–A.D. 10,000* (Ames: Iowa University Press).

8. M. Artzy, "Nomads of the Sea," in R. Hohlfelder and S. Swiny, eds., *Res Maritima 1994: Cyprus and the Eastern Mediterranean, Prehistory Through the Roman*

Period, Cyprus American Archaeological Research Institute, Monograph Series, vol. 1 (Atlanta: Scholars Press, 1997), 1–16; N. K. Sandars, *The Sea Peoples: Warriors of the Ancient Mediterranean, 1250–1150 B.C.* (London: Thames and Hudson, 1978).

9. See, for example, T. J. Abercrombie, "Arabia's Frankincense Trail," *National Geographic*, October 1985, 475–512.

10. The desert of Sinai (pronounced "seen-eye"), lying between Egypt and Canaan, was evidently named after the Semitic moon god, Sin, considered lord of the calendar and of wisdom (*The Columbia Encyclopedia* [New York: Columbia University Press, 1993], s.v. "Sinai").

11. The relationship between sedentary (rainfed or riverine) and nomadic (pastoral) societies traditionally has been depicted as one marked by unrelenting hostility, born of intense rivalry. In stressful times, it could indeed flare up into violent conflict. In "normal" times, however, the relationship was based on complementarity and mutual advantage. Pastoral societies could hardly have survived without some sort of modus vivendi with their neighboring sedentary societies, which naturally tended to be much more populous and dominant economically, politically, and militarily. The pastoralists relied on the villages and towns in the rainfed domain to buy their animal products and—in return—to supply the items they could not produce for themselves, such as grain and manufactured tools.

12. See, for example, W. F. Albright, *Yahweh and the Gods of Canaan: A Historical Analysis of Two Contrasting Faiths* (Winona Lake, Ind.: Eisenbraums, 1968).

13. T. Dothan and M. Dothan, *People of the Sea: The Search for the Philistines* (New York: Macmillan, 1992).

14. See, for example, J. D. Levenson, "From Temple to Synagogue," in B. Halpern and J. D. Levenson, eds., *Traditions in Transformation: Turning Points in Biblical Faith* (Winona Lake, Ind.: Eisenbrauns, 1981), 52–65.

15. Concerning the possible identities of the Bible's redactors, see R. E. Friedman, *Who Wrote the Bible?* (New York: Summit Books, 1987), and N. Naaman, *The Past that Shapes the Present: The Creation of Biblical Historiography in the Late First Temple Period and After the Downfall* (Jerusalem: Hess, 2002).

3. THE FIRST RIVERINE DOMAIN

1. Mesopotamia is the Greek name for the land between the twin rivers: the Tigris and Euphrates. The biblical name for that riverine land was Aram Naharayim, or "Aram of the two rivers." The name Aram referred to the Aramaean people, who inhabited parts of northern Mesopotamia and present-day Syria (itself named after the ancient Assyrians, whose center of power, Ashur, was located on the upper Tigris). The shifting kaleidoscope of history produced a confusion of names.

2. An alternative explanation for 'Ivrim, as those who came from across the Jordan River (rather than the Euphrates), is offered in F. Brown, S. R. Driver, and C. A. Briggs, *Hebrew and English Lexicon* (1906; reprint, Peabody, Mass.: Hendrickson, 1997).

3. The Mesopotamians, as well as the Egyptians, believed that the sky had been separated from the Earth by the winds.

4. In his history of the Persian wars, Herodotus (ca. 485–425 B.C.E.) reported that in Mesopotamia, grain yielded two- and three-hundredfold the amount of seed planted. This is undoubtedly an exaggeration; even a thirtyfold yield would be impressive.

5. According to Genesis 11:28–31, Ur-Kasdim (usually translated as "Ur of the Chaldeans") was the birthplace of Abram, son of Terah. It seems more plausible, however, that the family originated in northern, rather than southern, Mesopotamia. The reference to the Chaldeans is, in any case, an anachronism, since they did not enter the stage of history until many centuries after the presumed birth of Abraham.

6. "The Epic of Creation," in J. B. Pritchard, ed., *The Ancient Near East: An Anthology of Texts and Pictures* (Princeton, N.J.: Princeton University Press, 1958), 1:31–39; H. McCall, *Mesopotamian Myths* (London: British Museum Press, 1990).

7. The notion that evaporated water loses substantiality, that vapor (*hevel*) is nothingness, is expressed in a world-weary nihilistic lament: *havel havalim, hakol havel* (Ecclesiastes 1:2). The passage, generally translated as "vanity of vanities, all is vanity," implying "utter futility, all is futile," means literally "vapor of vapors, all is vapor."

8. Gihon (gusher) is also the name of the spring that provided water to ancient Jerusalem.

9. According to the textual analysis first offered by Julius Wellhausen, the first five books of the Bible, known as the Torah or Pentateuch, were composed from four different sources, which are designated E, which refers to God as Elohim; J, which refers to God as Jahweh (Yahweh); P, which is attributed to a priestly source; and D, which refers to the authorship of Deuteronomy. For details, see W. G. Plaut, "General Introduction to the Torah," in W. G. Plaut, ed., *The Torah: Modern Commentary* (New York: Union of American Hebrew Congregations, 1981), xviii–xxv, and R. E. Friedman, *Who Wrote the Bible?* (New York: Summit Books, 1987).

10. Recently, an alternative hypothesis has been offered regarding the Great Flood: it posits a marine, rather than a riverine, origin. It pertains to the sudden rise of the Black Sea, resulting from the penetration of the waters of the Mediterranean through the barrier of the Bosporus. See W. Ryan and W. Pitman, *Noah's Flood: The New Scientific Discoveries About the Event that Changed History* (New York: Simon and Schuster, 1998). However, that singular event has no relation to the periodic occurrence of catastrophic floods due to river overflows in southern Mesopotamia.

11. "Epic of Gilgamesh," in Pritchard, ed., *Ancient Near East*, 1:40–75.

12. "Epic of Gilgamesh," in Pritchard, ed., *Ancient Near East*, 1:40–75.

13. These two passages evidently originated from different sources (P and J), one referring to God as Elohim, and the other as Yahweh.

14. The "pitch" used to glue the seams of the boat and render it waterproof was probably asphalt, which oozed out of the ground in some parts of Mesopotamia. It was even found in the Dead Sea, which the Romans, much later, called Mare Asphaltum.

15. The olive twig figures in the biblical but not in the Mesopotamian version of the story. Olives grew typically in the rainfed highlands of Canaan, but were uncommon in the irrigated riverine valley of Mesopotamia. Thanks to the biblical story,

a twig of olive leaves still is a symbol of well-being, hence also of reconciliation and peace.

16. The description "Yahweh smelled the pleasing odor" echoes the *Epic of Gilgamesh*, in which the gods awaited the end of the flood so that humans could once again supply them with food. And when the flood did abate and humans offered sacrifices, "the gods smelled the sweet fragrance and they crowded like flies about the sacrificer" ("Epic of Gilgamesh," in Pritchard, ed., *Ancient Near East*, 1:70). A similar description is given in the *Iliad*: when Agamemnon and his men sacrificed animals on the shore, the aroma rose skyward to please the gods. For this and other cross-cultural parallels, see C. H. Gordon, *Before the Bible* (Hebrew) (Tel Aviv: Am Ored, 1966).

17. The rainbow, which in this passage is declared to be a permanent sign of God's guarantee, is not really a tangible or substantive object, but merely a fleetingly observed phenomenon resulting from the refraction of sunlight in a cloud of water droplets. Seen differently from different angles, it is very much in the eye of the beholder.

4. THE PASTORAL DOMAIN

1. Carried into Christianity, this collective role was personalized in the depiction of Jesus as Agnus Dei, the Lamb of God. Jesus's own words on the cross paraphrase Psalm 22:1, *Eli, Eli, lama 'azavthani* (My God, my God, why have You abandoned me), into Matthew 27:46, *lama zabachthani* (why have You sacrificed me). Those are, incidentally, the only words in the New Testament (written in Greek) that were retained in the original Hebrew (although the spoken language of the common folk at that time was mostly Aramaic, a language closely akin to Hebrew).

2. That boundary is dramatically evident in the very site of old Jerusalem, which sits right at the watershed divide. Here the average rainfall diminishes from about 24 inches (600 mm) a year to less than 8 inches (200 mm) within 1 mile (1 to 2 km)—a distance traversable in a fifteen-minute walk!

3. M. Evenari, L. Shanan, and N. Tadmor, *The Negev: The Challenge of a Desert* (Cambridge, Mass.: Harvard University Press, 1971); D. Hillel, *Negev: Land, Water, and Life in a Desert Environment* (New York: Praeger, 1982).

4. S. Ahituv, ed., *Studies in the Archaeology of Nomads in the Negev and Sinai* (Hebrew) (Jerusalem: Bialic Institute, 1998).

5. Pastoral activities persisted in these areas almost continuously (despite fluctuations in climate and shifts in political and military power) throughout the successive eras known as the Pottery Neolithic, Chalcolithic, Bronze, Iron, Babylonian, Persian, Hellenistic, Roman, Byzantine, and Arab Ages. Some pastoral societies have survived more or less intact until recent decades, but the onslaught of modernity has forced radical changes on their traditional way of life. Contemporary nomadic pastoralists have been regarded as the modern counterparts of the migratory herders of the past. There are differences, however. One involves the demarcation of international boundaries and the establishment of central political authorities that restrict movement and impose various regulations and modes of commerce. Another is the presence and availability of firearms, which the people in ancient times could not have imagined. See O. Bar Yosef, "The

Origins of Pastoral Societies in the Levant" (Hebrew), in Ahituv, ed., *Studies in the Archaeology of Nomads in Negev and Sinai*, 7–26.

6. A city named Ur (near present-day Urfa, in the part of the upper Euphrates watershed in southeastern Turkey) existed in northern Mesopotamia, and it (rather than the more famous Sumerian city in southern Mesopotamia) may have been the birthplace of the Hebrew clan.

7. Abraham's reported sojourn in Egypt may (or may not) have coincided with the time of the Hyksos (rulers of foreign lands), the Semitic pastoralists who invaded and ruled Egypt from about 1720 to 1550 B.C.E. However, Abraham's chronology—indeed, his very existence as a historical figure—is unprovable.

8. W. H. Quinn, "A Study of Southern Oscillation–Related Climatic Activity for A.D. 622–1900, Incorporating Nile River Flood Data," in H. F. Diaz and V. Markgraf, eds., *El Niño: Historical and Paleoclimatic Aspects of the Southern Oscillation* (Cambridge: Cambridge University Press, 1992), 119–149.

9. The same tale is told three times: twice with Abraham and Sarah (Genesis 12 and 20) and once with Isaac and Rebekah (Genesis 26). It seems that wife-sister marriages were not unusual in the ancient Near East during the second millennium B.C.E. Contracts found in ancient Nuzi (northern Iraq) indicate that a man might marry a woman and adopt her as his legal sister (or vice versa?). Such marriages may have been intended to preserve the purity of the lineage of an upper-class family. Brother–sister marriages were common in the royal families of Egypt. See E. A. Speiser, "The Wife-Sister Motif in the Patriarchal Narratives," in *Oriental and Biblical Studies: Collected Writings of E. A. Speiser*, ed. J. J. Finkelstein and M. Greenberg (Philadelphia: University of Pennsylvania Press, 1967), 62–82.

10. The custom of wives giving their handmaidens to their husbands to bear children for them (as described in Genesis regarding Abraham's wife, Sarah, and her handmaid Hagar, as well as Jacob's wives, Rachel and Leah, and their handmaids Bilhah and Zilpah) evidently was common in the ancient Near East. Records from Mari (a city on the upper Euphrates) and Nuzi (east of the Tigris) mention the practice. See A. Malamat, *Israel in Biblical Times* (Hebrew) (Jerusalem: Bialik Institute, 1983).

11. Abraham's willingness to banish Ishmael and favor Isaac is but one of many cases in the Bible in which a younger brother is granted (or assumes) primacy over his older brother or brothers. Other examples are God's preference for Abel over Cain, Jacob's assumption of Esau's blessing, Jacob's partiality for Joseph over his other sons, Ephraim (rather than Manasseh) receiving Jacob's main blessing, David being selected from among his brothers, and Solomon (rather than Adonijah) being chosen as David's successor.

12. A particularly gruesome example of the pagan practice of sacrificing the firstborn son in desperate circumstances is described in connection with the king of Moab, who was under siege by the combined forces of the northern and southern kings of the Israelites: "Seeing that the battle was going against him, the king of Moab led an attempt of seven hundred swordsmen to break a way through to the king of Edom, but they failed. So he took his firstborn son, who was to succeed him as king, and offered him up on the wall as a burnt offering. A great wrath came upon Israel, so they withdrew from him and went back to their own land" (2 Kings 3:26–28). The notion of reciprocity underlying ritual sacrifices of all sorts is elaborated in W. Burkert, *Creation of the Sacred: Tracks of Biology in Early Religions* (Cambridge, Mass.: Harvard University Press, 1996).

13. Curiously, a subsequent account reports that Abraham took another wife, named Keturah, and that she bore him six sons (Genesis 25:1–2). It also mentions that Abraham sired children with unnamed concubines. This narrative is inconsistent with the earlier description of Abraham's life. It would seem that the redactor of the final version of the Bible simply decided to include in Genesis several old stories that concerned a legendary ancient Patriarch, regardless of inconsistencies. Another implausible account is of Abraham as a warrior who vanquished the combined forces of four kings with his own private army of 318 men (Genesis 14:14–24), whereas in a later chapter (Genesis 20) he meekly submitted his wife to Abimelech.

14. Note the change in God's name in mid-story, from The Elohim (Genesis 22:9) to Yahweh (Genesis 22:11), suggesting the possibility that a "happy end" was inserted into the old story by a Yahwistic redactor.

15. The pagan practice of sacrificing the firstborn son to a voracious god in order to ensure subsequent fertility was modified by the Israelites so that it applied only to livestock: "Yahweh spoke unto Moses, saying 'Consecrate to Me every firstborn, man and beast, the first issue of every womb among the Israelites is Mine'" (Exodus 13:1). Consecrating did not require the actual sacrifice of the firstborn son, as in the pagan cults (and as almost committed by Abraham). Instead, the Israelites substituted a ritual—known as *pidyon haben* (redemption of the son)—by which the parents would redeem their child from Yahweh through prayer and acts of charity: "And when Yahweh has brought you into the land of the Canaanites, as He swore to you and your fathers, and has given it to you, you shall set apart for Yahweh every first issue of the womb; every male firstling that your cattle drop shall be to Yahweh. . . . And you must redeem every firstborn male among your children" (Exodus 13:11–13). The image of a child offered as sacrifice remained very powerful, however. It was applied metaphorically to the Jewish nation as a whole in times of persecution, and—of course—became a seminal theme in Christianity.

16. Mention of the Hittites in this context is puzzling. The Hittites were a powerful nation, of Indo-European origin, that flourished from 1600 to 1200 B.C.E. in the Anatolian plateau and were among the first peoples to smelt iron successfully (*The Columbia Encyclopedia* [New York: Columbia University Press, 1993], s.v. "Hittites"). However, they are not known to have lived in Canaan. In any case, Ephron (the name of the man from whom Abraham purchased the cave of Machpelah) is a Semitic (probably Canaanite) name. The name Machpelah, incidentally, seems to derive from the verb *kefel*, which means "doubling" or "multiplication," suggesting, perhaps, a wishful increase of good fortune.

17. The negotiation probably took place at the city gate, the traditional place of gathering for the townsfolk, where an agreed transaction could be witnessed by many and become a matter of public record. (Just inside the gate of a typical fortified city of the Bronze and Iron Ages was usually an open forum, where people of the town as well as travelers would exchange information, gossip, and goods.) Thus it was at the city gate that Boaz obtained the release of Ruth from Naomi's kinsman, who had the prior right to marry her (Ruth 4:1–10).

18. Compared with the high price of 400 shekels exacted from Abraham for the cave of Machpelah, the price received by Joseph's brothers from the Ishmaelites was only 20 pieces (probably shekels) of silver (Genesis 37:28). David paid just 50

shekels to Araunah the Jebusite for his threshing ground atop the mount (later to be called the Temple Mount) next to Jerusalem (2 Samuel 24:24). Jeremiah purchased a field in Anathoth for no more than 17 shekels (Jeremiah 32:9). King Omri of the northern kingdom of Israel bought the hill of Samaria (an area large enough for him to build an entire capital city) from a man named Shemer for just 2 talents (*kikrayim*) of silver (1 Kings 16:24). We do not know, however, the value of 1 *kikar* of silver relative to that of 1 shekel of silver.

19. Abraham's highly symbolic act of purchasing land was repeated by his grandson Jacob in the city of Shechem (Genesis 33:18–20).

20. Interestingly, the word "rival" was used in Roman law in a neutral sense in referring to neighbors along a *rivus* (stream). In an arid region, neighbors who draw water from the same stream tend all too often to become competitors. Hence the word "rival" has come to mean "adversary" in common parlance. That word is postbiblical, of course, but the sense it conveys is age-old.

21. *Be'er* means "well," and *shev'a* or *sheb'a* means both "seven" and "oath." Thus the name Beer-sheba implies both "well of seven" and "well of the oath." That oath was perhaps the first recorded peace treaty resolving a dispute over water.

22. The name Beer lahai-roi means, literally, "the well of the living [God] who sees me." It was so named by Hagar, when she first tried to escape from her mistress, Sarai (Genesis 16:13–14).

23. According to my observations, the Bedouin in the northern Negev traditionally sowed grain (barley or wheat as winter crops, sorghum as summer crop) at a rate of some 40 pounds per acre (50 kg per hectare), and applied no manure or fertilizers. In a good year, with a relative abundance of rain (say, 14 inches [350 mm]), they may obtain a grain crop of, say, 2200 pounds (1 ton) per acre (2500 kg per hectare) (that is, a fiftyfold yield). In an "average" year, with about 10 inches (250 mm), the yield is likely to be about half that in the same region. In a drought year, with rainfall below 8 inches (200 mm), the yield may be less than 800 pounds per acre (twenty times the planting rate). Thus Isaac's reported yield of a hundredfold seems highly exaggerated.

24. The rule of thumb in a semiarid region is that seasonal rainfall varies from half the perennial mean (average) to twice the mean. See D. Hillel, *Out of the Earth: Civilization and the Life of the Soil* (Berkeley: University of California Press, 1992), and *Rivers of Eden: The Struggle for Water and the Quest for Peace in the Middle East* (New York: Oxford University Press, 1994). Since a few anomalously rainy seasons tend to skew the mean, the mean is generally higher than the statistical median or the mode. In other words, there are more years with below-average rainfall than with above-average rainfall. To increase the chances of success, dryland farmers in such marginal areas generally select low-lying plots that receive increments of water (additional to direct rainfall) by the accumulation of runoff from adjacent slopes. That had long been the basis for pastoral–agricultural subsistence in the savanna-like plains on the fringes of the desert.

25. Mention of the Philistines in this context seems to be an anachronism. The southern coast of Canaan was occupied by the Philistines only toward the end of the second millennium B.C.E.—some centuries after the presumed time of the Hebrew Patriarchs. Abimelech is, in any case, a Semitic (probably Canaanite) name.

26. The relations between sedentary farmers and seminomadic pastoralists were, on the whole, mutually advantageous; but they were also fraught with tension. The two societies traded goods (for example, grain and fabricated tools for animal products). Farmers often allowed pastoralists to graze the stubble in their fields after the harvest, but only if the flocks were not too large and did not trample the fields excessively, stray onto growing crops, or consume too much water.

27. An example of an agreement obtained fraudulently involved the Gibeonites, who tricked the Israelites into believing that they came from far away (Joshua 9). A quite different example of a vow that was binding, even though it led to tragic consequences, was that of Jephthah, who inadvertently sealed the fate of his only daughter (Judges 11:30–40).

28. The name Leah may be interpreted to connote "weary" or "feeble."

29. N. Feinbrun-Dothan and A. Danin, *Analytical Flora of Eretz Israel* (Jerusalem: Keter, 1991).

30. The possible genetic basis for Jacob's preferential breeding is analyzed in Y. Feliks, *Agriculture in Eretz-Israel in the Periods of the Bible, the Misnah, and the Talmud: Halakkah [Doctrine] and Practice in Basic Agricultural Work* (Hebrew) (Jerusalem: Mas, 1990).

31. The river is unnamed, but the inclusion of the definite article (*ha*) in the appellation Ha-nahar suggests that it refers to the great river: the Euphrates. Elsewhere in the Bible, the Euphrates is repeatedly called Ha-nahar Ha-gadol, the Great River (Genesis 15:18; Deuteronomy 1:7, 11:24; Joshua 1:4, 24:2–3,14–15, to cite just a few examples).

It is interesting to speculate about how long it took Jacob to move his family and livestock from Haran to Canaan. Assuming that Haran is on the east side of the upper Euphrates, somewhere between its tributaries, the Balikh and Khabur Rivers, the distance from there to Gilead (on the east bank of the Jordan River) is about 300 miles (500 km) as the crow flies. The actual distance traversed while trekking flocks of sheep and goats must have been much greater (considering the need to find pasture and water en route). Since the biblical account suggests that Jacob's journey took place in the spring (as Laban was busy shearing his sheep at the time of Jacob's departure), there probably was a relative abundance of grass and water along the route. Hence the actual distance traveled may have been about 500 to 600 miles (800 to 1000 km). Having been a shepherd in the Negev and driven sheep across such irregular terrain, I estimate the average distance covered each day to have been no greater than 12 miles (about 20 km). If so, the entire journey could have taken as long as two months. The Bible mentions that Laban learned of the escape after three days and that he then gave chase for seven days before catching up with Jacob at Gilead. That seems much too short a time for Jacob to have covered so great a distance. So the biblical account seems to be quantitatively inaccurate. Still, such an event could have taken place, even if not precisely as described.

32. That ancient practice has been preserved symbolically in the still-prevalent custom of placing small stones on graves in Jewish cemeteries.

33. The word "bread" meant, in a generic sense, the principal staple food of the people. The main food of shepherds is, of course, meat, whereas that of farmers is what we normally call bread—that is, a baked loaf of dough made of grain. Sig-

nificantly, in Hebrew (which became the language of farmers in rainfed Canaan, akin to the Canaanite language), the word *lehem* means "baked bread," whereas in Arabic (which remained for much longer the language of pastoralists) *lahem*, the same word with a slightly different pronunciation, means "meat."

34. The "eleven children" evidently refer to Jacob's sons (not counting Benjamin, who was not yet born).

35. The name Hamor means, literally, "donkey." It may have been contrived by the biblical writer to disparage the man, or it may have been his real name—one of many characters in the Bible named after animals, including Caleb (dog), Huldah (rat), Nun (fish), Orev (raven), Shafan (hyrax), Tsipor (bird), Yael (ibex), Yonah (dove), and Zeev (wolf).

36. Whereas the daughters of farmers and city dwellers rarely ventured into the countryside by themselves, the daughters of pastoralists grew up in the great outdoors and were accustomed to being on their own, protected by the unwritten law of the pastoralists: the threat of revenge against the entire clan of whoever dared to harm them.

37. Nowhere in the biblical account of the rape of Dinah and its consequences is there any mention of what her own preferences may have been. The maiden's brothers were certain that they knew what was right, and that in asserting the family honor they were acting for her as well. But might she (perhaps like Helen in the *Iliad*) have reciprocated her captor's love and have preferred to remain with him? And what could have been her lot after her "liberation" from Shechem's house? Would she still be marriageable after having been "defiled"? And—if she were—whom could she marry? The men of the clan could choose spouses freely from among the local Canaanite women, for themselves and their sons, but not so the women.

The lesser status of women is exemplified by the absence of any reference to daughters that Jacob and his wives may have had, other than Dinah. A clan that had twelve sons was very likely to have had a few daughters as well. It may be that Dinah alone was mentioned by name only because of what happened to her and, consequently, to the clan and that other daughters were simply not considered noteworthy. Only in Genesis 46:7 is there a brief mention that Jacob had sons and daughters.

38. The name Beth-lehem means, literally, "house of bread." Implied in this name is "granary" or "breadbasket"—evidently a reference to the town as a center of grain production and storage. This passage is the first mention of Bethlehem, which many generations later featured prominently in the Hebrew Bible as the birthplace of David and, much later, in the New Testament as the birthplace of Jesus.

39. Rachel's death may have been a consequence of the hasty journey undertaken by the clan following the murderous raid on the city of Shechem by Simeon and Levi. No reason is given in the Bible why Jacob buried Leah, but not Rachel, in the ancestral cave of Machpelah. We can surmise, however, that since Rachel's death evidently preceded Isaac's (Genesis 35:27–29), Jacob could not yet have assumed that it was his right to bury her in the clan's tomb. More easily explainable is the fact that when the time came (some generations later) to bury Joseph's remains in Canaan, they were not interred in the same cave, but in an alternative family plot in Shechem (Joshua 24:32). After the settlement of Canaan by the twelve tribes and the division of the country among them, the cave of Machpelah became part of the estate of the tribe of Judah. Joshua, who had carried the remains of Joseph

from Egypt, was of the tribe of Ephraim (one of the two tribes descended from Joseph). Members of the tribe of Ephraim, which was in competition with the tribe of Judah for supremacy, must have preferred their separate burial place, within their own territorial estate.

A shrine dedicated to Rachel (called Rachel's Tomb) is still maintained as a place of pilgrimage and worship just outside the town of Bethlehem. As Jacob's beloved, Rachel is remembered in Jewish lore with much more affection than is her sister Leah, even though the Jews are supposed to be descendants primarily of Leah's son Judah.

40. A son's having intercourse with his father's concubines was a symbolic act of appropriating his status. Witness the audacious act of Absalom, who wrested the throne from his elderly father, David, and then—in a blatant demonstration of his prowess and authority—fornicated publicly with David's concubines (2 Samuel 16:20–22).

41. From his name comes the word "onanism," implying masturbation (or, more accurately, coitus interruptus).

42. The town of Timnah figured, many generations later, in another liaison between an Israelite man and a local woman. It was there that Samson married a Philistine woman (Judges 14), only to leave her after she had betrayed his confidence. His fascination with another Philistine woman, Delilah, eventually cost him his life (Judges 16).

43. Among David's progenitors was another daring woman, also of non-Israelite extraction—the Moabite Ruth. She married Perez's descendant Boaz and became David's great-grandmother. The most celebrated Jewish king and the founder of the Judean dynasty was thus the product of not one but two intermarriages!

44. The love–hate relationship between Joseph and his brothers is a dramatic topic that has been analyzed and elaborated by many authors. Perhaps the most noteworthy is Thomas Mann's monumental four-volume novel *Joseph and His Brothers*.

45. Jacob also introduced his father, Jacob, to Pharaoh. When asked, "How many are the years of your life?" Jacob answered, "The years of my sojourn [on Earth] are one hundred and thirty." Then he added: "Few and hard have been the years of my life." Indeed, the life of Jacob-Israel seems to have portended the entire subsequent history of the people to whom he bequeathed his name. From his place of origin in Canaan, he was forced into exile in Mesopotamia, later returned to his homeland, only to be forced to leave it again for another exile. It was a restless life of wandering and alienation and danger, of conflict and loss and recovery, of change and improvisation and living by his wits, of stubborn and resourceful survival.

46. That gesture symbolized intimacy and trust between men. The same scene occurred between Abraham and Eliezer just before the loyal servant left on his mission to find a wife for Isaac in Haran (Genesis 24:2–3).

5. THE SECOND RIVERINE DOMAIN

1. Herodotus, often called "the father of history," traveled throughout the ancient Levant. His writings combined historical and geographic information on Greece and its wars with Persia, Babylonia, Egypt, and Asia Minor.

2. So famously fertile was the black soil of the Nile floodplain that the Greeks considered it to be the mother lode of all substances and named the study of all mate-

rials (*khimia*) after it. That word has metamorphosed into the French *chimie*, Arabic *al-khemie* (alchemy), and English "chemistry."

3. Regarding the construction of the monuments in Egypt, see, for example, C. Hobson, *The World of the Pharaohs* (New York: Thames and Hudson, 1987).

4. The evident need for coordination in large riverine irrigation systems led K. A. Wittfogel to postulate that irrigation-based societies, which he called "hydraulic civilizations," necessarily tend to centralized control and even to despotic rule ("The Hydraulic Civilizations," in W. L. Thomas, ed., *Man's Role in Changing the Face of the Earth* [Chicago: University of Chicago Press, 1956], 152–164). Although his hypothesis now seems simplistic and has been much criticized, the concept contains a grain of plausibility, in Egypt more than in Mesopotamia.

5. "The Report of a Frontier Official," in J. B. Pritchard, ed., *The Ancient Near East: An Anthology of Texts and Pictures* (Princeton, N.J.: Princeton University Press, 1958), 1:183–184.

6. "The Tradition of Seven Lean Years in Egypt," in Pritchard, ed., *Ancient Near East*, 1:24–27.

7. Akhenaten moved his capital from Thebes to a new site, which he called Akhetaten. Excavations at that site, known by its present name, Tel el 'Amarna, revealed a treasure trove of clay tablets, representing the correspondence of the Egyptian court with states and cities throughout its far-flung empire in southwestern Asia. The tablets, inscribed in the Akkadian cuneiform that was at that time the language of international trade and communication, reveal much about the history of the region in the fourteenth century B.C.E.

8. S. Freud, *Moses and Monotheism* (New York: Knopf, 1939). For a thorough critique of Freud's treatment of the subject, see Y. H. Yerushalmi, *Freud's Moses: Judaism Terminable and Interminable* (New Haven, Conn.: Yale University Press, 1991).

9. D. B. Redford, *Akhenaten: The Heretic King* (Princeton, N.J.: Princeton University Press, 1984).

10. "Hymn to the Aton," in Pritchard, ed., *Ancient Near East*, 1:227–230.

11. Important items of trade in ancient Egypt were extracts of aromatic desert plants, including myrrh and other incense flavors, as well as medicinal herbs and such precious stones as lapis lazuli, turquoise, and malachite. Some of these commodities were carried from faraway locales (including southern Arabia, Eritrea, and even faraway Afghanistan), either by camel caravans traversing the desert or by ships sailing on the Red Sea.

12. This story is highly reminiscent of the "Tale of the Two Brothers," written on papyrus by the Egyptian scribe Inena, who lived around the last quarter of the thirteenth century B.C.E. That tale is a blend of mythology, folklore, and humor, apparently written to entertain rather than to serve any moral or religious purpose. The part of the story that bears a striking similarity to the encounter of Joseph and Potiphar's wife tells how a conscientious young man, after rebuffing the advances of his elder brother's wife, was falsely accused of attempting to rape her. See G. Hart, *Egyptian Myths* (London: British Museum Press, 1999), 75–78.

13. The reference to the "land of the Hebrews" is puzzling, since, according to the Bible, there was only one Hebrew clan in all of Canaan at the time of Joseph. So the story must have been written long after the purported event, when the land of Canaan had indeed become the land of the Hebrews. Another possibility is that

the word "Hebrews" was not specific to the clan of Israelites, but defined a wider assemblage of Semitic tribes living in and around Canaan. In the Amarna Letters to Akhenaten from the city-state rulers of Canaan and Syria, there are several references to a group of people called the Habiru, who appear to have been stateless. Among them were fighters who served local rulers as mercenaries. Indeed, some scholars have suggested that the Habiru mentioned in the Amarna Letters resembled the biblical description of David and his outlaw band of six hundred armed men (1 Samuel 27:2), who hired themselves out to King Achish of the Philistine city of Gath. See, for example, M. Greenberg, *The Hab/piru* (New Haven, Conn.: Yale University Press, 1995), and D. M. Rohl, *Pharaohs and Kings: A Biblical Quest* (New York: Crown, 1995). The similarity of name and situation suggests that the appellation Habiru is the same as the biblical term 'Ivrim (the plural of 'Ivri, which either stems from an eponymous ancestor, Ever, or designates someone who originated from across the river [presumably, the Euphrates]).

14. The interpretation of dreams was an accepted practice in ancient Egypt. The formulaic method for it is exemplified in the Egyptian book of dream interpretation, which provides interpretations to more than two hundred possible dreams (D. Alon, *A New Light on the Exodus from Egypt* [Hebrew] [Jerusalem: Ariel, 1999], 29), including the following:

> If a person sees himself in a dream being given white bread . . . good things will happen to him.
> If a person seems himself diving in the river, that is an omen that he shall be cleansed of all evil.
> If a person sees himself in a dream and his side is in pain . . . something is to be taken from him.
> If a man sees himself in a dream gazing at a deep well, it is an omen that he will be imprisoned.
> If a man sees himself in a dream looking at his own visage in a mirror, that is an omen that he shall take another woman.

15. As pointed out by D. B. Redford, certain elements of the story of Joseph reprise earlier motifs from Egyptian mythology (*Egypt, Canaan, and Israel in Ancient Times* [Princeton, N.J.: Princeton University Press, 1992]). He cites a story relating a series of disastrous events (failure of the Nile to flood, consequent famine). Another story is of the goddess Isis appearing to Pharaoh in a dream, after which a wise man is summoned to interpret the dream. The same motif, incidentally, appears in the Bible in the much later (Hellenistic era) Book of Daniel, in which a wise man is released from wrongful imprisonment and called on to interpret a king's portentous dream (Daniel 2, 5).

16. El Niño–La Niña cycles are large-scale weather phenomena apparently initiated in the equatorial Pacific and triggered by recurring (although randomly timed) oscillations of atmospheric low-pressure and high-pressure centers. These phenomena also involve fluctuations in sea-surface temperature, storm tracks, and rainfall amounts over regions subject to "teleconnections." See H. F. Diaz and V. Markgraf, eds., *El Niño: Historical and Paleoclimatic Aspects of the Southern Oscillation* (Cambridge: Cambridge University Press, 1992), and H. M. Glantz, ed., *Teleconnections Linking Worldwide Climate Anomalies* (Cambridge: Cambridge University Press, 1991).

17. Nilometers are gauging posts used by the ancient Egyptians to measure the level of the river during its flood stage, year after year. On the statistical correlation of Nilometer records with El Niño cycles, see W. H. Quinn, "A Study of Southern Oscillation–Related Climatic Activity for A.D. 622–1900, Incorporating Nile River Flood Data," in Diaz and Markgraf, eds., *El Niño*, 119–149.

18. Ancient Egyptian texts have frequent references to hunger, "years of misery," during which the Nile failed to rise and water the fields and orchards. A text whose setting is the reign of Djoser even tells of "seven years" of famine: "Those in the palace were in heart's affliction from a very great evil, since the Nile had not come in my time for a space of seven years. Grain was scant, fruits were dried up, and everything which they eat was short. Every man robbed his companion. They moved without going [ahead]. The infant was wailing; the youth was waiting; the heart of the old men was in sorrow, their legs were bent, crouching on the ground, their arms were folded. The courtiers were in need. The temples were shut up; the sanctuaries held air. Everything was found empty" ("Tradition of Seven Lean Years in Egypt," in Pritchard, ed., *Ancient Near East*, 1:24–27). Egypt apparently had a tradition of seven lean years, which, by a contractual arrangement between Pharaoh and a god, were to alternate with years of plenty.

19. There was one notable exception to the centralized power of the pharaoh, and it was the entrenched priesthood: "Only the land of the priests bought he not, for the priests had a portion from Pharaoh, and did eat their portion which Pharaoh gave them; wherefore they sold not their land" (Genesis 47:22). So powerful were the priests of Egypt that even the pharaoh had to grant them special privileges to placate them, lest they turn against him (as they did against the "heretical" Akhenaten).

20. For a discussion of this controversy, see, for example, P. Galpaz-Feller, *Exodus: Reality or Illusion?* (Hebrew) (Tel Aviv: Shoken, 2002).

21. In one papyrus text, Asiatic servants were listed by an original name, followed by an assigned Egyptian name (as in the case of Joseph, who was also given an Egyptian name by Pharaoh). One letter in a collection of letters, which apparently served as models for schoolboys, presents the form in which an official on the eastern frontier of Egypt (in Sinai) might report the passage of Asiatic tribes into the greener pastures of the Nile delta: "We have finished letting the nomadic tribes of Edom pass the Fortress of Mer-ne-Ptah Hotep-hir-Maat—life, prosperity, health! . . . to keep them and to keep their livestock alive" ("Report of a Frontier Official," in Pritchard, ed., *Ancient Near East*, 1:183–184).

22. Much of that territory was under Egyptian military and economic influence throughout most of the second millennium B.C.E.

23. The Bible does not specify the names of the pharaohs who reigned during the times of Joseph and Moses. Hence it is difficult to ascribe dates to the events described. The commonly supposed time of the Exodus is the thirteenth century B.C.E., during the reign of Rameses II; his father, Seti II; or his son Merneptah. There is no unanimity on this date, however, even among the historians and archaeologists who accept the Exodus as a real event. Alternative dates offered for the Exodus are the sixteenth and fifteenth centuries B.C.E., which would associate the presence of the Israelites in Egypt as welcome guests (initially) with the reign of the Semitic Hyksos (1760–1539 B.C.E.). In *New Light on the Exodus*

from Egypt, Alon sets the Exodus at the end of the third millennium B.C.E., much earlier than estimated by most other scholars.

24. Egyptian xenophobia is exemplified in the statement of the prophet Ifover: "Strangers are entering Egypt. . . . The children of nobility are being smashed against walls. . . . The slaves are now masters. . . . The corps of the king has been removed from its tomb by desperate people. . . . How is it that the army is commanded by strangers . . . who are showing Asiatics the way to Egypt" ("Asiatics in Egypt," in J. B. Pritchard, ed., *The Ancient Near East: An Anthology of Texts and Pictures* [Princeton, N.J.: Princeton University Press, 1975], 2:87–93).

25. Thus was expressed the ancient notion that only the males carry the identity of a nation and pose a danger to rival nations. That idea was shared in part by the Israelites. Therefore, all the sons of Jacob constituted the children of Israel, regardless of who their mothers were—whether "official" wives or concubines. Only gradually did Israelite (and later Jewish) tradition come to accept that a child's ethnic identity is determined by the mother, as the true identity of the father sometimes was uncertain.

26. Sargon the Great was king of Akkad (ca. 2340–2305 B.C.E.). The inscribed legend of his birth includes the following statement: "My changeling mother conceived me, in secret she bore me. She set me in a basket of rushes, with bitumen she sealed my lid. She cast me into the river which rose not (over) me. The river bore me up and carried me to Akki, the drawer of water . . . [who took me] as his son (and) reared me" ("The Legend of Sargon," in Pritchard, ed., *Ancient Near East*, 1:85–86).

27. As in the story of Gautama Buddha many centuries later, Moses the prince had to leave the palace and wander among common people in order to experience and understand the real world before he could attain enlightenment.

28. The name Reuel means, literally, "behold El," with El being the name of a Semitic god. The Midianites were evidently a Semitic pastoral tribe, not unlike the Amalekites, who apparently inhabited the territories of northeastern Sinai and the Negev.

29. No explanation is given in the Bible for the change of Moses's father-in-law's name from Reuel to Jethro. It is possible that Jethro was his regular name, whereas Reuel was his ecclesiastical name.

30. The word *horeb* means "dry." Moving far into the wilderness, Moses probably passed from the pastoral domain to the desert domain, where loneliness and silence reign, mirages appear, and mystical visions are occasionally experienced.

31. The word *hineni* (Here I am)—expressing submission to the divine calling—was uttered by several of those addressed directly by God, including Abraham (Genesis 22:1, 11), Jacob (Genesis 31:11, 46:2), Samuel (1 Samuel 3:4–8), and Isaiah (Isaiah 6:8).

32. The phrase "land flowing with milk and honey" is usually taken to have been a description of a particularly fertile agricultural land. It is more likely to have referred to a lush pastoral land. Milk was more typically a product of shepherds and goatherds than of farmers, whose main products were grain and fruits. Moreover, when speaking of honey, the Bible refers to wild honey found in the field (1 Samuel 14:25; Judges 14:8; Proverbs 25:18; Song of Songs 5:1). There seems to be no biblical reference to beekeeping.

33. *Eheyeh asher eheyeh* means, literally, "I shall be whatever I shall be," implying "I am what I am." The expression *eheyeh* itself evokes the special name of God:

Yahweh. That name was written in Biblical Hebrew (as words are written in most Semitic languages) without vowels, merely with the consonants YHWH or YHVH (known as the Tetragrammaton). In Orthodox Jewish tradition, the explicit name (Ha-Shem ha-mefurash) has long been considered too holy to be pronounced aloud, and the correct or original vowels are unknown.

34. For a thorough discussion of circumcision as a religious rite and of sacrificial offerings in general, see W. Burkert, *Creation of the Sacred: Tracks of Biology in Early Religions* (Cambridge, Mass.: Harvard University Press, 1996).

35. The practice of circumcision appears to have been prevalent in Egypt from early times. A stone monument found in Naga a-Dir carried an inscription dated to the twenty-third or twenty-second century B.C.E.: "I was beloved of my father and favored by my mother and dear to my brothers and sisters, when I was circumcised without injury, together with 120 others." A relief found in Saqqara (near Cairo) depicts the circumcision of a group of boys by Egyptian priests. See I. Shaw and P. Nicholson, *British Museum Dictionary of Ancient Egypt* (Cairo: American University in Cairo Press, 1998), 65. It brings to mind the mass circumcision of the Israelite men by Joshua (using flint blades) at the start of their campaign to conquer the promised land of Canaan (Joshua 5:2–9). The continued traditional use of a stone knife rather than a metal one suggests the great antiquity of the operation. See C. De Wit, "Le Circoncision chez les anciens Égyptiens," *Aegyptische Sprache und Altertumskunde* 99 (1972): 41–48, and J. M. Sasson, "Circumcision in the Ancient Near East," *Journal of Biblical Literature* 85 (1966): 473–476.

36. Digging wells alongside the Nile, the Egyptians could reach down to the water table and draw groundwater, which was fed by the river but cleansed of foul sediment by seepage through the soil (which thus serves as a "living filter"). Seepage through layers of sand and clay is a well-known and often-used practice for water filtration even today.

37. For an interesting discussion of the ecological basis for the scourges reportedly inflicted on Egypt, see Galpaz-Feller, *Exodus*.

38. A most vivid description of an invasion of locusts is given in Joel 1:1–4.

39. The confusion between the Reed Sea and the Red Sea apparently persists in subsequent books of the Bible, in which the actual Red Sea is erroneously referred to as Yam Soof (for example, 1 Kings 9:26). The ancient Israelites apparently did not have a special name for the then-faraway Red Sea. The Bible records only two brief periods during the monarchy in which the Israelites actually reached that body of water and attempted to sail on it: the time of Solomon (2 Chronicles 8:17–18) and the time of Jehoshaphat (2 Chronicles 20:36)

40. The story of the parting of the Sea of Reeds (Exodus 14:16–21, 15:8) had its precedent in Egyptian mythology, in a tale ascribed to the time of Pharaoh Senefru (ca. 2600 B.C.E.). See Alon, *New Light on the Exodus from Egypt*, citing I. M. Grintz, *Stories, Poems and Parables in Ancient Egypt's Literature* (Jerusalem: Keter, 1975).

41. What purports to be a vision of the future may actually have been written long after the settlement in Canaan, as a retrospective confirmation of Moses's prophecy.

6. THE DESERT DOMAIN

1. Lawless bands from the desert periodically raided and plundered pastoralists in the semiarid domain and occasionally even farmers in the rainfed domain; then they

hid in the wilderness, where they could evade constituted authority. An example of such banditry is described in 1 Samuel 25, which tells how David, long before he became king of Israel, fled from the wrathful King Saul and hid in the Judean Desert. There he gathered a group of desperados, outcasts from society ("everyone who was in straits and everyone who was in debt and who was desperate . . . and he became their leader" [1 Samuel 22:2]) and tried to subsist by offering "protection" to pastoralists and demanding (or extorting) goods in return. One of the pastoralists, Nabal, refused to provide the requested goods, so David was about to kill the man. He refrained from doing so only at the pleading of Nabal's wife, Abigail, who brought abundant provisions to David and his band. Conveniently, Nabal died shortly after, and David then took Abigail to be his wife.

2. According to E. Klein, the noun *midbar* originated from an infinitive meaning "to drive [livestock]," hence "open land whither flocks are driven" (*A Comprehensive Etymological Dictionary of the Hebrew Language* [Jerusalem: Carta, 1987], 317–318). The original Hebrews who lived in or along the desert distinguished among different types of habitats—depending on specific features of the landscape, degrees of aridity, and relative habitability or utility—and hence applied different terms to each.

3. According to W. F. Albright, the one-humped dromedary, native to the Arabian desert, probably was domesticated in the late second millennium B.C.E. and did not enter into common use until the last century of that millennium ("Abraham the Hebrew: A New Archaeological Interpretation," *Bulletin of the American School of Oriental Research* 163 [1961]: 36–54). If so, then the stories about the use of camels during the lifetimes of Abraham, Isaac, and Jacob must be anachronistic, inserted into the Bible by later writers or redactors. However, R. W. Bulliet has suggested that the beginning of domestication may well have occurred some centuries earlier than the date claimed by Albright (*The Camel and the Wheel* [New York: Columbia University Press, 1990]).

4. The importance of dyes, gems, spices, and aromatics in the religious rituals of the time is exemplified in the following passage: "And the rulers brought the onyx stones, and the stones to be set for the ephod, and for the breastplate [of the priest]; and the spice, and the oil for the light, and for the anointing oil, and for the sweet incense" (Exodus 35:26–28).

5. Another desert area that played an important role in the history of the Israelites is that along the eastern slopes of Canaan's central mountain range. These slopes, lying in the rain shadow of the westerly rain-bearing winds, are deeply incised by gorges that descend to the Jordan Rift Valley and the Dead Sea. This desert subregion (called the Samarian Desert in its northern section and the Judean Desert in its southern) is much smaller than the deserts of the Sinai Peninsula and the Negev, but no less rugged and much closer to the country's population center.

6. Comprehensive information on the geography and history of Sinai is in G. Gvirtzman, A. Shmueli, Y. Grados, I. Beit Arie, and M. Harel, eds., *Sinai* (Tel Aviv: Ministry of Defense, 1987).

7. The Nabateans were a Semitic nation that controlled the overland trade routes for spices, aromatics, and medicines across the deserts of Arabia, Edom, and the Negev from around the fourth century B.C.E. to the beginning of the second century C.E., at which time they came under the control of the Romans. Their capital was the "wine-red city" of Petra. For descriptions of the runoff-farming practices

of the Nabateans, see M. Evenari, L. Shanan, and N. Tadmor, *The Negev: The Challenge of a Desert* (Cambridge, Mass.: Harvard University Press, 1971), and D. Hillel, *Negev: Land, Water, and Life in a Desert Environment* (New York: Praeger, 1982).

8. For a description of the vegetation in the semiarid and arid parts of the region, see A. Danin, *Desert Vegetation of Israel and Sinai* (Jerusalem: Cana, 1983), and A. Danin and G. Orshan, eds., *Vegetation of Israel*, vol. 1, *Desert and Coastal Vegetation* (Leiden: Backhuys, 1999).

9. From Edward Palmer (an American explorer who toured Sinai in 1872) to M. Harel, *Travels in Sinai* (Tel Aviv: Am Oved, 1968); E. Anati, *Har Karkom in the Light of New Discoveries* (Capo di Ponte, Italy: Edizioni del Centro, 1993); and others too numerous to mention.

10. In *Travels in Sinai*, Harel suggested Mount Sin Bisher in western Sinai as the Mount Sinai of the Bible, while in *Har Karkom in the Light of New Discoveries*, Anati proposed Mount Karkom in the Negev. Others have suggested various mountains in northeastern as well as southern Sinai.

11. Wadi Tumeilat evidently was the course of an ancient distributary of the Nile that flowed eastward and spilled into the Great Bitter Lake. From there, it may have flowed southward toward the Gulf of Suez. During periods when that course was clogged, the Egyptians could have opened it by digging a canal that, in effect, enabled them to sail from the Mediterranean through an arm of the Nile and through Tumeilat all the way to the Gulf of Suez and beyond. That may well have been the route of the expedition ordered by Queen Hatshepsut to the land of Punt in eastern Africa, around 1497 B.C.E.

12. A similar feat of cleansing water by using salt to coagulate suspended pollutants was reported to have been performed centuries later by the prophet Elisha: "The men of the town said unto Elisha: 'Look, the town is a pleasant place to live in, as my lord can see; but the water is bad, and the land causes bereavement.' He responded, 'Bring me a new dish and put salt in it.' They brought it to him; he went to the spring and threw salt into it. And he said, 'Thus said Yahweh: I heal this water; no longer shall death and bereavement come from it!'" (2 Kings 2:19–22). A scientific explanation of how particles of clay and organic matter in an aqueous suspension can be flocculated (coagulated) by salts is given in D. Hillel, *Environmental Soil Physics* (San Diego, Calif.: Academic Press, 1998), 93–97.

13. On this occasion, Moses was instructed by Yahweh to strike the rock. On another occasion, he was commanded to speak to the rock, and when he nevertheless struck it, he was punished severely (Numbers 20:11).

14. The word "bread" (*lehem*) is used here in the generic sense to represent basic food.

15. This practice was prevalent until the early decades of the twentieth century, after which the number of birds diminished greatly as a result of overhunting. It is described in graphic detail in C. S. Jarvis, *Yesterday and Today in Sinai* (Edinburgh: Blackwood, 1933).

16. The expression in this passage regarding the utter destruction of Amalek is reminiscent of that regarding Israel itself on the stela that commemorates Merneptah's putative victory (ca. 1230 B.C.E.): "Israel is laid waste, his seed is not" ("Hymn of Victory of Mer-ne-ptah," in J. B. Pritchard, ed., *The Ancient Near East: An Anthology of Texts and Pictures* [Princeton, N.J.: Princeton University Press, 1958], 1:231).

It is also similar to the pronouncement in Jeremiah concerning the progeny of Esau: "I have bared Esau, . . . / He cannot hide. / His offspring is ravaged. . . . / He is no more" (Jeremiah 49:10). Such extreme manifestations of enmity and lust for revenge evidently were common in the ancient Near East and, sadly, are not uncommon even at present.

17. In this passage, the word *elohim* refers not to the particular God of the Israelites, but to various unnamed gods. Elsewhere in the Bible, whenever the term, although cast in the grammatical plural, is used in reference to the one God, it is capitalized as one of the proper names of God.

18. God's relationship with humanity is depicted in this exchange as a repeatedly frustrating process of trial and error. First he formed Adam and Eve as the pinnacles of creation, making them in his own image. Then they disappointed him and were banished from his garden. Their descendants turned out to be wicked, so he decided to destroy them and begin afresh with Noah and his progeny. But they, too, proved to be impudent and dangerous, and had to be scattered away from the Tower of Babel. Next God chose Abraham, but now his descendants also seemed to fail expectations. So, would God start all over again with Moses and his offspring? As reported, Moses—the most humble of men—refused the offer and the onus. Incidentally, although we know that Moses begat a son with Zipporah and had another wife (a Cushite) as well, the Bible does not tell us about the fate of that son or of Moses's subsequent descendants.

19. The Hebrew phrase *qaran 'or panaiv* (Exodus 34:29) means, figuratively, "the skin of his face radiated." However, the word *qaran* (radiated) derives from *qeren* (horn). The terms are evidently related because sunlight shining through clouds seems to send forth "horns" (or "beams" or "shafts") of light. Some translations of the Bible therefore rendered the phrase *qaran 'or panaiv*, literally, as "his face sprouted horns." This explains why Michelangelo's magnificent sculpture of Moses (which is in the church of St. Peter in Chains, in Rome) has horns protruding incongruously from his forehead. But this is merely one of numerous examples of how Biblical Hebrew could be mistranslated. As pointed out by D. L. Lieber, incidentally, horns were often associated with divinity in Mesopotamia and Canaan, where the gods were portrayed with horned helmets (*Etz Hayim: Torah and Commentary* [New York: Rabbinical Assembly, 2001], xvii–xxiii).

20. The Ten Commandments, first stated in Exodus, are repeated (with a few minor variations) in Deuteronomy 5:6–18.

21. The last word of this passage, translated into English as "rested," is *vayinafash* in Hebrew. It conveys a dual meaning: *nafash* means "rested," but it is related to the word *nefesh*, which means "soul." So *vayinafash* implies "restored His soul." Latent in this clause is the idea that the perfect rest is not just physical, but spiritual as well.

22. J. Roth, "Shabbat and the Holidays," in Lieber, ed., *Etz Hayim*, 1455–1459.

23. The shittim (acacia) was chosen to provide wood for the Tabernacle, because it is relatively prevalent in the deserts of the Negev and the Sinai Peninsula, growing in the streambeds (wadis). Several species are found there, and the particular one that may have been used by the Israelites during their sojourn in the desert is not known. Y. Feliks suggested that it was *Acacia albida*, which, however, grows mainly in relatively humid habitats rather than in the arid parts of the Sinai (*Species of Fruit Trees, Plants of the Bible and the Sages* [Hebrew] [Jerusalem: Mas, 1994]).

24. An idyllic vision of the Israelites in the desert was ascribed in the Bible to the pagan prophet Bala'am. He had been hired by the king of Moab to curse the Israelite intruders in his realm. Instead, Balaam was moved to praise them: "How fair are your tents, O Jacob, / your dwellings, O Israel; / like palm groves that stretch out, / like gardens beside a river, / like aloes planted by Yahweh" (Numbers 24:5–6). The existence of Balaam (although not at the time assumed in the Bible) was apparently confirmed by the discovery at Deir 'Alla, on the eastern side of the Jordan Valley, of an Aramaic-Ammonite inscription, dated to the seventh century B.C.E., on the plastered wall of a sanctuary. See J. A. Hackett, *The Balaam Text from Deir Alla*, Harvard Semitic Monographs, no. 31 (Chico, Calif.: Scholars Press, 1984).

25. The expression *vayelekh el nafsho* can mean either "he fled for his life" or "he went for his soul." The dual nuance is important: Elijah did not seek merely to save his life, but, indeed, to restore his soul.

26. This scene in the desert is reminiscent of the deliverance of Hagar by the timely appearance of an angel (Genesis 16:7–15).

27. It has long been assumed that Mount Horeb and Mount Sinai are one and the same. The word *horeb* means "arid" or "desolate." In a different context, it may also imply "ruined."

28. The "sound of thin silence" is my translation of the Hebrew *qol dmamah daqah*. The King James Version renders it "a thin small voice," and the translation by the Jewish Publication Society reads "a soft murmuring sound."

7. THE RAINFED DOMAIN

1. The Hebrew name of the valley, Yizr'a-El (sown by El [God]), celebrates its legendary fertility.

2. The past tense is used to describe the course of the Jordan River because its natural regimen was changed drastically by diversions carried out in the second half of the twentieth century.

3. A thorough treatment of the country's climate is in Y. Goldreich, *The Climate of Israel: Observations, Research, and Applications* (Ramat Gan: Bar Ilan University Press, 1998).

4. Some of the species of trees that grow in the country are evergreen, including pine (*Pinus halepensis, P. pinea*), tamarisk (*Tamarix aphylla*), sycamore (*Ficus sycamorus*), common oak (*Quercus calliprinos*), olive (*Olea europaea*), and carob (*Ceratonia siliqua*). Other species of trees undergo an annual period of dormancy, such as Tabor oak (*Quercus ithaburensis*), terebinth (*Pistacia palaestina, P. atlantica*), and almond (*Prunus dulcis*). Two of the most impressive of these trees are the oak and the terebinth, which can attain large size and great age. They are often mentioned in the Bible as symbols of strength and permanence. See A. Shmida and D. Darom, *Handbook of Trees and Bushes of Israel* (Jerusalem: Keter, 1992).

5. Six sources of water existed in the region and are mentioned in the Bible:

   1. *Gevim*, or hollows in the exposed bedrock at the bottom of a streambed, which retain water for some weeks after a flash flood: "make this streambed full of trenches . . . and the stream shall be filled with water, and you

shall drink" (2 Kings 3:16) and "The nobles sent their lads for water: / they come to the pits, / and find no water; / their vessels return empty. / They are ashamed and confounded, / and cover their heads, / because of the ground which is cracked, / for there had been no rain in the land" (Jeremiah 14:3).

2. *Borot*, or cisterns, chambers hewn into the bedrock or otherwise dug into the ground and lined with stone, which collect surface runoff during rainstorms. The walls of cisterns generally were plastered with lime to seal the cracks and prevent the seepage of the stored water through the porous rock. Cisterns that were not properly sealed were called broken cisterns (Jeremiah 2:13).

3. Shallow wells dug into the source of groundwater (nowadays called an aquifer), generally in the vicinity of intermittent streams, where groundwater may be found at a depth of no more than a few feet. When drawn out, the water in such wells may be gradually replenished by seepage from the surrounding strata as long as they remain saturated: "As a cistern wells with her waters" (Jeremiah 6:7). Shallow wells are not reliable as perennial sources of water, however, since they tend to dry up after a season of drought.

4. Deep wells, typically dug to a depth of 30 feet (10 m) and occasionally to a depth of 160 feet (50 m) or more, which tap into a deep, perennial aquifer. Such wells require communal action to dig, maintain, and regulate, and they provide a reliable, long-term supply of water for an entire village or town. Indeed, the well of each town became the center of communal life, an important meeting place but occasionally also a focus of rivalry.

5. Springs, where water from shallow aquifers emerges spontaneously and flows onto the surface. Numerous springs occur throughout the country, but most are rather meager and subject to fluctuations, as they are affected by the rainfall of the previous winter. Springs may even appear in the desert—for example, the oases of Kadesh Barnea, which served the Israelites in Sinai, and Ein Geddi in the Judean Desert. The most copious of the country's springs are those that feed the upper Jordan River and the Yarkon River (in the coastal plain). A locally important spring is the Gihon or Shiloah, which originally was the main water supply of Jerusalem. As the city grew in population, it needed auxiliary sources of water, which were obtained by means of digging cisterns and, later, diverting and channeling distant springs.

6. Perennial rivers and lakes, which, however, occur in only certain parts of the country (mainly in the north, in the upper Jordan Valley) and served the needs of local settlements.

6. J. N. Tubb has speculated that the name Canaan derives from a Semitic word that implies "subdued" (*Canaanites* [London: British Museum Press, 1998]). Indeed, that sliver of land between sea and desert has always been a sort of no-man's-land between the imperial power centers on either side of it. Hence it has been conquered and subdued repeatedly during its tumultuous history.

7. The Israelites' settlement of Canaan is generally considered to have taken place in the late thirteenth or early twelfth century B.C.E. Egypt's earlier control of

territories in southwestern Asia (including Canaan) had weakened by then. The resulting power vacuum may well have been an important factor in the array of circumstances that made possible the Israelites' settlement. However, the conventional chronology has been challenged in P. James, *Centuries of Darkness: A Challenge to the Conventional Chronology of Old World Archaeology* (London: Cape, 1991).

8. Again, Joshua's act was described as a miracle. As with the Israelites' crossing of the Sea of Reeds (Exodus 14:22), however, a mundane explanation can be offered: as the river meanders through the Lower Jordan Valley, it scours the soft marly deposits of that valley. From time to time, especially when the river is in spate, the swirling flow undermines blocks of earth that may collapse into the river and temporarily dam its flow.

9. I. Finkelstein and N. A. Silberman, *The Bible Unearthed: Archaeology's New Vision of Ancient Israel and the Origin of Its Sacred Texts* (New York: Free Press, 2001), following the earlier conjecture by G. E. Mendenhall, "The Hebrew Conquest of Palestine," *Biblical Archaeology* 25 (1962): 66–87, doubt that the Israelites entered the country from the outside and raise the possibility that they were of indigenous Canaanite origin. A diametrically opposed view is presented in A. Zertal, *A Nation Is Born: The Altar of Mount Ebal and the Origin of Israel* (Tel Aviv: Yedioth, 2000).

10. The Book of Deuteronomy evidently was written long after the settlement of Canaan. See R. E. Friedman, *Who Wrote the Bible?* (New York: Summit Books, 1987). Still, its message is presented as though it were to be delivered to a nation yet in the desert, preparing to enter the promised land.

11. The sharply different appearances of the mountainsides may have led the Samaritans, who continued to worship Yahweh in spite of their separation from Judea at the time of the Second Temple, to build their temple (rivaling that in Jerusalem) on Mount Gerizim. The remaining Samaritans continue to worship there even today. Curiously, however, it was on Mount Ebal that Joshua reportedly built his altar (Joshua 8:30).

12. The scarcity of available land is reflected in the following passage: "The Amorites pressed the Danites into the hill country, they would not let them come down to the plain" (Judges 1:34).

13. Since Jacob had twelve sons, each of whom begot a tribe named after himself, there should have been twelve tribes. But the two sons of Joseph, Ephraim and Manasseh, became the progenitors of separate tribes. Hence there were really thirteen tribes. The country was divided into just twelve territories, though, since the members of one tribe, Levi, were assigned ecclesiastical duties as assistants to the priesthood throughout the territories of all the other tribes, so the Levites remained landless. The priests, known as the Kohanim, being direct descendants of the first priest, Aaron (Moses's brother), were also of the tribe of Levi. After the centralization of religious rites during the monarchy, the Levites and Kohanim in the provincial areas either moved to the capital or assimilated into the local populations.

14. Clearing the forests in the hill district required much work and risked land degradation. The dense growth of trees and shrubs that once covered the slopes had protected the soil naturally. Removing that cover and pulverizing the ground by cultivation, as well as by allowing livestock to trample it, made the soil vulnerable

to accelerated erosion, which could be controlled only partially by the painstaking construction of walled terraces.

15. Evidence that many of the Israelites accepted the Canaanite gods is also seen in some of the names borne by even a few of the leaders. In the time of the Judges, for instance, there was Shamgar, the son of Anath, who smote the Philistines with an ox goad. His mother, Anath, had the name of a particularly gory pagan goddess who features prominently in the records of ancient Ugarit (on the Mediterranean coast of present-day Syria). Even Gideon, the Judge who, with a small band of Israelites, drove away the raiding Midianites and Amalekites, had a second, Canaanite name: Jerubaal, which may have meant "fearer of Ba'al."

16. This event is reminiscent of what may have happened to the Egyptian charioteers at the Sea of Reeds (Exodus 14:25). The story seems aimed at confirming a pattern (or at least a possibility) of divine intervention to save the Israelites in times of extreme peril.

17. The story of Yael and Sisera (Judges 4:17–21) is a twist on the theme of Bedouin hospitality. Alone in a tent, her husband being out with the flocks, Yael was vulnerable to attack by passing strangers. But she was a woman of valor who knew what to do when necessary.

18. The Flood lasted for forty days (Genesis 7:17); Isaac was forty years old when he married Rebekah (Genesis 25:20); Esau was forty years old when he married Judith (Genesis 26:34); the Israelites sojourned in Sinai for forty years (Exodus 16:35; Numbers 14:33); Moses stayed on Mount Sinai for forty days and nights (Deuteronomy 9:25); and the land was peaceful for forty years (Judges 3:11, 8:28, 13:1). Forty-year periods appear in 1 Samuel 4:18; 2 Samuel 5:4, 15:7; 1 Kings 2:11, 11:42; Ezekiel 29:11, 13; Amos 2:10, 5:25; Psalm 95:10; 1 Chronicles 29:27; and 2 Chronicles 9:30.

19. The work of establishing a vineyard on terraced land included clearing the area of stones, digging up the plot to eradicate the weeds and loosen the soil for better penetration of roots, building a fence to prevent incursions by domestic and wild animals, erecting a tower to guard against intruders, planting a superior variety of grape, and fashioning a winepress to extract the juice (which is to be fermented). No explicit mention is made here of the hewing of a cistern to supply water for supplementary irrigation, but we know that farm units in the hill district needed a cistern or access to a spring for a guaranteed supply of water. All this hard labor was generally done by hand. For example, the tilling probably was done by mattock (Isaiah 7:25), as many of the terraced plots were typically too small and narrow to permit plowing by means of an animal-drawn plow.

20. During a famine that followed a prolonged drought, people resorted to desperate measures in the effort to survive. It was at such a time that Elisha bade his servants to prepare pottage from edible wild plants. But as they gathered the plants, one of them evidently made a bad choice and threw wild gourds into the pot. As they began eating the pottage, they cried out: "There is death in the pot!" Elisha, however, did not throw away the pottage; he merely neutralized the offensive taste by adding meal (2 Kings 4:38–41). Of the wild herbs growing in the country, some are indeed edible (for example, mallow [*Malva sylvestris*], which the Arab villagers and shepherds call *khubeiza* [bread]), whereas others are distasteful or even poisonous (for instance, the wild watermelon-gourd [*Citrullus colocynthis*]). See Shmida and Darom, *Handbook of Trees and Bushes of Israel.*

21. The wheat harvest normally takes place in May, after the usual end of the rainy season (October to April). However, a rainstorm in May, while unusual, does occur from time to time.

22. The difference between the semidemocratic regime in Israel, a vestige of the formerly loose and unregimented tribal tradition, and the rigidly absolute monarchy in the Phoenician city-states is exemplified in the story of King Ahab, Queen Jezebel, and the vineyard of Naboth (1 Kings 21). In Israel, it was possible for a commoner to refuse the king's request to purchase his vineyard. But to Jezebel, a Phoenician, such a refusal was an intolerable violation of the king's prerogative, so she contrived to have the recalcitrant Naboth killed and to turn over his coveted property to her husband, the king. But then the prophet Elijah intervened and condemned the king with the famous words: "Would you murder and also take possession? Thus said Yahweh: In the very place where the dogs lapped up Naboth's blood, the dogs will lap up your blood too!" (1 Kings 21:19). Elijah's audacity, like that of the prophet Nathan, in his rebuke to King David (2 Samuel 12), would never have been tolerated in any other monarchical state. When King Rehoboam, the successor to King Solomon, refused to consider the grievances of his subjects and lower the tax burden imposed on them, they split his kingdom and established a rival kingdom over the greater part of the Land of Israel.

23. A depiction of Yahweh in terms reminiscent of the Canaanite imagery of Ba'al was given in the prophesy of Isaiah: "Behold, Yahweh rides upon a swift cloud" (Isaiah 19:1). The word *ba'al*, incidentally, has been retained as a word in Hebrew and Arabic, referring to rainfed farming.

24. Pouring water on the hands of a mentor or master was a sign of a pupil's or servant's obeisance and loyalty. The use of that metaphor suggests that Elisha was heir to Elijah's well-known powers as Yahweh's prophet, believed to be able to invoke supernatural miracles.

25. The very name of the district Edom (and of the people who dwelt there, the Edomites) probably derived from the word *adom* (red).

26. The filling in of wells and cisterns by adversaries of those who dug them is an age-old method of warfare in semiarid regions. It is meant to drive away competitors where grazing land and drinking water are in short supply (Genesis 26:15).

27. A very different version of the conflict between Israel and Moab was provided by the Moabites. An important inscription, known as the Moabite Stone, was discovered in 1868 and was taken to the Louvre. The contents of the inscription point toward the end of King Mesha's reign, perhaps around 830 B.C.E.: "I Mesha, son of Chemosh, king of Moab . . . triumphed over all my adversaries. . . . As for Omri, king of Israel, he humbled Moab many years, for Chemosh [the god of Moab] was angry at his land. And his son followed him and also said, 'I will humble Moab.' . . . But I have triumphed over him and over his house, while Israel hath perished for ever! . . . I fought against the [Israelite] town [Ataroth] and took it and slew all the people of the town as satiation (intoxication) for Chemosh and Moab. . . . And Chemosh said to me, 'Go, take Nebo from Israel!' So I went by night and fought against it from the break of dawn until noon, taking it and slaying all, seven thousand men, boys, women, girls and maid-servants, for I had devoted them to destruction for (the god) Ashtar-Chemosh. And I took from there the [. . .] of Yahweh, dragging them before Chemosh. . . . And the king of

Israel . . . was fighting against me, but Chemosh drove him out before me. . . . It was I (who) built Qarhoh . . . the wall of the citadel . . . its gates . . . and its towers and . . . the king's house, and I made both its reservoirs for water inside the town. And there was no cistern inside the town of Qarhoh, so I said to all the people, 'Let each of you make a cistern for himself in his house!' And I cut beams for Qarhoh with Israelite captives" ("The Moabite Stone," in J. B. Pritchard, ed., *The Ancient Near East: An Anthology of Texts and Pictures* [Princeton, N.J.: Princeton University Press, 1958], 1:209–210).

    The fate of the Moabite Stone is an object lesson in the importance and difficulty of preserving historical and archaeological records. It was first shown by an Arab guide to a Christian missionary in 1868, and it was intact. When the Bedouin who owned the stone realized that it had excited wide interest, they broke it into fragments, probably assuming that selling the large stone piece by piece might bring them more money. Fortunately, however, a man named Charles Claremont-Ganneau had managed to make an imprint of the lettering on the stone before its destruction. So although we do not have the stone itself, we do know what was inscribed on it. Who can tell how many important records of antiquity have been lost or destroyed without being so fortuitously transcribed?

8. THE MARITIME DOMAIN

1. Trade in timber was very important because both Mesopotamia and Egypt, the great centers of civilization and population in the ancient Near East, lacked their own sources of the long, strong, and durable logs of wood needed for the construction of large buildings, including palaces and temples. A mythological description of how the Mesopotamians obtained the cedars of Lebanon is related in the *Epic of Gilgamesh*: Gilgamesh, the king of Uruk, journeyed with his formerly wild companion Enkidu to the cedar mountain, where "Gilgamesh felled a cedar. Then Huwawa, [the ogre] guardian of the forest, was enraged. . . . Enikdu killed the ogre, and Gilgamesh cut down the trees" ("Epic of Gilgamesh," in J. B. Pritchard, ed., *The Ancient Near East: An Anthology of Texts and Pictures* [Princeton, N.J.: Princeton University Press, 1958], 1:40–75).

2. The city of Ugarit, near the Mediterranean coast opposite Cyprus, was occupied from about 3500 to about 1200 B.C.E., when it was destroyed by the Sea Peoples. For some centuries, especially during the mid-second millennium B.C.E., it was a thriving center of trade on the route from Mesopotamia (along the Euphrates River) to Anatolia. Many records have been discovered at the site of Ugarit, mainly in the Akkadian language and written in cuneiform on clay tablets, using numerous word and syllable signs. Some of the records, however, were written in an original kind of alphabet, with about thirty letters. Among these texts were literary expositions of the culture's myths, some of which resembled a few of the stories in the Hebrew Bible.

3. B. Oded, "Israel's Neighbors," in J. Barton, ed., *The Biblical World* (London: Routledge, 2002), 492–525.

4. The noun *ba'al* means, literally, "master" or "owner." By extension, it designates "husband." It is also used to describe rainfed (unirrigated) agriculture. As a verb, the word refers to the performance of the sexual act by a male. Thus the name of the storm god Ba'al implies great power and mastery. Various manifestations of

the storm god were also worshiped in some European cultures, under the names Zeus, Iovis, Jupiter (Io-pater [father Io]), and Thor.

5. A. J. Brody, *"Each Man Cried Out to His God": Specialized Religion of Canaanite and Phoenician Seafarers* (Atlanta: Scholars Press, 1998).

6. Other indications of the Israelites wariness toward the sea are in Isaiah 5:6, Jonah 1:13, and Psalm 77:17.

7. M. S. Smith, *The Origins of Biblical Monotheism: Israel's Polytheistic Background and the Ugaritic Texts* (New York: Oxford University Press, 2001), 33–40.

8. T. Dothan and M. Dothan, *People of the Sea: The Search for the Philistines* (New York: Macmillan, 1992); Oded, "Israel's Neighbors."

9. Amos 9:7 and Jeremiah 47:4 attributed the origin of the Philistines to Caphtor, but the location of Caphtor is not known—perhaps the island of Crete or the coast of Anatolia

10. M. Artzy, "Nomads of the Sea," in R. Hohlfelder and S. Swiny, eds., *Res Maritima 1994: Cyprus and the Eastern Mediterranean, Prehistory Through the Roman Period*, Cyprus American Archaeological Research Institute, Monograph Series, vol. 1 (Atlanta: Scholars Press, 1997), 1–16, and "Routes, Trade, Boats and 'Nomads of the Sea,'" in S. Gitin, A. Mazar, and E. Stern, eds., *Mediterranean Peoples in Transition: Thirteenth to Early Tenth Centuries B.C.E.* (Jerusalem: Israel Exploration Society, 1998), 439–448.

11. The reliefs and inscriptions of the mortuary temple at Medinet Habu are of special interest as records that document the great movements of peoples across and around the eastern Mediterranean, migrations that challenged the dominance of Egypt in the region. In the pictorial carvings, various categories of foreign peoples are distinguished by features and garb. Typical representations are recognizable as characteristic of the Hittites, Syro-Canaanites, Africans, Libyans, and various groups of Sea Peoples. See N. K. Sandars, *The Sea Peoples: Warriors of the Ancient Mediterranean, 1250–1150 B.C.* (London: Thames and Hudson, 1978); J. Finegan, *Archaeological History of the Ancient Middle East* (Bolder, Colo.: Westview Press, 1979); and D. B. Redford, *Egypt, Canaan, and Israel in Ancient Times* (Princeton, N.J.: Princeton University Press, 1992).

12. The traditional plow of the region, used in some parts of the Middle East until recently, was made of wood, and the spike that penetrated and grooved the soil was covered with sheet metal for strength, hardness, and resistance to abrasion. The word translated as "coulter" was rendered as "sickle" in the Septuagint (the translation of the Hebrew Bible into Greek, ca. 250 B.C.E.). The sickle was originally fitted with sharp chips of flint, which served as cutting blades by which to harvest grain and which were disposable once they had lost their edge. Later, it was made of hardened metal that could be sharpened repeatedly.

13. The principal constituent of bronze is copper. Ores of copper were found in many places in the ancient Near East, including Sinai, Anatolia, Transjordan, and even the Negev. When smelted, the ore provided elemental copper, which is a soft metal. So the production of high-quality bronze from copper depended on the availability of tin, which was used, in preference to lead or arsenic, to alloy with copper. Limited quantities of tin apparently were found in the Zagros Mountains and in Spain, but the shortage persisted. In some cases, lead was used as a substitute for tin, but the resulting bronze alloy was of lesser quality.

The advent of iron technology did not immediately signal the end of the use of bronze, and there was no sharp transition from the so-called Bronze Age to the

Iron Age (generally reckoned to have begun around 1200 B.C.E.). When Samson was caught by the Philistines, he was bound in copper fetters (Judges 16:21). When David (ca. 1000 B.C.E.) vanquished the Aramaeans of Zobah and Damascus, he took a vast amount of bronze (2 Samuel 8:8). Then King To'i of Hamath sent his son to greet David and offer him objects of silver, gold, and bronze (2 Samuel 8:9–10). David's successor, Solomon, who built the Temple and an ornate palace in Jerusalem, enlisted the skill of a master craftsman in bronze named Hiram (1 Kings 7:14). The same story, retold in the Book of Chronicles, relates that Huram [*sic*] was an artisan in iron as well as in bronze, gold, silver, precious stones, and wood (2 Chronicles 2:14).

14. One of the many ironies of history is the geographic reversal that has occurred in modern times. The putative descendants of the Israelites, in the state of Israel, are arrayed mainly along the coast, whereas the people named after the Philistines (the Palestinians) are in large part (except in the Gaza Strip) settled in the hill districts of biblical Samaria and Judea.

15. The Hebrew word *yagur* (here translated as "linger") can also mean "shall fear." Thus the sentence *Dan, lamah yagur oniot?* may imply "Dan, why does he fear ships?" The ambiguity may have been intentional. This is but one example of how difficult it is for a translation to capture the nuances of an original text.

16. The story of Samson's heroic exploits takes place entirely in a Philistine milieu, and it is reminiscent of a Mycenaean saga. (His feud with the Philistines, for example, was provoked by a dispute over a woman, much as was the anger of Achilles during the Trojan War. There are other similarities as well.)

    The entire tribe of Dan, indeed, may have been related to the tribes of Sea Peoples known variously as the Denyen, Danaans, Dannuna, and Shardona, as well as to one of the designations for the Greeks in the *Iliad*: the Danaoi. See Y. Yadin, "And Dan, Why Did He Remain in Ships?" *American Journal of Biblical Archaeology* 1 (1968): 9–23, and Dothan and Dothan, *People of the Sea*, 215–219. Curiously, the tribe of Dan is missing from the extensive genealogy of the Israelites listed in 1 Chronicles 1–9.

17. The name Dagon may also be related to the Semitic word *dagan*, which means "grain," so the name may actually suggest "god of grain." Indeed, when the Philistines settled on the land, they became primarily farmers. However, the identification of Dagon with a symbolic fish seems more consistent with the tradition of what was originally a maritime culture. In *Sea Peoples*, Sandars cites contradictory evidence suggesting that the Philistines were people of the land from the start, not one of the Sea Peoples.

18. The strategic importance of Beth-Shean is its location at the eastern edge of the Valley of Jezreel, astride the major route connecting the Mediterranean coast with the Jordan Valley and, beyond it, Transjordan to the east and Aram (Damascus) to the northeast.

19. The Philistines, incidentally, although vanquished by the Israelites and gradually eliminated from the subsequent history of the Near East, eventually had their symbolic revenge. Nearly a millennium later, in the wake of two major Jewish rebellions against the conquering Romans, both ruthlessly crushed, the victors, after having boasted "Judaea capta" on their coins, sought to erase the association of the country with the Jews completely. They spitefully renamed the entire country Palestinae, thus resurrecting the memory of the Philistines, who had

never occupied much more than a small fraction of the country, mainly along the southern coastal plain.

20. Eilat, Elat, Elath, or Ilat means, literally, "goddess [of]." A coin minted in ancient Tyre depicts a goddess standing in a galley, representing the ship's divine guardian. Her name, inscribed in Phoenician, reads "Ilat Tsur" (goddess of Tyre). See Brody, "*Each Man Cried Out to His God.*"

21. The Phoenician colonies had Semitic (Canaanite–Hebrew) names. Carthage and Cartagena are derived from the word *karta* or *keret* (city), and the name Malta implies "haven" (refuge from a stormy sea).

22. A. Yardeni, *The Book of Hebrew Script* (Jerusalem: Carta, 1991).

23. Solomon's naval expedition to Ophir may have inspired (or been inspired by) the reported visit to Jerusalem of the queen of Sheba (1 Kings 10:1–13), who may have come from the same region. However, the veracity and chronology of both reported events have not been established.

24. This venture is reminiscent of the much earlier expedition sent from Egypt during the reign of Queen Hatshepsut (1503–1482 B.C.E.) to the "Land of Punt," also believed to have been somewhere on the Horn of Africa or the southern coast of the Arabian Peninsula.

25. It was Jezebel who conspired to kill Naboth, whose vineyard King Ahab coveted. She was reviled by the writers of the Bible for "her countless harlotries and sorceries" (2 Kings 9:22) and eventually met a very cruel fate (2 Kings 9:30–33). Her daughter Athaliah met a similar end (2 Kings 11:20).

26. Joppa, or Jaffa (alternative transliterations of its Hebrew name, Yaffo), was one of the main seaports in ancient Canaan. It is partly shielded from the waves of the Mediterranean Sea by an offshore rock ("Andromeda's rock" of Greek mythological fame) that constitutes a sort of breakwater and a haven for seafarers (although it can be hazardous to bypass in a stormy sea). Jaffa remained the principal seaport of Palestine until the early part of the twentieth century and became famous in recent times as the port of export for "Jaffa oranges."

27. Several instances of casting lots to divine God's will are given in the Bible. See, for example, the case of Jonathan (1 Samuel 14:41–42).

28. During a windstorm, the sailors of the period could furl the sails of their boat (lest they be torn by the gusting winds or cause the ship to be smashed against the protruding rocks of the seacoast). They would then rely on their own rowing to approach the shore slowly and safely. If the wind was very strong, however, the sailors may not have been able to save themselves and their ship.

29. As the story progresses, the designated deity changed from Yahweh to Yahweh-Elohim to Elohim. One may speculate that the later part of the story was tacked onto the earlier part by an editor. (Admittedly, that is pure speculation.)

30. The plant, with the Hebrew name *kikayon*, has ever since been a symbol of the ephemeral nature of human life and endeavor. It is generally identified with the castor-bean shrub (*Ricinus communis*), which has long been used in the Middle East as a source of oil, mainly for medicinal purposes. It grows very quickly if planted in favorable conditions of soil and water and may attain a height of 7 to 13 feet (2 to 4 m). Its large leaves provide shade, but in dry conditions they wilt and shrivel. The seeds are encased in a smooth cover that contains toxins, but the fleshy interior is nonpoisonous and contains high-quality oil. See A. Shmida and D. Darom, *Handbook of Trees and Bushes of Israel* (Jerusalem: Keter, 1992).

9. THE URBAN DOMAIN

1. An often expressed idea is that Israel's distinctive consciousness originated in its primal desert experience and that the revelation of monotheism occurred necessarily in the vast, silent, clear, and immutable environment of the desert. See A. Alt, "The Settlement of the Israelites in Palestine," in *Essays on Old Testament History and Religion* (Garden City, N.Y.: Doubleday, 1967), 112–120, and H. A. Frankfort, *Kingship and the Gods: A Study of Ancient Near Eastern Religion as the Integration of Society and Nature* (Chicago: University of Chicago Press, 1978). While it is true that prophets such as Moses and Elijah are reported in the Bible to have found their inspiration in the desert, it may also be significant that they came to the desert from the outside. Desert dwellers were less likely to be inspired by the place, as they were engaged daily in a difficult struggle to subsist in an extremely austere environment. Outsiders who sojourn in the desert for a limited time are more likely to be struck—even awe-struck—by its grandeur.

2. K. M. Kenyon, *Archaeology in the Holy Land* (London: Benn, 1979).

3. Although Joshua 10 describes the defeat of the king of Jerusalem and his allies, the city itself was not captured at that time: "And as for the Jebusites, the inhabitants of Jerusalem, the children of Judah could not drive them out; so the Jebusites dwelt with the children of Judah at Jerusalem unto this day" (Joshua 15:63). The last sentence suggests that this passage was written before David captured the city. It contradicts the statement that the men of the tribe of Judah "fought against Jerusalem, and took it, and smote it with the edge of the sword, and set the city on fire" (Judges 1:8).

4. The identification of Mount Moriah with Jerusalem has no basis at all in the story of Abraham and Isaac in Genesis 22, in which the place of sacrifice is described as an uninhabited wilderness. Elsewhere in Genesis (14:18–20), Abraham is reported to have been greeted by King Melchizedek of the city of Salem, which must have been at a different location from Mount Moriah. Salem was later identified along with Zion as Jerusalem: "Elohim has made Himself known to Judah, / His name is great in Israel; / Salem became His abode; / Zion his den" (Psalm 76:3). Only at the time of Solomon was the association made between the Temple Mount of Jerusalem and Mount Moriah: "Then Solomon began to build the House of Yahweh in Jerusalem on Mount Moriah . . . at the threshing floor of Ornan the Jebusite" (2 Chronicles 3:1).

5. The construction of the Temple by Solomon, so elaborately described in the Bible (1 Kings 5–7) and long accepted universally as a historical fact, has become a matter of controversy in recent years. Some scholars even contend that there was no Israelite kingdom before the ninth century B.C.E. and, therefore, that the biblical account of King Solomon is purely mythical. See N. P. Lemche, *The Israelites in History and Tradition* (Louisville, Ky.: Westminster John Knox Press, 1998). These contentions are rebutted by, among others, W. G. Dever, *What Did the Biblical Writers Know and When Did They Know It?* (Grand Rapids, Mich.: Eerdmans, 2001).

6. The injunction to worship Yahweh in one place (rather than in the various places where he had been worshiped before) was confirmed in Deuteronomy 12:11, 14:23, 16:2, 6, 11, and 26:2.

7. The name Urushalima or its Hebrew equivalent, Yerushalayim, was only much later interpreted to signify City of Peace ('Ir-shalom). (That designation expresses the triumph of hope over reality, for—ironically—no city on Earth has seen and suffered more strife and bloodshed than Jerusalem.) Over the centuries, Jerusalem acquired various other epithets, including City of the Jebusites before its conquest by David, City of David and Zion after the conquest, and, much later, City of Justice, Holy City, and City of God. The Arabs call it Urshalim, but add the honorific El Quds (the holy). Arab legend (not mentioned in the Qur'an) has it that Mohammad ascended to heaven from atop the Temple Mount, now known in Arabic as Haram esh-Sharif, or Noble Sanctuary.

8. The measured discharge of the spring is about 260 to 390 cubic yards (200 to 300 cubic m) a day, which would have been sufficient to supply the needs in antiquity of a population numbering several thousand—if the water were used entirely for human needs. In fact, some of the water may have been allocated for agriculture (for example, the growing of vegetable and fruit crops in the vale of Qidron). So before the development of water-tight cisterns, the original population of the Canaanite settlement probably did not exceed one or two thousand.

9. During the Bronze Age, the Gihon served as the principal source of water for Jerusalem, and its output limited the size of the city's population. After lime mortar was developed for plastering porous or fissured walls, in the Iron Age, the people of Jerusalem relied for their water supply mainly on the collection of run-off water from roofs and paved surfaces during winter rains and its storage in numerous underground cisterns. During the Second Temple period, however, as the city's population grew to exceed a hundred thousand, great hydraulic works were undertaken—especially by King Herod—to convey additional water from distant springs into Jerusalem.

10. Before besieging Jerusalem, Sennacherib systematically destroyed most of the fortified cities of Judah. Most famous is his siege and destruction of Lachish in 701 B.C.E., vividly depicted in a relief that is reproduced in P. James, *Centuries of Darkness: A Challenge to the Conventional Chronology of Old World Archaeology* (London: Cape, 1991), 176–177. That city was rebuilt, but again destroyed, this time by King Nebuchadnezzar of Babylonia, in 587 B.C.E.

11. Adventurous visitors can now wade through the tunnel and observe the work of the ancient excavators. It is one of the most impressive hydraulic projects of antiquity. The tunnel, incidentally, is not straight but sinuous, probably because the excavators followed natural fissures in the karstic limestone bedrock. See D. Gill, "Subterranean Water Works of Biblical Jerusalem," *Science* 254 (1991): 1467–1471, and R. Reich and E. Shukron, "Channel II in the City of David," in C. Ohlig, Y. Peleg, and T. Tsuk, eds., *Cura Aquarum in Israel* (Norderstedt: DWhG, 2002), 1–6.

12. Sennacherib's own testimony, recorded in his inscription in Nineveh was quite different: "As for Hezekiah the Jew, he did not submit to my yoke. I laid siege to 46 of his strong cities . . . and conquered them. . . . I drove out of them 200,150 people. . . . Himself I made prisoner in Jerusalem . . . like a bird in a cage. . . . Hezekiah himself, whom the terror-inspiring splendor of my lordship had overwhelmed . . . did send me later, to Nineveh, my lordly city, together with 30 talents of gold, 899 talents of silver" (J. R. Bartlett, ed., *Archaeology and Biblical Interpretation* [London: Routledge, 1997], 80).

13. The site of Megiddo is strategically located at a major crossroads linking the north–south highway from Egypt to Assyria (through what is known today as the 'Arah Pass) and the east–west highway from the Mediterranean coast (by way of the Valleys of Jezreel and Beisän). Many major battles were fought there, including that of Deborah and Barak against the Canaanite commander Sisera (Judges 4–5). According to Revelation 16:16, Megiddo is also destined to be the site of the final battle of Armageddon between the forces of good and evil. The name Armageddon is derived from the Hebrew name Har Megiddo (Mount Megiddo).

14. R. E. Friedman, *Who Wrote the Bible?* (New York: Summit Books, 1987); N. Naaman, "The Kingdom of Judah Under Josiah," *Tel Aviv* 18 (1991): 3–71, and *The Past that Shapes the Present: The Creation of Biblical Historiography in the Late First Temple Period and After the Downfall* (Jerusalem: Hess, 2002); B. M. Metzger and M. D. Coogan, eds., *The Oxford Companion to the Bible* (New York: Oxford University Press, 1993).

15. According to Friedman, in *Who Wrote the Bible?*, Deuteronomy was most probably written by the prophet Jeremiah, who himself was of the priestly cast (although from the provincial town of Anathoth). Deuteronomy also differs from the other books of the Pentateuch in its emphasis on the moral and humanistic spirit in the formulation of its laws. A detailed analysis of these aspect is given in *Encyclopaedia Judaica* (Jerusalem: Keter, 1972), s.v. "Deuteronomy."

16. I. Finkelstein and N. A. Silberman, *The Bible Unearthed: Archaeology's New Vision of Ancient Israel and the Origin of Its Sacred Texts* (New York: Free Press, 2001).

10. THE EXILE DOMAIN

1. The tenuous and continually shifting chain of military-geopolitical enmities and alliances was encapsulated by K. Kenyon: "Between the Hebrew kingdoms and Assyria lay the Aramaic kingdom of Damascus. When Assyria was weak, Damascus was apt to be a thorn in the flesh of Israel. When Assyria was threatening Damascus, Israel was freed from pressure and could recover her lost possessions. When Israel was at grips with Damascus, Judah would free herself from Israelite control. When Judah was suffering at the hands of Israel, Edom could revolt from her, and when Israel in turn was weak, the other kingdoms east of the Jordan could likewise break away, or attack in their turn. And so the train of events went on, with now one country and now another in the ascendant" (*Archaeology in the Holy Land* [London: Benn, 1979], 287).

2. Generations later—after the destruction of Jerusalem, the exile of the Judeans to Babylon, and the return of some of them to Jerusalem—the Samaritans, who apparently retained a memory of the historical tradition of the Israelites and a sense of kinship with the Judeans, wished to join in the reconstruction of the city. By then, however, the religious leaders of the Judeans had evolved a puritanical and exclusive culture of their own and so they rejected the Samaritans as alien heathens. Thereafter, the relationship between the Samaritans and the Judeans was marked by hostility. Throughout the period of the Second Temple, the Judeans considered the Samaritans, who occupied the region known to this day as Samaria (which separated the main Jewish population centers of Judea and Galilee), to be enemies. That is why the story of the "Good Samaritan" (Luke

10:29–37) was such an unusual one, the point being that *even a Samaritan* can be good. A detailed comparison between the Jewish and the Samaritan versions of the Torah is in A. Sadaqa and R. Sadaqa, *Jewish and Samaritan Versions of the Pentateuch* (Tel Aviv: Mas, 1961.)

3. The northern kingdom of Israel, which was destroyed by the Assyrians (2 Kings 17), had included the tribes of Ephraim, Manasseh, Reuben, Gad, Issachar, Zebulun, Asher, Naphtali, Dan, and Benjamin. Over the centuries since that event, numerous legends have been told concerning the fate of the ten "Lost Tribes," speculating that they had survived as a community. Some writers have suggested that they migrated to Central Asia, India, Burma, the British Isles, or even North America.

4. One of those "fortified towns" was Lachish, whose capture by Sennacherib is illustrated dramatically in the reliefs from his palace at Nineveh, now displayed at the British Museum. See Kenyon, *Archaeology in the Holy Land*, 297. The relief provides a vivid portrayal of the military action up to the storming of the city and the carrying off of captives into exile. The accompanying text describes the siege and conquest of "forty-six of the strong cities, walled forts" of "Hezekiah the Jew" and the expulsion of their populations and plunder of their flocks and herds. Excavations at Lachish have revealed much evidence of that fierce battle. See D. Ussishkin, "The Destruction of Lachish by Sennacherib and the Dating of the Royal Judean Storage Jars," *Tel Aviv* 4 (1977): 28–60.

5. The phrase "grass of the roofs" refers to the custom of the time to cover roofs with a thin layer of thatch and soil in order to insulate houses against cold and rain in winter and heat in summer. Any grass that germinated in that shallow substrate would quickly wilt during a spell of hot and dry weather.

6. The name Mattaniah means "gift of Yah[weh]," whereas the name Zedekiah means "justice of Yah[weh]."

7. According to 2 Kings 24:14, the number exiled in the first wave was ten thousand, while eight thousand are specified in 2 Kings 24:16. That initial deportation was followed by a second expulsion of an unspecified number after the destruction of the Temple (2 Kings 25:11). However, Jeremiah 52:28–30 lists a total of 4600 in both expulsions. Although their numbers were limited, the captives included the leaders, and hence the administration and trade of the country was disrupted and the economy could no longer support the populated towns of the former kingdom. See Kenyon, *Archaeology in the Holy Land*. An unknown number of the Judeans who were opposed to the Babylonians then migrated to Egypt (even taking the prophet Jeremiah with them, against his will) and evidently established an exile community there.

8. Some of the locations in which the Judeans were settled are mentioned in the Books of Ezekiel and Ezra: Tel-Aviv (Ezekiel 3:15); Tel-Melah, Tel-Hersha, Cherub, Addan, and Immer (Ezra 2:59); and Casiphia (Ezra 8:17). The prefix *tel* (mound) attached to the names of some of those settlements suggests that they were built on the ruins of earlier towns.

9. The military aspect of the Judeans' allegiance to Persia was exemplified by the presence of a Jewish contingent in the frontier outpost of Aswan in Upper Egypt (guarding the southernmost limit of the Persian Empire) in the late sixth century B.C.E.

10. There is an obvious dichotomy between Isaiah 1–39 and Isaiah 40–66. The differences in style and in historical context have been ascribed to the juxtaposition of two scrolls, the first composed during the First Temple period and the

second—called Deutero-Isaiah—written during the Babylonian exile. See, among other sources, *Encyclopaedia Judaica* (Jerusalem: Keter, 1972), s.v. "Isaiah."

11. The fundamental tenet or message derived from the Bible was that Yahweh's redemption would always be conditional and that the people's security and prosperity would forever depend on their behavior, individually and collectively.

12. In some ways, the cultural and theological creativity of the community in Babylonia surpassed even that of the community in Judea during and after the Second Temple period. The leading centers of learning in Sura, Pumbadita, and Nehardea gave rise to the great collection of works known as the Babylonian Talmud, which has ever since been recognized as one of the pillars of Orthodox Judaism. See, for example, A. Eban, *Heritage: Civilization and the Jews* (New York: Summit Books, 1984).

13. E. Yamauchi, "The Eastern Jewish Diaspora Under the Babylonians," in M. W. Chavalas and K. L. Younger, eds., *Mesopotamia and the Bible* (Grand Rapids, Mich.: Baker Academic, 2002), 356–377.

14. Yamauchi, "Eastern Jewish Diaspora Under the Babylonians," 356–377.

15. R. E. Friedman, *Who Wrote the Bible?* (New York: Summit Books, 1987), 223–225; D. N. Freedman, *The Unity of the Hebrew Bible* (Ann Arbor: University of Michigan Press, 1993).

16. Even the ancient Hebrew (Canaanite) script was abandoned in favor of the Aramaic script, which remains in use, despite an attempt in the second century C.E. to revive the original script. See A. Yardeni, *The Book of Hebrew Script* (Jerusalem: Carta, 1991).

17. The root *babel* or *bavel*, contained in the name Zerubbabel, means, literally, "gate of [the god] El" (*bab-el*).

18. When Manasseh, grandson of the high priest Eliashib, was deposed by Nehemiah because of intermarriage in his family (Nehemiah 13:28), Sanballat, governor of Samaria, built for him a temple on Mount Gerizim to rival the Temple in Jerusalem. Thus the schism between the Samaritans and the Jews acquired a religious as well as an ethnic character.

19. As pointed out by Freedman, the idyllic Book of Ruth, describing the acceptance of a Moabite woman who became an ancestor of King David, has been interpreted as a plea for tolerance against the severe ethnocentric rules laid down by Ezra and enforced by Nehemiah (*Unity of the Hebrew Bible*, 8).

11. THE OVERARCHING UNITY

1. For a different view, in support of the notion of a unique revelation, see the monumental work by Y. Kaufman, *The Religion of Israel* (Chicago: University of Chicago Press, 1960).

2. A rhetorical question, posed in a number of places in the Bible, "Who is like among the gods, O Yahweh?" (Exodus 15:11), presupposes that other gods exist, but that they are inferior to Yahweh. The image of Yahweh as an enthroned king, surrounded by a retinue of lesser divine figures, is found in many passages, including 1 Kings 22, Isaiah 6, and Daniel 7.

3. R. Patai, *The Hebrew Goddess* (Detroit: Wayne State University Press, 1990); S. Scham, "The Lost Goddess of Israel," *Archaeology* 58 (2005): 36–40.

4. The inscription was discovered at a site named Kuntilet 'Ajrud, on the ancient desert road to Eilat (near the present boundary between the western Negev and

eastern Sinai). The outpost apparently was manned by an Israelite contingent, which left several telltale inscriptions there. One of them stated: "May you be blessed by Yahweh of Samaria and his Ashera." Underneath the text is a sketch of three figures, two of whom appear to represent Yahweh and Ashera. See Z. Meshel, "Kuntilat 'Ajrud, an Israelite Site on the Sinai Border" (Hebrew), *Qadmoniot* 9 (1976): 119–124.

5. M. S. Smith speculates that Yahweh and Ashera were worshiped as a divine couple (a form of "ditheism") in Israel as well as in Judah (*The Origins of Biblical Monotheism: Israel's Polytheistic Background and the Ugaritic Texts* [New York: Oxford University Press, 2001], 47). The repeated prophetic condemnations of Ashera (who, in Canaanite tradition, was associated with Ba'al) can be regarded as indirect evidence that such worship was prevalent (as otherwise there would have been no need for such condemnations).

6. Interestingly, the commonly used Hebrew words for "justice" and "charity" derive from the same root. *Tsedeq* (justice) is assigned the masculine gender; *tsedaqah* (charity), the feminine gender. The two concepts, indeed, complement each other.

7. The metaphor of God as father or shepherd is often invoked. Less prevalent is the metaphor of God as intimate suitor or husband (Jeremiah 2:2; Ezekiel 16:6–14).

8. The idea and ideal of theodicy—the justification of all divine acts—was never unanimous, however. The two books of the Bible that explicitly cast doubt on the purposefulness of life and the beneficence of God are Ecclesiastes and Job. Both were included in the canon, but only after the somewhat incongruous addition of finales affirming the very faith that they called into question.

9. This point is made again and again in the Book of Judges, which describes an almost regular pattern: the Israelites did what was evil in the sight of Yahweh. Consequently, he gave them into the hand of an enemy. They then cried out in distress to Yahweh, who raised up a deliverer from among them. After the enemy was subdued or expelled, the Israelites enjoyed a period of tranquility (during which each "sat under his vine and fig tree," an agricultural metaphor indeed!) before they transgressed again, and the cycle was repeated.

10. The impulse toward individual and communal self-examination, acknowledgment of sins, and repentance is exemplified by the ten Days of Awe, beginning with Rosh Hashanah (Head of the Year) and culminating with Yom Kippur (Day of Atonement)—the most solemn holidays at the beginning of the Jewish annual calendar.

11. Only the Book of Job seems to contradict the assumption of guilt as the cause of misfortune and, in its stead, to advocate the stoic and fatalistic acceptance of whatever happens in human life, on the counter-assumption that God's deeds are beyond human understanding.

EPILOGUE

1. Shittim is the name of the acacia tree that grows in the semiarid pastoral domain and in the streambeds of the desert domain. It is the scraggly and thorny tree from whose timber the Israelites fashioned the Ark of the Covenant in the wilderness of Sinai (Exodus 25:10).

APPENDIX 1. ON THE HISTORICAL VALIDITY
OF THE BIBLE

1. If the story of David is untrue, composed merely as a paean to glorify the putative founder of the Judean dynasty, why would the writer have included such embarrassing details as David's leadership of an outlaw band engaged in extortion (1 Samuel 25); David's alliance with Achish, king of Philistia, who was an enemy of Israel (1 Samuel 27); and David's immoral affair with Bathsheba and his atrocious act of arranging the death of her husband (2 Samuel 11)?

2. This position was popularized in W. Keller, *The Bible as History: A Confirmtion of the Book of Books* (New York: Morrow, 1956). The book has been a consistent best-seller since its publication.

3. A case in point is the biblical account of the capture of Jericho and the destruction of the city's walls (Joshua 5:13–6:27). The date of the Israelites' invasion has long been thought to be around 1200 B.C.E. However, excavations in Jericho have revealed no trace of the described event at that time. See P. James, *Centuries of Darkness: A Challenge to the Conventional Chronology of Old World Archaeology* (London: Cape, 1991).

4. J. R. Bartlett stated: "History is the representation of the past as it is understood in the present and in the light of the concerns of the present. The author or authors of the Penateuch . . . were concerned above all with the life and preservation of a community. . . . Hence the emphasis on the common ancestry of the Israelites" (*The Bible: Faith and Evidence: A Critical Enquiry into the Nature of Biblical History* [London: British Museum, 1990], 42)

5. An example of this is the stone structure discovered on the slope of Mount Ebal, near Shechem (present-day Nablus). Archaeologist A. Zertal has suggested that it is the remains of the altar erected by Joshua on that mountain (Deuteronomy 27; Joshua 24) (*A Nation Is Born: The Altar of Mount Ebal and the Origin of Israel* [Tel Aviv: Yedioth, 2000]). Other archaeologists have discounted this possibility, believing that the structure was more likely to have been a guardhouse. See A. Kempinski, "'Joshua's Altar': An Iron Age Watchtower," *Biblical Archaeology Review* 12 (1986): 42–49, and A. Mazar, "Iron Age I and II Towers at Giloh and the Israelite Settlement," *Israel Exploration Journal* 40 (1990): 77–101.

6. A. Biran, *Biblical Dan* (Jerusalem: Israel Exploration Society, 1994).

7. N. C. Habel, *Literary Criticism of the Old Testament* (Philadelphia: Fortress Press, 1971).

8. Spinoza paid dearly for his skepticism. He was excommunicated by the orthodox Jewish community of Amsterdam and shunned by his coreligionists to the end of his life.

9. R. E. Friedman, *Who Wrote the Bible?* (New York: Summit Books, 1987).

10. It was then, according to N. Naaman, that the "Deuteronomic historiography" was compiled, consisting of the Books of Deuteronomy, Joshua, Judges, 1 and 2 Samuel, and 1 and 2 Kings (*The Past that Shapes the Present: The Creation of Biblical Historiography in the Late First Temple Period and After the Downfall* [Jerusalem: Hess, 2002]).

11. For example, I. Finkelstein and N. A. Silberman, *The Bible Unearthed: Archaeology's New Vision of Ancient Israel and the Origin of Its Sacred Texts* (New York: Free Press, 2001). Other scholars, though, point to the remarkable similarity in

detail between some of the stories of the Patriarchs and documents found at such sites in Syria as Nuzi, Mari, and Emar, dating from the second millennium B.C.E. The similarities pertain to personal names (Abraham, Jacob), social customs, and legal practices (for example, the wife-sister episode in Genesis 12, 20, and 26). See L. I. Levine, "Biblical Life and Perspectives: Biblical Archaeology," in D. L. Lieber, ed., *Etz Hayim: Torah and Commentary* (New York: Rabbinical Assembly, 2001), 1339–1344.

12. R. W. Bulliet, *The Camel and the Wheel* (New York: Columbia University Press, 1990).

13. Documents discovered at Nuzi, Mari, and Emar even include such social norms and family arrangements as barren wives submitting their handmaids to their husbands and a firstborn son selling his birthright to his younger brother, which appear to be very similar to those of the Patriarchs. According to biblical scholar W. F. Albright: "Abraham, Isaac, and Jacob . . . now appear as true children of their age, bearing the same names, moving about over the same territory . . . practicing the same customs as their contemporaries" (*From the Stone Age to Christianity: Monotheism and the Historical Process* [Baltimore: Johns Hopkins University Press, 1967], 10–11). See also A. Malamat, *Israel in Biblical Times* (Hebrew) (Jerusalem: Bialik Institute, 1983).

14. In Genesis, the ancestral experience of nomadic (or, rather, seminomadic) pastoralism is compressed into three generations: those of Abraham, Isaac, and Jacob. But each of them may represent the abstraction of several lifetimes. This is especially so with Abraham, whose story is replete with strange twists and turns and inconsistencies: at one time, he was a mighty warrior who defeated the combined forces of four kings (Genesis 14), but then meekly submitted his wife to the king of the small town of Gerar (Genesis 20); he achieved old age (Genesis 24) after having had one wife, one concubine, and two sons, but then married another woman and begat as many as six sons (Genesis 25). It seems that a number of ancestral stories were combined into one in this, as well as in other, narratives.

15. The semiarid and arid lands of southwestern Asia always have been prone to the periodic occurrences of dry seasons, and sometimes to a succession of dry seasons. Egypt, too, would suffer from occasional droughts. However, the weather regimen that determines the water supplies in Canaan differs fundamentally from the regimen that determines the water supply in Egypt. The two water-supply regimens do not rise and fall in tandem, and may actually display an opposite, or seesaw, pattern. See W. H. Quinn, "A Study of Southern Oscillation–Related Climatic Activity for A.D. 622–1900, Incorporating Nile River Flood Data," in H. F. Diaz and V. Markgraf, eds., *El Niño: Historical and Paleoclimatic Aspects of the Southern Oscillation* (Cambridge: Cambridge University Press, 1992), 119–149.

16. M. S. Smith, *The Origins of Biblical Monotheism: Israel's Polytheistic Background and the Ugaritic Texts* (New York: Oxford University Press, 2001); "Asiatics in Egyptian Household Service," in J. B. Pritchard, ed., *The Ancient Near East: An Anthology of Texts and Pictures* (Princeton, N.J.: Princeton University Press, 1975), 2:87–89.

17. Bartlett, *Bible: Faith and Evidence.*

18. "Hymn of Victory of Merneptah," in J. B. Pritchard, ed., *The Ancient Near East: An Anthology of Texts and Pictures* (Princeton, N.J.: Princeton University Press, 1958), 1:231; Bartlett, *Bible: Faith and Evidence,* 50, 54.

19. J. Bright, *A History of Israel* (Philadelphia: Westminster Press, 1972). However, D. B. Redford expressed a completely different view, opining that the biblical story of Joseph is a novella created during the seventh and sixth centuries B.C.E. to set the theme of Israel's descent into Egypt, that the sojourn there and the Exodus are Israelite adaptations of earlier Canaanite traditions, and that there is no reason to believe that the entire story has any basis in fact (*Egypt, Canaan, and Israel in Ancient Times* [Princeton, N.J.: Princeton University Press, 1992], 429).

20. The tales in the Book of Joshua of the massacres of Canaanites by the conquering Israelites (Joshua 6:21, 8:22, 10:20, 30–40, 11:8, 11, 14) are, in effect, refuted in the Book of Judges. That book, pertaining to a later period, attests that the Canaanite cities remained strong. Far from being vanquished early, the Canaanites continued for some generations to harass and subjugate the Israelites. Excavations at Hazor, however, have revealed that the city suffered massive destruction at the end of the thirteenth century B.C.E., about the time of Joshua's alleged conquest. See A. Mazar, "On the Connection Between the Archaeological Research and the Writing of the History of Early Israel" (Hebrew), *Cathedra* 100 (2001): 65–88.

21. In *Bible Unearthed*, Finkelstein and Silberman note that the inhabitants of the Early Iron Age settlements in the central hills of Canaan evidently avoided consuming pork, a fact for which they offer no explanation. One possible reason is that the settlers were of nomadic pastoral origin and, likely, Israelites.

22. I. Finkelstein does not attribute the Israelite settlement in the central hill district of Canaan to invaders from across the Jordan River, but mainly to people who were there all along, although living a different lifestyle (*The Archaeology of Israelite Settlement* [Jerusalem: Israel Exploration Society, 1988]).

23. The difficulty of verifying biblical history is illustrated by the recent controversy regarding an event that reportedly occurred after the death of King Solomon (believed to have ruled the united kingdom of Israel from 970 to 930 B.C.E.). Early in the reign of his son Rehoboam, the kingdom was invaded by the Egyptian pharaoh Sheshonq I, called Shishaq in the Bible (1 Kings 14:25; 2 Chronicles 12:2–9). His army sacked several Israelite strongholds and exacted a heavy tribute from Jerusalem. A team of Israeli and Dutch archaeologists and physicists tested the organic remains from a layer indicating violent destruction in a mound called Tel Rehov, in northern Israel. Using C14 isotope measurements, they dated the stratum to the late tenth century B.C.E., thus apparently corroborating the biblical account of the destructive conquest by Sheshonq. However, this chronology is disputed by another team of archaeologists and physicists in Israel. See H. J. Bruins, J. van der Plicht, and A. Mazar, "C14 Dates from Tel Rehov: Iron Age Chronology, Pharaohs, and Hebrew Kings," *Science* 300 (2003): 315, and the comments and counter-comments in *Science* 302 (2003): 508. In 2005, archaeologist Eilat Mazar discovered the remains of a monumental structure of the tenth century B.C.E. in Jerusalem that she believes may have been the palace of King David. Other archaeologists are skeptical of her claim. And so the controversy continues.

24. The earliest documentary evidence referring to a biblical figure seems to be the stela of King Mesha of Moab (in present-day southern Jordan), which describes events corresponding to the reigns of Omri and Ahab in the mid-ninth century B.C.E.

25. In *Past that Shapes the Present*, Naaman considers that the systematic writing of Hebrew texts began only after literacy became fairly common, after the eighth century B.C.E.

26. By comparison, the histories of Egypt and Mesopotamia had to be reconstructed by modern scholars from disjointed records. Moreover, while the main concern of the Egyptian and Mesopotamian scribes was to glorify their kings' exploits, the authors of the Bible were much more balanced in their approach. They recounted the failures, as well as the successes, of the kings, alternating flattering accounts with critical (even scathing) ones. See James, *Centuries of Darkness*.

27. So inspiring and timeless is this verse that it was inscribed, three millennia after it first was written, on the Liberty Bell in Philadelphia.

APPENDIX 2. PERCEPTIONS OF HUMANITY'S ROLE ON GOD'S EARTH

1. L. White, "The Historical Roots of Our Ecological Crisis," *Science* 155 (1966): 1203.

2. The expression "to serve and preserve it" is my translation of the Hebrew *l'ovdah ul'shomra* (Genesis 2:15), usually rendered "to dress it and keep it" (King James Version), "to till it and keep it" (Revised Standard Version), or "to till it and tend it" (Jewish Publication Society).

3. The word *adamah* is also related to the words *adom* (red) and *dam* (blood). Indeed, the typical soil of the hill districts of Canaan (Judea, Samaria, and Galilee) has, in modern times, been given the name *terra rossa*, for its deep red color. The soil type is prevalent all around the Mediterranean basin.

4. The ancient Hebrew association of humans with soil is echoed in the Latin name for man, *Homo*, derived from *humus*, the living soil. So powerful a metaphor suggests that older civilizations perceived a profound truth about humanity's relationship to the Earth, which modern humans may tend to forget.

5. A little later, there is mention of the beginning of metallurgy, in the account of Tubal-Cain, "an instructor of every artificer in bronze and iron" (Genesis 4:22). The invention of baked bricks is linked with a "plain in the land of Shinar" (Genesis 11:2–3), likely a reference to Sumer. Memories of shipbuilding and viticulture are associated with the story of Noah (Genesis 9:20).

APPENDIX 3. SELECTED PASSAGES REGARDING THE SEVEN DOMAINS

1. H. N. Moldenke and A. L. Moldenke, *Plants of the Bible* (New York: Ronald Press, 1952); A. Alon, *The Natural History of the Land of the Bible* (Jerusalem: Steimatzky, 1969); Y. Feliks, *Nature and Man in the Bible* (Jerusalem: Sancino Press, 1981); Y. Hashulami, *The Scriptures in the Light of Natural Phenomena* (Tel Aviv: Milo, 1996).

2. This pastoral metaphor defines the function of the king in a nation with strong pastoral traditions.

3. Grass symbolizes the brevity and uncertainty of human life.

4. The threat conjures up the perennial fear of shepherds, who must tread through the grass and the bushes. The word *lahash* (whisper) is an onomatopoeic expression of the sound made by a snake, and it is a variant on the Hebrew word for "snake" (*nahash*).

5. Hosea expressed his nostalgia for the simplicity and idyllic purity of the shepherds' life in earlier times.

6. This is, incidentally, one of several mentions of earthquakes—a not unusual occurrence, given the inherent geologic instability of the country.

7. Zephaniah yearned for the renewal of the ancient pastoral life, for a return to the halcyon days of the Patriarchs, to relive the experience of the wilderness and to reaffirm the original love between God and his people.

8. Zechariah likened a leader who abandons his allegiance to Yahweh to an irresponsible shepherd.

9. "Still waters" are the dream of every dweller in an arid land, where water is either sparse or comes in the form of sudden, violent flash floods of short duration. The "valley of the shadow of death" refers to a narrow ravine between sheer cliffs, from which there is no escape against an unexpected flood or lurking marauders.

10. Isaiah alluded to the dust storms and dust devils that swirl in the desert.

11. A streambed in the desert may seem dry and lifeless, but will teem with life after the sudden appearance of water.

12. There are many names for rain in Hebrew (including *delef, geshem, matar, mimtar, nezel, raviv, sagrir,* and *seirim*). The special name for early rain is *yoreh*. It usually occurs in October or November, and it determines the onset of the wet season, which lasts from autumn to spring, during which winter crops are grown. If the rain comes too soon, it is likely to be followed by a dry spell, so seedlings that germinate early may desiccate before the main rains arrive to sustain them. If the *yoreh* is too scanty or arrives too late, the cold weather of winter may hinder early growth, so plants may enter the dry season before they have had the chance to grow and mature sufficiently. The concluding rain of the wet season is called *malkosh*, and it usually arrives in April, at grain-filling and -maturing time. If the *malkosh* is too early or too late, it can limit yields as well as hinder farm operations such as harvesting, threshing, and preparing for the planting of summer crops.

13. Jeremiah expressed the notion that God above holds reservoirs (treasuries) of water, fire, and wind that he can unleash at will.

14. The olive tree is the source of the prized and sanctified olive oil, hence the specter of breaking and burning it is so shocking.

15. In the prophecy of Hosea, the beneficence of God is exemplified by the dependable rain.

16. In contrast, the moral weakness of humanity is compared to the failed rain. Dew is often mentioned, and great benefits are attributed to it, even though the actual contribution of dew to crops in an arid region is generally marginal. Dew, incidentally, does not "fall," as does rain. It condenses on cold surfaces, usually at night, and by mid-morning, most of it usually has evaporated.

17. Hosea provided a series of metaphors drawn from farming in the rainfed domain.

18. In rainfed farming, the seemingly random temporal and spatial distribution (as well as the quantity) of rain can be of crucial importance. Although the early season may be moist, a dry spell in the late season, during the critical grain-filling stage just before ripening, can greatly reduce yields.

19. The roar is of the waves breaking against the rocky shore.

20. Nahum used an alliterative expression, apparently describing the panic and chaos of a city under siege: *Buqa umvuqa umvulaqa, velev namess ufiq birkayim vehalhala bekhol motnayim ufnei kulam qibtzu farur.*

21. There follows a sequence of actions taken by the people of a besieged city.

22. The regularity of migratory birds is contrasted with the waywardness of the people.

23. Micah offered a striking image of a volcanic eruption.

24. Channels in arid lands conduct water infrequently and unpredictably, but a flash flood quickly passes away and the channel turns dry when water is needed most.

BIBLIOGRAPHY

Abercrombie, T. J. "Arabia's Frankincense Trail." *National Geographic*, October 1985, 475–512.

Abramsky, S. *Ancient Towns in Israel*. Jerusalem: World Zionist Organization, 1963.

Adams, R. McC. *Heartland of Cities: Surveys of Ancient Settlements and Land Use on the Central Floodplain of the Euphrates*. Chicago: University of Chicago Press, 1981.

Adams, R. McC. "Soil and Water as Critical Factors in the History of the Fertile Crescent." In R. S. Baker, G. W. Gee, and C. Rosenzweig, eds., *Soil and Water Science: Key to Understanding Our Global Environment*, 11–14. Proceedings of a Symposium of the Soil Science Society of America Held in Honor of Daniel Hillel. Madison, Wis.: Soil Science Society of America, 1994.

Ahad Haam [Asher Ginzberg]. "Moses." *Hashelah* 13 (1904): 342–347.

Aharoni, Y. "The Horned Altar of Beer Sheba." *Biblical Archaeology* 37 (1979): 2–6.

Aharoni, Y. *The Land of the Bible: A Historical Geography*. Philadelphia: Westminster Press, 1979.

Aharoni, Y., M. Avi-Yonah, A. F. Rainey, and Z. Zafrai. *The Macmillan Bible Atlas*. Jerusalem: Carta, 1993.

Ahituv, S., ed. *Studies in the Archaeology of Nomads in the Negev and Sinai* (Hebrew). Jerusalem: Bialic Institute, 1998.

Ahituv, S., and B. A. Levine, eds. *Avraham Malamat Volume*. Jerusalem: Israel Exploration Society, 1993.

Ahlström, G. W. *The History of Ancient Palestine from the Palaeolithic Period to Alexander's Conquest*. Sheffield, Eng.: Sheffield Academic Press, 1993.

Albert, J., M. Bernhardsson, and R. Kenna, eds. *Transformation of Middle Eastern Natural Environments: Legacies and Lessons*. New Haven, Conn.: Yale University School of Forestry and Environmental Studies, 1998.

Albright, W. F. "Abraham the Hebrew: A New Archaeological Interpretation." *Bulletin of the American School of Oriental Research* 163 (1961): 36–54.

Albright, W. F. "The Babylonian Matter in the Predeuteronomic Primeval History in Gen. 1–11." *Journal of Biblical Literature* 58 (1939): 9–103.

Albright, W. F. *From the Stone Age to Christianity: Monotheism and the Historical Process.* Baltimore: Johns Hopkins University Press, 1967.

Albright, W. F. "The Gezer Calendar." *Bulletin of the American School of Oriental Research* 92 (1943): 16–26.

Albright, W. F. *The Proto-Sinaic Inscriptions and Their Decipherment.* Cambridge, Mass.: Harvard University Press, 1966.

Albright, W. F. *Yahweh and the Gods of Canaan: A Historical Analysis of Two Contrasting Faiths.* Winona Lake, Ind.: Eisenbrauns, 1968.

Allan, J. P. *Genesis in Egypt: The Philosophy of Ancient Egyptian Creation Accounts.* Yale Egyptological Studies, vol. 2. New Haven, Conn.: Yale University Press, 1988.

Alon, A. *The Natural History of the Land of the Bible.* Jerusalem: Steimatzky, 1969.

Alon, D. *A New Light on the Exodus from Egypt* (Hebrew). Jerusalem: Ariel, 1999.

Alt, A. "The Settlement of the Israelites in Palestine." In *Essays on Old Testament History and Religion*, 112–120. Garden City, N.Y.: Doubleday, 1967.

Alter, R. *The Art of Biblical Narrative.* New York: Basic Books, 1981.

Alter, R. *The World of Biblical Literature.* New York: Basic Books, 1992.

Alter, R., and F. Kermode, eds. *The Literary Guide to the Bible.* Cambridge, Mass.: Harvard University Press, 1987.

Amit, Y. *Revealed and Hidden in the Bible* (Hebrew). Tel Aviv: Yediot Aharonot, 2003.

Anati, E. *Har Karkom in the Light of New Discoveries.* Capo di Ponte, Italy: Edizioni del Centro, 1993.

Anati, E. "The Rock Art of the Negev Desert." *Near Eastern Archaeology* 62 (1999): 22–34.

Anderson, B. W., ed. *Creation in the Old Testament.* Philadelphia: Fortress Press, 1984.

Artzy, M. "Incense, Camels, and Collared Rim Jars: Desert Trade Routes and Maritime Outlets in the Second Millennium." *Oxford Journal of Archaeology* 13 (1994): 121–127.

Artzy, M. "Nomads of the Sea." In R. Hohlfelder and S. Swiny, eds., *Res Maritima 1994: Cyprus and the Eastern Mediterranean, Prehistory Through the Roman Period,* 1–16. Cyprus American Archaeological Research Institute, Monograph Series, vol. 1. Atlanta: Scholars Press, 1997.

Artzy, M. "Routes, Trade, Boats and 'Nomads of the Sea.'" In S. Gitin, A. Mazar, and E. Stern, eds., *Mediterranean Peoples in Transition: Thirteenth to Early Tenth Centuries B.C.E.,* 439–448. Jerusalem: Israel Exploration Society, 1998.

Artzy, M., and D. Hillel. "A Defense of the Theory of Progressive Salinization in Ancient Mosopotamia." *Geoarchaeology* 3 (1988): 235–238.

Assman, J. *Moses der Ägypter.* Munich: Hanzer, 1998.

Aubet, M. E. *The Phoenicians and the West: Politics, Colonies, and Trade.* Cambridge: Cambridge University Press, 2001.

Austin, R. C. *Hope for the Land: Nature in the Bible.* Atlanta: John Knox Press, 1988.

Avidag, N. *Discovering Jerusalem.* Oxford: Blackwell, 1980.

Avitsur, S. *Daily Life in Eretz Israel in the XIX Century.* Tel Aviv: Am Hassefer, 1972.

Avitsur, S. *Man and His Work: Historical Atlas of Tools and Workshops in the Holy Land.* Jerusalem: Carta and Israel Exploration Society, 1976.

Avi-Yonah, M., and E. G. Kraeling. *Our Living Bible*. Chicago: Biblical Publications, 1972.

Bailey, C., ed. *Notes on the Bedouins: Aspects of Bedouin Culture and Folklore in Sinai and the Negev* (Hebrew). Sde Boker: Field Studies School, 1975.

Baker, J. A. "Biblical Attitudes to Nature." In H. Montefiore, ed., *Man and Nature*, 56–67. London: Collins, 1975.

Baly, D. *The Geography of the Bible*. New York: Harper & Row, 1974.

Bare, G. *Plants and Animals of the Bible*. London: United Bible Study, 1969.

Baron, S. *The Desert Locust*. New York: Scribner, 1972.

Barr, J. *The Semantics of Biblical Language*. Oxford: Oxford University Press, 1961.

Bar-Tanah, A. *Fundamental Issues in Judaism*. Tel Aviv: University of Tel Aviv Press, 1985.

Bartlett, J. R. *The Bible: Faith and Evidence: A Critical Enquiry into the Nature of Biblical History*. London: British Museum, 1990.

Bartlett, J. R., ed. *Archaeology and Biblical Interpretation*. London: Routledge, 1997.

Barton, J., ed. *The Biblical World*. London: Routledge, 1990.

Bar Yoseph, O. "The Origins of Pastoral Societies in the Levant" (Hebrew). In S. Ahituv, ed., *Studies in the Archaeology of Nomads in Negev and Sinai*, 7–26. Jerusalem: Bialik Institute, 1998.

Ben Dov, M. *In the Shadow of the Temple: The Discovery of Ancient Jerusalem*. New York: Harper & Row, 1985

Ben Gurion, D. *Biblical Reflections*. Tel Aviv: Davidson-Itai, 1976.

Ben Tor, A., ed. *The Archaeology of Ancient Israel*. New Haven, Conn.: Yale University Press, 1992.

Ben Yehuda, E. *A Complete Dictionary of Ancient and Modern Hebrew*. New York: Yoseloff, 1960.

Ben Yoseph, J. "The Climate in Eretz Israel During Biblical Times." *Hebrew Studies* 26 (1985): 225–239.

Berquist, J. L. *Judaism in Persia's Shadow: A Social and Historical Approach*. Minneapolis: Fortress Press, 1995.

Berry, T., and T. Clarke. *Befriending the Earth: A Theology of Reconciliation Between Humans and the Earth*. Mystic, Conn.: Twenty-third, 1991.

Biale, D. *Eros and the Jews: From Biblical Israel to Contemporary America*. New York: Basic Books, 1992.

Bienkowski, P., and A. Millard, eds. *Dictionary of the Ancient Near East*. Philadelphia: University of Pennsylvania Press, 2000.

Bietak, M. *Avaris: The Capital of the Hyksos*. London: British Museum Press, 1996.

Biran, A. *Biblical Dan*. Jerusalem: Israel Exploration Society, 1994.

Biran, A., and J. Naveh. "An Aramaic Stele Fragment from Tel Dan." *Israel Exploration Journal* 43 (1993): 81–98.

Bodenheimer, F. S. *Animals and Man in Bible Lands*. Leiden: Brill, 1960.

Borowski, O. *Agriculture in Iron Age Israel*. Winona Lake, Ind.: Eisenbrauns, 1987.

Bright, J. *A History of Israel*. Philadelphia: Westminster Press, 1972.

Brody, A. J. *"Each Man Cried Out to His God": Specialized Religion of Canaanite and Phoenician Seafarers*. Atlanta: Scholars Press, 1998.

Bronowski, J. *A Sense of the Future: Essays in Natural Philosophy*. Cambridge, Mass.: MIT Press, 1977.

Brower, D. J., and E. Teeter. *Egypt and the Egyptians*. Cambridge: Cambridge University Press, 1999.

Brown, F., S. R. Driver, and C. A. Briggs. *Hebrew and English Lexicon*. 1906. Reprint, Peabody, Mass.: Hendrickson, 1997.

Brown, W. P. *The Ethos of the Cosmos: The Genesis of Moral Imagination in the Bible*. Grand Rapids, Mich.: Eerdmans, 1999.

Bruins, H. J., J. van der Plicht, and A. Mazar. "¹⁴C Dates from Tel Rehov: Iron-Age Chronology, Pharaohs, and Hebrew Kings." *Science* 300 (2003): 315–318.

Buber, M. *On the Bible*. Syracuse, N.Y.: Syracuse University Press, 2000.

Budge, E. A. W. *From Fetish to God in Ancient Egypt*. Oxford: Oxford University Press, 1934.

Bulfinch, T. *The Age of Fable*. New York: Harper & Row, 1966.

Bulliet, R. W. *The Camel and the Wheel*. New York: Columbia University Press, 1990.

Burkert, W. *Creation of the Sacred: Tracks of Biology in Early Religions*. Cambridge, Mass.: Harvard University Press, 1996.

Bustenay, O. "Israel's Neighbors." In J. Barton, ed., *The Biblical World*, 492–525. London: Routledge, 1990.

Butzer, K. W. "Agricultural Origins in the Near East as a Geographical Problem." In S. Struever, ed., *Prehistoric Agriculture*, 79–92. Garden City, N.Y.: Natural History Press, 1971.

Butzer, K. W. *Early Hydraulic Civilization in Egypt: A Study in Cultural Ecology*. Chicago: University of Chicago Press, 1976.

Butzer, K. W. *Environment and Archeology: An Ecological Approach to Prehistory*. Chicago: Aldine Atherton, 1971.

Cahill, T. *The Gifts of the Jews: How a Tribe of Desert Nomads Changed the Way Everyone Thinks and Feels*. New York: Random House, 1998.

Calvocoressi, P. *Who's Who in the Bible*. London: Penguin Books, 1987.

Carne, J. *Syria, the Holy Land, and Asia Minor, Illustrated*. London: Fisher, 1842.

Carroll, J. E., P. Brockelman, and M. Westfall, eds. *The Greening of Faiths: God, the Environment, and the Good Life*. Hanover, N.H.: University Press of New England, 1997.

Cassuto, M. D. "Gezer: The Gezer Calendar." In M. D. Cassuto, ed., *Biblical Encyclopedia*, 2:471–474. Jerusalem: Bialik Institute, 1954.

Cassuto, U. *The Goddess Anath: Canaanite Epics of the Patriarchal Age* (Hebrew). Jerusalem: Bialik Institute, 1965.

Chomsky, W. *Hebrew: The Eternal Language*. Philadelphia: Jewish Publication Society, 1957.

Clare, J. D., and H. Wansbrough. *The Bible Alive: Witness the Great Events of the Bible*. London: HarperCollins, 1993.

Clutton-Brock, J. *Horse Power: A History of the Horse and Donkey in Human Societies*. London: Natural History Museum, 1992.

Clutton-Brock, J. *A Natural History of Domesticated Mammals*. Cambridge: Cambridge University Press, 1999.

Cohon, S. S. "The Name of God: A Study in Rabbinic Theology." In *Annual* 23, pt. 1:579–604. Cincinnati: Hebrew Union College, 1951.

Coogan, M. D. *Stories from Ancient Canaan*. Philadelphia: Westminster Press, 1978.

Crane, E. *The Archaeology of Beekeeping*. London: Duckworth, 1983.

Cross, F. M. *Canaanite Myth and Hebrew Epic: Essays in the History of the Religion of Israel*. Cambridge, Mass.: Harvard University Press, 1973.

Dalley, S, trans. *Myths from Mesopotamia: Creation, the Flood, Gilgamesh, and Others.* Oxford: Oxford University Press, 1989.

Danin, A. *Desert Vegetation of Israel and Sinai.* Jerusalem: Cana, 1983.

Danin, A. "Man and the Natural Environment." In T. E. Levy, ed., *The Archaeology of Society in the Holy Land,* 24–30. New York: Facts on File, 1995.

Danin, A. "The Vegetation and Flora of Sinai." In G. Gvirtsman, A. Shmueli, Y. Grados, I. Beit Arie, and M. Harel, eds., *Sinai,* 46–56. Tel Aviv: Ministry of Defense, 1987.

Danin, A. "Vegetation Units of Israel in Satellite Imaging at a Scale of 1:750,000" (Hebrew). In B. Z. Kedar and A. Danin, eds., *Remote-Sensing Aerial Photography and Satellite Imaging as Investigative Tools in Israel Studies,* 20–30. Jerusalem: Yad Ben-Zvi, 2000.

Danin, A., and G. Orshan, eds. *Vegetation of Israel.* Vol. 1, *Desert and Coastal Vegetation.* Leiden: Backhuys, 1999.

David, A. R. *The Ancient Egyptians: Religious Beliefs and Practices.* London: Routledge and Kegan Paul, 1982.

Davis, J. D., and H. S. Gehman. *Westminster Dictionary of the Bible.* Philadelphia: Westminster Press, 1944.

Day, J. *God's Conflict with the Dragon and the Sea: Echoes of a Canaanite Myth in the Old Testament.* Cambridge: Cambridge University Press, 1985.

Day, J. *The Gods of Canaan and the Old Testament.* Sheffield, Eng.: JSOT Press, 1999.

Dearman, J. A., ed. *Studies in the Mesha Inscription and Moab.* Atlanta: Scholars Press, 1989.

Dennett, D. C. "Appraising Grace: What Evolutionary Good Is God?" *Sciences* 37 (1997): 39–45.

De Vaux, R. *The Early History of Israel.* Philadelphia: Westminster Press, 1978.

Dever, W. G. *What Did the Biblical Writers Know and When Did They Know It?* Grand Rapids, Mich.: Eerdmans, 2001.

De Wit, C. "Le Circoncision chez les anciens Égyptiens." *Aegyptische Sprache und Altertumskunde* 99 (1972): 41–48.

Diaz, H. F., and V. Markgraf, eds. *El Niño: Historical and Paleoclimatic Aspects of the Southern Oscillation.* Cambridge: Cambridge University Press, 1992.

Dor, M. *Animal Life in the Time of the Bible, Mishna, and Talmud.* Tel Aviv: Daftal-Graphor, 1997.

Dothan, T. "Ekron of the Philistines: Where They Came From, How They Settled Down, and the Place They Worshipped." *Biblical Archaeology Review* 16 (1990): 26–36.

Dothan, T., and M. Dothan. *People of the Sea: The Search for the Philistines.* New York: Macmillan, 1992.

Drews, R. *Early Riders: The Beginnings of Mounted Warfare in Asia and Europe.* London: Routledge, 2004.

Driver, G. R. *Semitic Writing from Pictograph to Alphabet.* London: Oxford University Press, 1976.

Duvshani, M. *The Fundamentals of Prophesy* (Hebrew). Tel Aviv: Yavneh, 1965.

Eban, A. *Heritage: Civilization and the Jews.* New York: Summit Books, 1984.

Edelstein, G., and Y. Gat. "Terraces Around Jerusalem" (Hebrew). *Israel—Land and Nature* 6 (1981): 72–78.

Efal, I., ed. *The History of Eretz Israel: Israel and Judah in the Biblical Period* (Hebrew). Jerusalem: Keter, 1984.

Einstein, A. *Ideas and Opinions*. New York: Crown, Wings Books, 1954.

Eissenfeldt, O. *The Old Testament: An Introduction*. New York: Harper & Row, 1965.

El 'Aref, A. *Bedouin Tribes in the District of Beersheba*. 1937. Reprint, Jerusalem: Ariel, 2000.

Eliade, M., ed. *The Encyclopedia of Religion*. 16 vols. New York: Macmillan, 1987.

Elitzur, A. C. *Into the Holy of Holies: Psychoanalytic Insights into the Bible and Judaism* (Hebrew). Tel Aviv: Yarom, 1988.

Encyclopaedia Hebraica. 32 vols. Jerusalem: Encyclopaedia, 1949–1981.

Encyclopaedia Judaica. 16 vols. Jerusalem: Keter, 1972.

Encyclopedia of the Bible. 8 vols. Jerusalem: Bialic Institute, 1971.

Etzion, Y. *The Lost Bible* (Hebrew). Jerusalem: Shoken, 1992.

Evenari, M., L. Shanan, and N. Tadmor. *The Negev: The Challenge of a Desert*. Cambridge, Mass.: Harvard University Press, 1971.

Fauna and Flora of the Bible. New York: United Bible Society, 1980.

Feinbrun-Dothan, N., and A. Danin. *Analytical Flora of Eretz Israel*. Jerusalem: Keter, 1991.

Feliks, J. *Agriculture in Palestine in the Period of the Mishna and Talmud* (Hebrew). Jerusalem: Magnes Press, 1963.

Feliks, Y. *Agriculture in Eretz-Israel in the Periods of the Bible, the Misnah, and the Talmud: Halakkah [Doctrine] and Practice in Basic Agricultural Work* (Hebrew). Jerusalem: Mas, 1990.

Feliks, Y. *Nature and Man in the Bible*. Jerusalem: Sancino Press, 1981.

Feliks, Y. *Species of Fruit Trees, Plants of the Bible and the Sages* (Hebrew). Jerusalem: Mas, 1994.

Finegan, J. *Archaeological History of the Ancient Middle East*. Bolder, Colo.: Westview Press, 1979.

Finkelstein, I. *The Archaeology of Israelite Settlement*. Jerusalem: Israel Exploration Society, 1988.

Finkelstein, I. "The Great Transformation: The 'Conquest' of the Highlands Frontiers and the Rise of the Territorial States." In T. E. Levy, ed., *The Archaeology of Society in the Holy Land*, 349–365. New York: Facts on File, 1995.

Finkelstein, I. "The Philistines in the Bible: A Late-Monarchic Perspective." *Journal for the Study of the Old Testament* 27 (2002): 131–167.

Finkelstein, I. "State Formation in Israel and Judah." *Near Eastern Archaeology* 62 (1999): 35–52.

Finkelstein, I., and N. Na'aman, eds. *From Nomadism to Monarchy: Archaeological and Historical Aspects of Early Israel*. Jerusalem: Yad Ben-Zvi, 1994.

Finkelstein, I., and N. A. Silberman. "Archaeology and Scripture at the Beginning of the Third Millennium: A View from the Center" (Hebrew). *Cathedra* 100 (2001): 47–64.

Finkelstein, I., and N. A. Silberman. *The Bible Unearthed: Archaeology's New Vision of Ancient Israel and the Origin of Its Sacred Texts*. New York: Free Press, 2001.

Flexner, A. *Lectures on the Geology of Israel*. Tel Aviv: Ministry of Defense, 1989.

Flusser, D. *Jewish Sources in Early Christianity* (Hebrew). Tel Aviv: Ministry of Defense, 1980.

Foster, B. R., ed. *The Epic of Gilgamesh: A New Translation, Analogues, Criticism*. New York: Norton, 2000.

Frank, H. T., ed. *Hammond's Atlas of the Bible Lands*. Maplewood, N.J.: Hammond, 1997.

Frank, H. T., C. W. Swain, and C. Canby. *The Bible Through the Ages*. Cleveland: World, 1967.

Frankfort, H. A. *Kingship and the Gods: A Study of Ancient Near Eastern Religion as the Integration of Society and Nature*. Chicago: University of Chicago Press, 1978.

Frazer, J. G. *Folklore in the Old Testament: Studies in Comparative Religion, Legend, and Law*. New York: Tudor, 1923.

Freedman, D. N. *The Unity of the Hebrew Bible*. Ann Arbor: University of Michigan Press, 1993.

Freud, S. *Moses and Monotheism*. New York: Knopf, 1939.

Frick, F. S. *The Formation of the State in Ancient Israel*. Social World of Biblical Antiquity, no. 4. Sheffield, Eng.: JSOT Press, 1985.

Friedman, D. *To Kill and Inherit* (Hebrew). Tel Aviv: Dvir, 2000.

Friedman, R. E. *The Hidden Book in the Bible*. San Francisco: Harper, 1998.

Friedman, R. E. *Who Wrote the Bible?* New York: Summit Books, 1987.

Frumkin, A., A Shimron, and J. Rosenbaum. "Radiometric Dating of the Siloam Tunnel, Jerusalem." *Nature* 425 (2003): 169–171.

Frye, N. *The Great Code: The Bible and Literature*. San Diego, Calif.: Harcourt Brace Jovanovich, 1983.

Frymer-Kensky, T. *In the Wake of the Goddesses: Women, Culture, and the Biblical Transformation of Pagan Myth*. New York: Free Press, 1992.

Galil, J. "An Ancient Technique for Ripening Sycamore Fruit in East Mediterranean Countries." *Economic Botany* 22 (1968): 179–190.

Galpaz-Feller, P. *Exodus: Reality or Illusion?* (Hebrew). Jerusalem: Shoken, 2002.

Gardner, G. *Invoking the Spirit: Religion and Spirituality in the Quest for a Sustainable World*. Washington, D.C.: Worldwatch Institute, 2002.

Gates, C. Ancient Cities: *The Archaeology of Urban Life in the Ancient Near East and Egypt, Greece, and Rome*. London: Routledge, 2004.

Geller, S. A. *Sacred Enigmas: Literary Religion in the Hebrew Bible*. London: Routledge, 1996.

Gesenius, W. *A Hebrew and English Lexicon of the Old Testament*. Oxford: Clarendon Press, 1906.

Gilead, I. "Pastoralists and Farmers in the Northern Negev During the Chalcolithic Age" (Hebrew). In S. Ahituv, ed., *Studies in the Archaeology of Nomads in the Negev and Sinai*, 43–58. Jerusalem: Bialik Institute, 1998.

Gill, D. "Subterranean Water Works of Biblical Jerusalem." *Science* 254 (1991): 1467–1471.

Gleick, J. *Chaos: Making a New Science*. New York: Viking Books, 1987.

Glueck, N. *Deities and Dolphins*. New York: Farrar, Straus and Giroux, 1965.

Glueck, N. *The Other Side of the Jordan*. Cambridge, Mass.: American School of Oriental Research, 1970.

Glueck, N. *The River Jordan*. Philadelphia: Westminster Press, 1946.

Glueck, N. *Rivers in the Desert: A History of the Negev*. New York: Norton, 1959.

Goldreich, Y. *The Climate of Israel: Observations, Research, and Applications*. Ramat Gan: Bar Ilan University Press, 1998.

Gordis, R. "The Earth Is the Lord's: Ecology in the Jewish Tradition." *Midstream* 31 (1985): 19–23.

Gordon, C. H. *Before the Bible* (Hebrew). Tel Aviv: Am Ored, 1966.

Gordon, C. H. "Biblical Customs and the Nuzi Tables." In E. F. Campbell and D. N. Freedman, eds., *The Biblical Archaeologist Reader*, 2:20–30. Garden City, N.Y.: Doubleday, 1964.

Gordon, C. H., and G. A. Rendoburg. *The Bible and the Ancient Near East.* New York: Norton, 1997.

Gottwald, N. K. *The Tribes of Yahweh: A Sociology of the Religion of Liberated Israel.* London: SCM Press, 1980.

Gould, S. J. *Rocks of Ages: Science and Religion in the Fullness of Life.* New York: Ballantine Books, 1999.

Gould, S. J. *Time's Arrow, Time's Cycle: Myth and Metaphor in the Discovery of Geological Time.* Cambridge, Mass: Harvard University Press, 1987.

Graves, R., and R. Patai. *Hebrew Myths: The Books of Genesis.* New York: Greenwich House, 1983.

Green, A. R. W. *The Storm-God in the Ancient Near East.* Winona Lake, Ind.: Eisenbrauns, 2003.

Greenberg, M. *The Hab/piru.* New Haven, Conn.: Yale University Press, 1995.

Greenberg, M. *Understanding Exodus.* New York: Behrman House, 1969.

Grintz, I. M. *Stories, Poems and Parables in Ancient Egypt's Literature.* Jerusalem: Keter, 1975.

Gvirtsman, G., A. Shmueli, Y. Grados, I. Beit Arie, and M. Harel, eds. *Sinai.* Tel Aviv: Ministry of Defense, 1987.

Habel, N. C. *Literary Criticism of the Old Testament.* Philadelphia: Fortress Press, 1971.

Hackett, J. A. *The Balaam Text from Deir 'Alla.* Harvard Semitic Monographs, no. 31. Chico, Calif.: Scholars Press, 1984.

Halpern, B. *The First Historians: The Hebrew Bible and History.* University Park: Pennsylvania State University Press, 1996.

Halpern, B., and J. D. Levenson, eds. *Traditions in Transformation: Turning Points in Biblical Faith.* Winona Lake, Ind.: Eisenbrauns, 1981.

Haran, M., ed. *The World of the Bible* (Hebrew). Tel Aviv: Davidson-Itai, 1993.

Harden, D. *The Phoenicians.* London: British Museum Press, 1971.

Harel, M. *Historical Geography of the Land of Israel.* Tel Aviv: Am Oved, 2002.

Harel, M. *Landscape, Nature, and Man in the Scriptures.* Tel Aviv: Am Oved, 1984.

Harel, M. *Travels in Sinai.* Tel Aviv: Am Oved, 1968.

Hareuveni, N. *Desert and Herder in the Legacy of Israel* (Hebrew). Lod: Neot Kudimim, 1991.

Hareuveni, N. *Ecology in the Bible.* Kiryat Ono: Neot Kedumim, 1974.

Hareuveni, N. *Nature and Landscape in the Legacy of Israel* (Hebrew). Kiryat Ono: Neot Kedumim, 1980.

Hareuveni, N. *Shrub and Tree in the Legacy of Israel* (Hebrew). Kiryat Ono: Neot Kedumim, 1984.

Harlan, J. R. *The Living Fields: Our Agricultural Heritage.* Cambridge: Cambridge University Press, 1995.

Harlan, J. R., and D. Zohary. "Distribution of Wild Wheats and Barley." *Science* 153 (1966): 1074–1080.

Hart, G. *Egyptian Myths.* London: British Museum Press, 1999.

Hashulami, Y. *The Bible in Light of Natural Phenomena.* Tel Aviv: Milo, 1996.

Hastings, J., ed. *Dictionary of the Bible.* New York: Scribner, 1963.

Hatziminaoglou, Y., and J. Boyazoglu. "The Goat in Ancient Civilizations: From the Fertile Crescent to the Aegean Sea." *Small Ruminant Research* 51 (2004): 123–129.

Hay, L. S. "What Really Happened at the Sea of Reeds?" *Journal of Biblical Literature* 83 (1964): 397–403.

Heidel, A. *The Babylonian Genesis*. Chicago: University of Chicago Press, 1951.

Herzog, Z. "The *Tanakh*: No Findings on the Ground." *Haaretz Weekly Magazine*, 29 October 1999, 30–40.

Heschel, A. J. *God in Search of Man: A Philosophy of Judaism*. New York: Harper & Row, 1955.

Hetzron, R., ed. *The Semitic Languages*. London: Routledge, 1997.

Hiebert, T. *The Yahwist's Landscape: Nature and Religion in Early Israel*. New York: Oxford University Press, 1996.

Hillel, D. *Environmental Soil Physics*. San Diego, Calif.: Academic Press, 1998.

Hillel, D. *Negev: Land, Water, and Life in a Desert Environment*. New York: Praeger, 1982.

Hillel, D. *Out of the Earth: Civilization and the Life of the Soil*. Berkeley: University of California Press, 1992.

Hillel, D. *Rivers of Eden: The Struggle for Water and the Quest for Peace in the Middle East*. New York: Oxford University Press, 1994.

Hillel, D., and N. Tadmor. "Water Regime and Vegetation in the Negev Highlands." *Ecology* 43 (1962): 33–41.

Hillers, D. R. *Covenant: The History of a Biblical Idea*. Baltimore: Johns Hopkins University Press, 1969.

Hjelm, I. *The Samaritans and Early Judaism*. Sheffield, Eng.: Sheffield Academic Press, 1998.

Hobbs, J. J. *Bedouin Life in the Egyptian Wilderness*. Cairo: American University Press, 1990.

Hobson, C. *The World of the Pharaohs*. New York: Thames and Hudson, 1987.

Hodder, I. "Women and Men at Catalhoyuk." *Scientific American*, January 2004, 76–83.

Hoffman, J. M. *In the Beginning: A Short History of the Hebrew Language*. New York: New York University Press, 2004.

Hopf, M. "Plant Remains and Early Farming in Jericho." In P. J. Ucko and G. W. Dimbleby, eds., *The Domestication and Exploitation of Plants and Animals*, 97–110. Chicago: Aldine, 1969.

Hopkins, D. C. *The Highlands of Canaan: Agricultural Life in the Early Iron Age*. Sheffield, Eng.: JSOT Press, 1985.

Hopkins, D. C. "Life on the Land: The Subsistence Struggles of Early Israel." *Biblical Archaeologist* 50 (1987): 178–191.

Hostetter, E. C. *An Elementary Grammar of Biblical Hebrew*. Sheffield, Eng.: Sheffield Academic Press, 2000.

Houtman, C. *Der Himmel im Alten Testament: Israels Weltbild und Weltanschauung*. Leiden: Brill, 1993.

Howells, V. *A Naturalist in Palestine*. London: Melrose, 1956.

Hughes, J. D. *Ecology in Ancient Civilizations*. Albuquerque: University of New Mexico Press, 1975.

Hurlbut, J. L. *Manual of Biblical Geography*. Chicago: Rand, McNally, 1884.

Hütterman, A. "The Ecological Message of the Torah: A Biologist's Interpretation of the Mosaic Law." *Israel Journal of Botany* 40 (1991): 183–195.

Hütterman, A. "Die ökologische Botschaft der Thora." *Naturwissenschaften* 80 (1993): 147–156.

Hyams, E. *Plants in the Service of Man: 10,000 Years of Domestication*. Philadelphia: Lippincott, 1971.

Jackson, F. "The Coevolutionary Relationship of Humans and Domesticated Plants." *Yearbook of Physical Anthropology* 39 (1996):161–176.

Jacobsen, T. *Salinity and Irrigated Agriculture in Antiquity*. Malibu, Calif.: Undena, 1982.

Jacobsen, T. *The Treasures of Darkness: A History of Mesopotamian Religion*. New Haven, Conn.: Yale University Press, 1976.

Jacobsen, T., and R. McC. Adams. "Silt and Salt in Mesopotamia." *Science* 128 (1958): 1251–1258.

Jaffe, S. "How People Faced the Drought in the Land of Israel." *Israel Meteorological Research Papers* 5 (1994): 46–48.

James, P. *Centuries of Darkness: A Challenge to the Conventional Chronology of Old World Archaeology*. London: Cape, 1991.

James, T. G. H. *An Introduction to Ancient Egypt*. New York: Harper & Row, 1979.

Japhet, S. "Was the History of the People of Israel Invented During the Persian Period?" (Hebrew). *Cathedra* 100 (2001): 110–112.

Jarman, H. N. "The Origins of Wheat and Barley Cultivation." In E. S. Higgs, ed., *Papers in Economic History*, 15–26. Cambridge: Cambridge University Press, 1972.

Jarvis, C. S. *Yesterday and Today in Sinai*. Edinburgh: Blackwood, 1933.

Jean, G. *Writing: The Story of Alphabets and Scripts*. London: Thames and Hudson, 1992.

Joines, K. R. "The Bronze Serpent in the Israelite Cult." *Journal of Biblical Literature* 81 (1968): 245–256.

Jordan, M. *Encyclopedia of Gods*. New York: Facts on File, 1993.

Judaism and Ecology. New York: Hadassah and Shomrei Adamah, 1993.

Kadmon, R., and A. Danin. "Distribution of Plant Species in Israel in Relation to Spatial Variation in Rainfall." *Journal of Vegetation Science* 10 (1999): 421–432.

Kadmon, R., and A. Danin. "Floristic Variation in Israel: A GIS Analysis." *Flora* 192 (1977): 341–345.

Kaufman, Y. *The Religion of Israel*. Chicago: University of Chicago Press, 1960.

Kautzsch, E., ed. *Gesenius' Hebrew Grammar*. Oxford: Oxford University Press, 1910.

Kay, J. "Concepts of Nature in the Hebrew Bible." *Environmental Ethics* 10 (1988): 309–327.

Kedar, Y. *The Ancient Agriculture in the Negev* (Hebrew). Jerusalem: Bialik Institute, 1967.

Keel, O. *The Song of Songs: A Continental Commentary*. Minneapolis: Fortress Press, 1994.

Keel, O., and C. Uehlinger. *Gods, Goddesses, and the Images of God in Ancient Israel*. Minneapolis: Fortress Press, 1996.

Keller, W. *The Bible as History: A Confirmation of the Book of Books*. New York: Morrow, 1956.

Kempinski, A. "'Joshua's Altar': An Iron Age Watchtower." *Biblical Archaeology Review* 12 (1986): 42–49.

Kempinski, A. "When History Sleeps, Theology Arises: A Note on Joshua 8:30–35 and the Archaeology of the 'Settlement Period.'" In S. Ahituv and B. A. Levine, eds.,

Avraham Malamat Volume, 175–183. Jerusalem: Jerusalem Exploration Society, 1993.

Kenyon, K. M. *Archaeology in the Holy Land*. London: Benn, 1979.

Khalid, F. M., and J. O'Brien, eds. *Israel and Ecology*. London: Cassell, 1994.

King, E. *Plants of the Holy Scriptures, with a Check-list of Plants that Are Mentioned in the Bible*. New York: New York Botanical Garden, 1948.

Kirsch, J. *The Harlot by the Side of the Road: Forbidden Tales of the Bible*. New York: Ballantine Books, 1997.

Kislev, M. E., E. Weiss, and A. Hartmann. "Impetus for Sowing and the Beginning of Agriculture." *Proceedings of the National Academy of Science* 10 (2004): 2692–2695.

Klein, E. *A Comprehensive Etymological Dictionary of the Hebrew Language*. Jerusalem: Carta, 1987.

Klingbeil, M. *Yahweh Fighting from Heaven: God as a Warrior and as a God of Heaven in the Hebrew Psalter and Ancient Near Eastern Iconography*. Fribourg: University Press, 1999.

Kloos, C. *Yhwh's Combat with the Sea: A Canaanite Tradition in the Religion of Ancient Israel*. Amsterdam: Leiden, 1986.

Knapp, J. *The History and Culture of Ancient Western Asia and Egypt*. Chicago: Dorsey Press, 1997.

Kramer, S. N., ed. *Mythologies of the Ancient World*. New York: Anchor Books, 1961.

Kugel, J. L. *The Bible as It Was*. New York: Free Press, 1997.

Kugel, J. L. *The God of Old: Inside the Lost World of the Bible*. New York: Free Press, 2003.

Kugel, J. L., and R. A. Greer. *Early Biblical Interpretation*. Philadelphia: Westminster Press, 1986.

Kurtz, P., ed. *Science and Religion: Are They Compatible?* Amherst, N.Y.: Prometheus Books, 2003.

Lambdin, T. O. "Egyptian Loan Words in the Old Testament." *Journal of the American Oriental Society* 73 (1953): 145–155.

Lamberg-Karlovsky, C. C., and J. A. Sabloff. *Ancient Civilizations: The Near East and Mesopotamia*. Prospect Heights, Ill.: Waveland Press, 1995.

Lawrence, T. E. *Seven Pillars of Wisdom*. Garden City, N.Y.: Doubleday Doran, 1935.

Laymon, C. M., ed. *The Interpreters One-Volume Commentary on the Bible*. Nashville, Tenn.: Abingdon Press, 1971.

Leick, G. *Mesopotamia: The Invention of the City*. London: Penguin Books, 1999.

Lemche, N. P. *The Canaanites and Their Land: The Tradition of the Canaanites*. Sheffield, Eng.: Sheffield Academic Press, 1995.

Lemche, N. P. *The Israelites in History and Tradition*. Louisville, Ky.: Westminster John Knox Press, 1998.

Levenson, J. D. "From Temple to Synagogue." In B. Halpern and J. D. Levenson, eds., *Traditions in Transformation: Turning Points in Biblical Faith*, 52–65. Winona Lake, Ind.: Eisenbrauns, 1981.

Levin, L., and A. Mazar, eds. *The Controversy over the Historical Truth in the Bible* (Hebrew). Jerusalem: Yad Ben-Zvi and Merkaz Dinur, 2001.

Levy, T. E., ed. *The Archaeology of Society in the Holy Land*. New York: Facts on File, 1995.

Lieber, D. L., ed. *Etz Hayim: Torah and Commentary*. New York: Rabbinical Assembly, 2001.

Liebermann, Y. *Biblical Statistics* (Hebrew). Tel Aviv: Cherikover, 1995.

Lloyd, S. *The Archaeology of Mesopotamia: From Old Stone Age to the Persian Conquest.* London: Thames and Hudson, 1978.

Luria, B. Z. *Antiquity of the Hebrews: Studies in the Patriarchal Era* (Hebrew). Jerusalem: Kiryat Sepher, 1977.

Maas, A. J. "Jehovah (Yahweh)." In *The Catholic Encyclopedia.* Available at: www.newadvent.org/cathen.

MacKay, A. I. *Farming and Gardening in the Bible.* Old Tappan, N.J.: Rodale Press, 1950.

Mackenzie, D. A. *Egyptian Myths and Legends.* New York: Gramercy Books, 1978.

Malamat, A. *Israel in Biblical Times* (Hebrew). Jerusalem: Bialik Institute, 1983.

Malamat, A. "The Kingdom of Judah Between Egypt and Babylon: A Small State Within a Grand Power Confrontation." In W. Classen, ed., *Text and Context*, 77–98. Sheffield, Eng.: JSOT Press, 1983.

Mazar, A. "Iron Age I and II Towers at Giloh and the Israelite Settlement." *Israel Exploration Journal* 40 (1990): 77–101.

Mazar, A. "On the Connection Between the Archaeological Research and the Writing of the History of Early Israel" (Hebrew). *Cathedra* 100 (2001): 65–88.

Mazar, B. *Canaan and Israel* (Hebrew). Jerusalem: Bialik Institute, 1980.

Mazar, B. "The Early Israelite Settlement in the Hill Country." *Bulletin of the American School of Oriental Research* 241 (1981): 75–85.

McCall, H. *Mesopotamian Myths.* London: British Museum Press, 1999.

McKay, A. L. *Farming and Gardening in the Bible.* Emmaus, Pa.: Rodale Press, 1950.

McKibben, B. *The Comforting Whirlwind: God, Job, and the Scale of Creation.* Grand Rapids, Mich.: Eerdmans, 1994.

Mendenhall, G. E. "Ancient Israel's Hyphenated History." In D. N. Freedman and D. F. Graf, eds., *Palestine in Transition*, 91–103. Sheffield, Eng.: Almond Press, 1983.

Mendelhall, G. E. "The Hebrew Conquest of Palestine." *Biblical Archaeology* 25 (1962): 66–87.

Mendenhall, G. E. *The Tenth Generation: The Origins of Biblical Tradition.* Baltimore: Johns Hopkins University Press, 1973.

Meshel, Z. "Kuntilat 'Ajrud, an Israelite Site on the Sinai Border" (Hebrew). *Qadmoniot* 9 (1976): 119–124.

Meshel, Z., and I. Finkelstein, eds. *Sinai in Antiquity: Researches in the History and Archaeology of the Peninsula* (Hebrew). Tel Aviv: Hakkibutz Hamenchad, 1980.

Metzger, B. M., and M. D. Coogan, eds. *The Oxford Companion to the Bible.* New York: Oxford University Press, 1993.

Meyerson, P. *The Ancient Agricultural Regime of Nessana and the Central Negev.* London: Colt Archaeological Institute, 1960.

Miller, J. M., and J. H. Hayes. *A History of Ancient Israel and Judah.* Philadelphia: Westminster Press, 1991.

Moldenke, H. N., and A. L. Moldenke. *Plants of the Bible.* New York: Ronald Press, 1952.

Moorey, P. R. S. *A Century of Biblical Archaeology.* Cambridge: Lutterworth Press, 1991.

Moses, C. *Science in Ancient Mesopotamia.* New York: Grolier, 1998

Naaman, N. "*Habiru* and Hebrews: The Transfer of a Social Order to the Literary Sphere." *Journal of Near Eastern Studies* 45 (1986): 271–288.

Naaman, N. "The Kingdom of Judah Under Josiah." *Tel Aviv* 18 (1991): 3–71.

Naaman, N. *The Past that Shapes the Present: The Creation of Biblical Historiography in the Late First Temple Period and After the Downfall*. Jerusalem: Hess, 2002.

Nakhai, B. A. *Archaeology and the Religions of Canaan and Israel*. Boston: American School of Oriental Research, 2001.

Naveh, J. *The Beginning of the Alphabet*. Jerusalem: Carta, 1989.

Naveh, J. *Early History of the Alphabet: An Introduction to West Semitic Epigraphy and Palaeography*. Leiden: Brill, 1982.

Naveh, Z. *Ecology of Man and Landscape* (Hebrew). Haifa: Gestlit, 1981.

Ness, Y., and E. Shturm, eds. *Questions About God* (Hebrew). Or Yehuda: Hed Arzi, 1998.

Niditch, S. *Folklore and the Hebrew Bible*. Minneapolis: Fortress Press, 1993.

Nir, D. *Geomorphology of Eretz Israel* (Hebrew). Jerusalem: Akademon, 1989.

Noth, M. *The History of Israel*. New York: Harper & Row, 1960.

Oats, J. *Babylon*. London: Thames and Hudson, 1986.

Oded, B. "Ancient Israel's Eastern Neighbors." In *Biblical Encyclopedia*. Jerusalem: Keter, 1985.

Oded, B. *The History of Israel in the Period of the First Temple* (Hebrew). Tel Aviv: Open University, 1983.

Oded, B. "Israel's Neighbors." In J. Barton, ed., *The Biblical World*, 492–525. London: Routledge, 2002.

Ohlig, C., Y. Peleg, and T. Tsuk, eds. *Cura Aquarum in Israel*. Norderstedt: DWhG, 2002.

Oppenheim, A. *Ancient Mesopotamia: Portrait of a Dead Civilization*. Chicago: University of Chicago Press, 1977.

Oppenheimer, A., A. Kasher, and U. Rapaport, eds. *Man and Land in Ancient Eretz Israel* (Hebrew). Jerusalem: Yad Ben-Zvi, 1986.

Orion, E., ed. *The Bible as a Meeting of Cultures on the Axis of Time* (Hebrew). Sdeh Boqer: Environmental Education, 1995.

Orlinsky, H. M. *Understanding the Bible Through History and Archaeology*. New York: Ktav, 1972.

Orni, E., and E. Efrat. *Geography of Israel*. Philadelphia: Jewish Publication Society, 1973.

Osborne, C., ed. *The Israelites*. New York: Time-Life Books, 1968.

Patai, R. *The Hebrew Goddess*. Detroit: Wayne State University Press, 1990.

Plaut, W. G. "General Introduction to the Torah." In W. G. Plaut, ed., *The Torah: Modern Commentary*, 1–10. New York: Union of American Hebrew Congregations, 1981.

Ponting, C. *A Green History of the World: The Environment and the Collapse of Great Civilizations*. New York: Penguin Books, 1993.

Porter, J. R. *The Illustrated Guide to the Bible*. New York: Oxford University Press, 1995.

Porter, S. E., and R. S. Hess, eds. *Translating the Bible: Problems and Prospects*. Sheffield, Eng.: Sheffield Academic Press, 1996.

Postgate, J. N., and M. A. Powell, eds. "Irrigation and Cultivation in Mesopotamia, Part 1" [special issue]. *Bulletin on Sumerian Culture* 4 (1988).

Postgate, J. N., and M. A. Powell, eds. "Irrigation and Cultivation in Mesopotamia, Part 2" [special issue]. *Bulletin on Sumerian Culture* 5 (1990).

Pritchard, J. B., ed. *The Ancient Near East: An Anthology of Texts and Pictures.* 2 vols. Princeton, N.J.: Princeton University Press, 1958, 1975.

Pritchard, J. B., ed. *The Times Atlas of the Bible.* London: Times Books, 1987.

Quinn, W. H. "A Study of Southern Oscillation–Related Climatic Activity for A.D. 622–1900, Incorporating Nile River Flood Data." In H. F. Diaz and V. Markgraf, eds., *El Niño: Historical and Paleoclimatic Aspects of the Southern Oscillation,* 119–149. Cambridge: Cambridge University Press, 1992.

Rabin, C. *Semitic Languages: An Introduction.* Jerusalem: Bialik, 1991.

Rasmussen, C. G. *Atlas of the Bible.* Grand Rapids, Mich.: Zondervan, 1989.

Ravitsky, R., ed. *Readers of Genesis: Israel's Women Write About the Women of Genesis* (Hebrew). Tel Aviv: Yedioth, 2001.

Reader, J. *Man on Earth.* London: Collins, 1988.

Redford, D. B. *Akhenaten: The Heretic King.* Princeton, N.J.: Princeton University Press, 1984.

Redford, D. B. *Egypt, Canaan, and Israel in Ancient Times.* Princeton, N.J.: Princeton University Press, 1992.

Redman, C. L. *The Rise of Civilization: From Early Farmers to Urban Society in the Ancient Near East.* San Francisco: Freeman, 1978.

Reich, R., G. Avni, and T. Winter. *The Jerusalem Archaeological Park.* Jerusalem: Israel Antiquities Authority, 1999.

Reifenberg, A. *The Soils of Palestine: Studies in Soil Formation and Land Utilisation in the Mediterranean.* London: Murray, 1938.

Reifenberg., A. *The Struggle Between the Desert and the Sown: Rise and Fall of Agriculture in the Levant.* Jerusalem: Jewish Agency, 1955.

Reik, T. *Myth and Guilt: The Crime and Punishment of Mankind.* New York: Grosset and Dunlap, 1970.

Richardson, M. E. J. *Hammurabi's Laws: Text, Translation, and Glossary.* Sheffield, Eng.: Sheffield Academic Press, 2000.

Roberts, D. *Holy Land Lithographs.* London: Day, 1855.

Robinson, A. *The Story of Writing.* New York: Thames and Hudson, 1999.

Rockefeller, S. C., and J. C. Elder, eds. *Spirit and Nature: Why the Environment Is a Religious Issue.* Boston: Beacon Press, 1992.

Rogerson, J., and P. R. Davies. "Was the Siloam Tunnel Built by Hezekiah?" *Biblical Archaeologist* 59 (1997): 138–149.

Rohl, D. M. *Pharaohs and Kings: A Biblical Quest.* New York: Crown, 1995.

Römer, T. C. *The Deuteronomistic History: A Social Science Commentary.* Sheffield, Eng.: Sheffield Academic Press, 2000.

Ron, Z. "Agricultural Terraces in the Judean Mountains." *Israel Exploration Journal* 16 (1966): 33–49.

Rose, A., ed. *Judaism and Ecology.* London: Cassell, 1990.

Rosen, S. A. "The Archaeology of Pastoral Nomads" (Hebrew). In S. Ahituv, ed., *Studies in the Archaeology of Nomads in the Negev and Sinai,* 27–42. Jerusalem: Bialic Institute, 1998.

Rubinstein, E. *Contemporary Hebrew and Ancient Hebrew.* Tel Aviv: Ministry of Defense, 1989.

Ruether, R. R. *Gaia and God: An Ecofeminist Theology of Earth Healing.* San Francisco: Harper, 1994.

Ryan, W., and W. Pitman. *Noah's Flood: The New Scientific Discoveries About the Event that Changed History.* New York: Simon and Schuster, 1998.

Sadaqa, A., and R. Sadaqa. *Jewish and Samaritan Versions of the Pentateuch.* Tel Aviv: Mas, 1961.

Saenz-Badillas, A. *A History of the Hebrew Language.* Cambridge: Cambridge University Press, 1993.

Saggs, H. W. F. *Civilization Before Greece and Rome.* New Haven, Conn.: Yale University Press, 1989.

Saggs, H. W. F. *The Greatness that Was Babylon* (Hebrew). Tel Aviv: Friedman, 1972.

Said, R. *The River Nile: Geology, Hydrology, and Utilization.* Oxford: Pergamon Press, 1993.

Sandars, N. K. *The Sea Peoples: Warriors of the Ancient Mediterranean, 1250–1150 B.C.* London: Thames and Hudson, 1978.

Sandmel, S. *The Hebrew Scriptures: An Introduction to Their Literature and Religious Ideas.* New York: Oxford University Press, 1978.

Sarna, N. M. *Exploring Exodus.* New York: Shocken Books, 1986.

Sarna, N. M. *On the Book of Psalms: Exploring the Prayers of Ancient Israel.* New York: Shocken Books, 1993.

Sasson, J. M. "Circumcision in the Ancient Near East." *Journal of Biblical Literature* 85 (1966): 473–476.

Scham, S. "The Lost Goddess of Israel." *Archaeology* 58 (2005): 36–40.

Schama, S. *Landscape and Memory.* New York: Knopf, 1995.

Schniedewind, W. M. *How the Bible Became a Book: The Textualization of Ancient Israel.* New York: Cambridge University Press, 2004.

Shanks, H. "Face to Face: Biblical Minimalists Meet Their Challengers." *Biblical Archaeology Review* 23–24 (1997): 26–42, 66.

Shanks, H., ed. *Ancient Israel: A Short History from Abraham to the Roman Destruction of the Temple.* Washington, D.C.: Biblical Archaeological Society, 1988.

Shavit, Y., and M. Eran. *The War of the Tablets: Defense of the Bible in the Nineteenth Century and the Controversy Regarding Babylonia and the Bible.* Tel Aviv: Am Oved, 2003.

Shaw, I., and P. Nicholson. *British Museum Dictionary of Ancient Egypt.* Cairo: American University in Cairo Press, 1998.

Shifra, S., and J. Klein, eds. *In Those Distant Days: Anthology of Mesopotamian Literature in Hebrew.* Tel Aviv: Am Oved, 1996.

Shiloh, Y. "Underground Water Systems in the Land of Israel in the Iron Age." In A. Kempinski and R. Reich, eds., *The Architecture of Ancient Israel: From the Prehistoric to the Persian Period,* 55–75. Jerusalem: Israel Exploration Society, 1992.

Shmida, A., and D. Darom. *Handbook of Trees and Bushes of Israel.* Jerusalem: Keter, 1992.

Simkins, R. A. *Creator and Creation: Nature in the Worldview of Ancient Israel.* Peabody, Mass.: Hendrickson, 1994.

Simmonds, N. W., ed. *Evolution of Crop Plants.* London: Longman, 1976.

Smith, B. *The Westminster Concise Bible Dictionary.* Philadelphia: Westminster Press, 1981.

Smith, G. A. *The Historical Geography of the Holy Land.* New York: Harper & Row, 1972.

Smith, M. S. *The Origins of Biblical Monotheism: Israel's Polytheistic Background and the Ugaritic Texts.* New York: Oxford University Press, 2001.

Smith, M. S. "Ugaritic Studies and the Hebrew Bible, 1968–1998 (with an Excursus on Judean Monotheism and the Ugaritic Texts)." In A. Lemaire and M. Soebo, eds., *Congress Volume Oslo 1998,* 112–120. Leiden: Brill, 2000.

Smith, R. B., J. Foster, N. Kouchoukos, P. A. Gluhosky, R. Young, and E. DePauw. "Spatial Analysis of Climate, Landscape, and Hydology in the Middle East: Modeling and Remote Sensing." Paper, Department of Geology and Geophysics, Yale University, 1999.

Smith, W. S. *Animals and Birds of the Bible.* Nashville, Tenn.: Abingdon Press, 1974.

Speiser, E. A. *Genesis: A New Translation with Introduction and Commentary.* Anchor Bible, vol. 1. New York: Doubleday, 1964.

Speiser, E. A. *Oriental and Biblical Studies: Collected Writings of E. A. Speiser.* Edited by J. J. Finkelstein and M. Greenberg. Philadelphia: University of Pennsylvania Press, 1967.

Sperling, S. D. *The Original Torah: The Political Intent of the Bible Writers.* New York: New York University Press, 1998.

Spong, J. S. *Rescuing the Bible from Fundamentalism.* San Francisco: Harper, 1992.

Spring, D., and E. Spring, eds. *Ecology and Religion in History.* New York: Harper & Row, 1974.

Stager, L. E. "Archaeology, Ecology, and Social History: Background Themes to the Song of Deborah." In J. A. Emerton, ed., *Congress Proceedings,* 221–234. Leiden: Brill, 1988.

Stager, L. E. "The Archaeology of the Family in Ancient Israel." *Bulletin of the American School of Oriental Research* 260 (1985): 1–35.

Stager, L. E. "Merneptah, Israel, and the Sea Peoples: New Light on an Old Relief." *Israel Exploration Journal* 18 (1985): 56–64.

Stager, L. E. "The Song of Deborah: Why Some Tribes Answered the Call and Some Did Not." *Biblical Archaeology Review* 15 (1989): 50–64.

Stein, D. E., ed. *A Garden of Choice Fruit: 200 Classic Jewish Quotes on Human Beings and the Environment.* Wyncote, Pa.: Shomrei Adamah / Keepers of the Earth, 1991.

Steinsaltz, A. *Figures in the Bible* (Hebrew). Tel Aviv: Ministry of Defense, 1980.

Thompson, T. L. *The Historicity of the Patriarchal Narratives: The Quest for the Historical Abraham.* Berlin: de Gruyter, 1974.

Thompson, T. L. *The Mythic Past: Biblical Archaeology and the Myth of Israel.* London: Cape, 1999.

Tov, E. *Textual Criticism of the Hebrew Bible.* Minneapolis: Fortress Press, 1992.

Tristram, H. B. *The Fauna and Flora of Palestine.* London: Palestine Exploration Fund, 1884.

Tristram, H. B. *The Natural History of the Bible.* London: Society for Promoting Christian Knowledge, 1873.

Tsiper, E. *The Canaanite Was Then in the Land* (Hebrew). Tel Aviv: Museum of Eretz Israel, 1990.

Tubb, J. N. *Canaanites.* London: British Museum Press, 1998.

Tucker, M. E. "Worldly Wonder: Religions Enter an Ecological Phase." *E: The Environmental Magazine,* November–December 2002, 12–14.

Tucker, M. E., and J. A. Grim, eds. *Worldviews and Ecology: Religion, Philosophy, and the Environment.* Maryknoll, N.Y.: Orbis Books, 1994.

Ussishkin, D. "The Destruction of Lachish by Sennacherib and the Dating of the Royal Judean Storage Jars." *Tel Aviv* 4 (1977): 28–60.

Van der Woude, A. S., ed. *The World of the Old Testament.* Grand Rapids, Mich.: Eerdmans, 1989.

Van Goudoever, J. *Biblical Calendars.* Leiden: Brill, 1959.

van Seters, J. *Abraham in History and Tradition.* New Haven, Conn., Yale University Press, 1975.

van Seters, J. *In Search of History: Historiography in the Ancient World and the Origins of Biblical History.* New Haven, Conn.: Yale University Press, 1983.

Vasey, D. E. *Ecological History of Agriculture, 10,000 B.C.–A.D. 10,000.* Ames: Iowa State University Press, 1992.

von Dassow, E. "What the Canaanite Cuneiformists Wrote: Review Article." *Israel Exploration Journal* 53 (2003): 196–217.

von Soden, W. *The Ancient Orient: An Introduction to the Study of the Ancient Near East.* Grand Rapids, Mich.: Eerdmans, 1985.

Walker, W. *All the Plants of the Bible.* New York: Harper, 1957.

Ward, M., ed. *Biblical Studies in Contemporary Thought.* Sommerville, Mass.: Greene, Hadden, 1975.

Waskow, A., ed. *Torah of the Earth: Exploring 4,000 Years of Ecology in Jewish Thought.* 2 vols. Woodstock, Vt.: Jewish Lights, 2000.

Weinberg, W. *The History of Hebrew Plene Spelling.* Cincinnati: Hebrew Union College Press, 1985.

Weinfeld, M. *The Promise of the Land: The Inheritance of the Land of Canaan by the Israelites.* Taubman Lectures in Jewish Studies. Berkeley: University of California Press, 1993.

Weinstein, J. "Exodus and Archaeological Reality." In E. S. Frerichs and H. H. Lesko, eds., *Exodus: The Egyptian Evidence,* 87–103. Winona Lake, Ind.: Eisenbrauns, 1997.

Wellhausen, J. *Prolegomena to the History of Ancient Israel.* 1885. Reprint, New York: Meridian Books, 1957.

Westman, H. *The Structure of Biblical Myths: The Ontogenesis of the Psyche.* Dallas: Spring, 1983.

White, L. "The Historical Roots of Our Ecological Crisis." *Science* 155 (1966): 1203–1207.

Whitewell, W. M. "Insects in the Bible." In M. C. Tenney, ed., *Zondervan Pictorial Bible Dictionary,* 70–94. Grand Rapids, Mich.: Zondervan, 1963.

Wilkinson, R. H. *The Complete Gods and Goddesses of Ancient Egypt.* London: Thames and Hudson, 2003.

Wilkinson, T. J. *Archaeological Landscapes of the Near East.* Tucson: University of Arizona Press, 2003.

Williams, J. G. *Understanding the Old Testament.* Woodbury, N.Y.: Educational Series, 1972.

Wilson, C. W., ed. *Picturesque Palestine, Sinai and Egypt.* 4 vols. London: Virtue, 1880–1884.

Wittfogel, K. A. "The Hydraulic Civilizations." In W. L. Thomas, ed., *Man's Role in Changing the Face of the Earth,* 152–164. Chicago: University of Chicago Press, 1956.

Wittfogel, K. A. *Oriental Despotism.* New Haven, Conn.: Yale University Press, 1957.

Woolley, C. L., and T. E. Lawrence. *The Wilderness of Zin.* Winona Lake, Ind.: Eisenbrauns, 2003.

Wright, G. E., ed. *The Bible and the Ancient Near East.* Garden City, N.Y.: Doubleday, 1965.

Wurthwein, E. *The Text of the Old Testament.* Grand Rapids, Mich.: Eerdmans, 1995.

Yadin, Y. "And Dan, Why Did He Remain in Ships?" *American Journal of Biblical Archaeology* 1 (1968): 9–23.

Yadin, Y. *The Art of Warfare in the Lands of the Bible.* Ramat Gan: International, 1963.

Yadin, Y. "Is the Biblical Account of the Israelite Conquest of Canaan Historically Reliable?" *Biblical Archaeology Review* 8 (1982): 16–23.

Yagil, R. *The Camel in Today's World: A Handbook on Camel Management.* Tel Aviv and Bonn: German-Israel Fund for Research and International Development, 1994.

Yahalom, M. *The Book of God: A Secular Bible* (Hebrew). Tel Aviv: Dor, 2000.

Yahav, D. *Biblical Idioms and Proverbs.* Tel Aviv: Tammuz, 1988.

Yamauchi, E. "The Eastern Jewish Diaspora Under the Babylonians." In M. W. Chavalas and K. L. Younger, eds., *Mesopotamia and the Bible,* 356–377. Grand Rapids, Mich.: Baker Academic, 2002.

Yardeni, A. *The Book of Hebrew Script.* Jerusalem: Carta, 1991.

Yerushalmi, Y. H. *Freud's Moses: Judaism Terminable and Interminable.* New Haven, Conn.: Yale University Press, 1991.

Young, D. A. *The Biblical Flood: A Case Study of the Church's Response to Extrabiblical Evidence.* Grand Rapids, Mich.: Eerdmans, 1995.

Younger, K. L. "Deportation of the Israelites." *Journal of Biblical Literature* 117 (1998): 201–227.

Yurco, F. "Merneptah's Canaanite Campaign and Israel's Origin." In E. S. Frerichs and H. H. Lesko, eds., *Exodus: The Egyptian Evidence,* 57–75. Winona Lake, Ind.: Eisenbrauns, 1997.

Zaharoni, I., ed. *Derekh Eretz: Man and Nature* (Hebrew). Tel Aviv: Ministry of Defense, 1985.

Zakovitch, Y., and A. Shinan. *That's Not What the Good Book Says* (Hebrew). Tel Aviv: Yedioth Ahronoth, 2004.

Zertal, A. *A Nation Is Born: The Altar of Mount Ebal and the Origin of Israel.* Tel Aviv: Yedioth, 2000.

Zohary, D., and M. Hopf. *Domestication of Plants in the Old World.* Oxford: Clarendon Press, 1988.

Zohary, D., and M. Hopf. "Domestication of Pulses in the Old World." *Science* 182 (1973): 887–894.

Zohary, D., and P. Spiegel-Roy. "Beginning of Fruit Growing in the Old World." *Science* 187 (1975): 319–327.

Zohary, M. *Plants of the Bible.* Cambridge: Cambridge University Press, 1982.

Zohary, M. *Introduction to the Geobotany of Palestine* (Hebrew). Tel Aviv: Sifriat Poalim, 1994.

Zornberg, A. G. *The Beginning of Desire: Reflections on Genesis.* New York: Doubleday, 1995.

Zugorodsky, M. *The Labor of Our Ancestors: Hebrews According to the Scriptures.* Tel Aviv: Twersky, 1949.

PERMISSIONS AND CREDITS

INDEX

Numbers in italics refer to pages on which illustrations appear.